D1764999

90 0900823 5

Interface engineering of natural fibre composites for maximum performance

Related titles:

Polymer–carbon nanotube composites: preparation, properties and applications
(ISBN 978-1-84569-761-7)
This important book reviews the use of carbon nanotubes as reinforcements in a
polymer matrix, creating a new class of nanocomposites with useful properties and
wide potential applications. The book discusses matrix materials for polymer–carbon
nanotube composites, different processing techniques, interfacial and other properties
as well as some of the important applications of this new group of materials.

Creep and fatigue in polymer matrix composites
(ISBN 978-1-84569-656-6)
This authoritative book reviews the experimental and theoretical approaches to creep
and fatigue of polymer matrix composites in the context of durability. The first group
of chapters focuses on viscoelastic and viscoplastic modelling. The book then reviews
stress corrosion, creep rupture and damage interaction before concluding with chapters
on fatigue modelling and monitoring.

Fatigue life prediction of composites and composite structures
(ISBN 978-1-84569-525-5)
This important book addresses the highly topical subject of fatigue life prediction
of composites and composite structures. Fatigue is the progressive and localised
structural damage that occurs when a material is subjected to cyclic loading. The use
of composites is growing in structural applications and they are replacing traditional
materials, primarily metals. Many of the composites being used have only recently
been developed and there are uncertainties about the long term performance of these
composites and how they will perform under cyclic fatigue loadings. The book will
provide a comprehensive review of fatigue damage and fatigue life modelling.

Details of these and other Woodhead Publishing materials books can be obtained by:

- visiting our web site at www.woodheadpublishing.com
- contacting Customer Services (e-mail: sales@woodheadpublishing.com; fax:
  +44 (0) 1223 832819; tel.: +44 (0) 1223 499140; address: Woodhead Publishing
  Limited, 80 High Street, Sawston, Cambridge CB22 3HJ, UK)

If you would like to receive information on forthcoming titles, please send your address
details to: Francis Dodds (address, tel. and fax as above; e-mail: francis.dodds@
woodheadpublishing.com). Please confirm which subject areas you are interested in.

Interface engineering of natural fibre composites for maximum performance

Edited by

Nikolaos E. Zafeiropoulos

WOODHEAD
PUBLISHING

Oxford   Cambridge   Philadelphia   New Delhi

Published by Woodhead Publishing Limited,
80 High Street, Sawston, Cambridge CB22 3HJ, UK
www.woodheadpublishing.com

Woodhead Publishing, 1518 Walnut Street, Suite 1100, Philadelphia, PA 19102-3406, USA

Woodhead Publishing India Private Limited, G-2, Vardaan House, 7/28 Ansari Road,
Daryaganj, New Delhi – 110002, India
www.woodheadpublishingindia.com

First published 2011, Woodhead Publishing Limited
© Woodhead Publishing Limited, 2011
The authors have asserted their moral rights.

British Library Cataloguing in Publication Data
A catalogue record for this book is available from the British Library.

ISBN 978-1-84569-742-6 (print)
ISBN 978-0-85709-228-1 (online)

The publisher's policy is to use permanent paper from mills that operate a sustainable forestry policy, and which has been manufactured from pulp which is processed using acid-free and elemental chlorine-free practices. Furthermore, the publisher ensures that the text paper and cover board used have met acceptable environmental accreditation standards.

Typeset by Replika Press Pvt Ltd, India
Printed by TJI Digital, Padstow, Cornwall, UK

Contents

K.-Y. Lee and A. Bismarck, Imperial College London, UK

M. Misra, S. S. Ahankari and A. K. Mohanty, University of Guelph, Canada and A. D. Ngo, University of Quebec, Canada

C. Scarponi, Sapienza University of Rome, Italy

Contributor contact details

(* = main contact)

Editor

N. Zafeiropoulos
Department of Materials Science
and Engineering
University of Ioannina
PO Box 1186
Ioannina, GR-45 110
Greece

Email: nzafirop@cc.uoi.gr

Chapter 1

Alessandro Gandini*
University of Aveiro
CICECO and Chemistry
Department
3810-193 Aveiro
Portugal

Email: agandini@ua.pt

Mohamed Naceur Belgacem
Polytechnic Institute of Grenoble
(Pagora)
Papermaking Process Engineering
Laboratory (LGP2)
UMR CNRS 5518
Domaine universitaire – 461 rue de
la papeterie – BP 65
F-38402 St Martin d'Hères
France

Email: naceur.belgacem@pagora.
grenoble-inp.fr

Chapter 2

Nektaria-Marianthi Barkoula
Centre for Materials Research
and School of Engineering and
Materials Science
Queen Mary University of London
Mile End Road
London E1 4NS
UK
and
Department of Materials Science
and Engineering
University of Ioannina
PO Box 1186
Ioannina, GR-45 110
Greece

Email: nbarkoul@cc.uoi.gr

Ton Peijs*
Centre for Materials Research
    and School of Engineering and
    Materials Science
Queen Mary University of London
Mile End Road
London E1 4NS
UK
and
Eindhoven Polymer Laboratories
Eindhoven University of
    Technology
PO Box 513
5600 MB Eindhoven
Netherlands

Email: t.peijs@qmul.ac.uk

Chapter 3

A. Dufresne
Federal University of Rio de
    Janeiro
Department of Metallurgical
    Engineering and Materials
Coppe
Rio de Janeiro
Brazil

Email: alain.dufresne@pagora.grenoble-
    inp.fr

Chapter 4

M. A. Mosiewicki*, N. E.
    Marcovich and M. I. Aranguren
Institute of Materials Science and
    Technology (INTEMA)
University of Mar del Plata –
    National Research Council
    (CONICET)
J. B. Justo 4302
7600 Mar del Plata
Argentina

Email: mirna@fi.mdp.edu.ar

Chapter 5

A. Arbelaiz and I. Mondragon*
Materials and Technology Group
Department of Chemical
    Engineering and Environmental
    Engineering
University of the Basque Country
Plaza Europa
20018 Donostia-San Sebastián
Spain

Email: iapmoegi@sc.ehu.es
    inaki.mondragon@ehu.es

Chapter 6

K. L. Pickering
School of Engineering
Faculty of Science and Engineering
Waikato University
Private Bag 3105
Hamilton 3240
New Zealand

Email: klp@waikato.ac.nz

Chapter 7

K. K. C. Ho and A. Bismarck*
Department of Chemical
    Engineering
Polymer and Composite
    Engineering (PaCE) Group
Imperial College London
South Kensington Campus
London SW7 2AZ
UK

Email: a.bismarck@imperial.ac.uk

Chapter 8

P. J. Herrera-Franco* and A.
    Valadez-González
Materials Unit
Centro de Investigación Científica
    de Yucatan, A. C.
Calle 43 No. 130, Col. Chuburná
    C. P. 97200
Mérida, Yucatán
Mexico

Email: pherrera@cicy.mx

Chapter 9

K. R. Rajisha
Department of Chemistry
CMS College
Kottayam
Kerala
India

B. Deepa and L. A. Pothan
Department of Chemistry
Bishop Moore College
Mavelikara-609101
Kerala
India

Email: lapothan@gmail.com

S. Thomas*
Centre for Nanoscience and
    Nanotechnology
Mahatma Gandhi University
Priyadarsini Hills
Kottayam-686560
Kerala
India

Email: sabupolymer@yahoo.com

Chapter 10

K.-Y. Lee and A. Bismarck*
Polymer and Composite
    Engineering (PaCE) Group
Department of Chemical
    Engineering
Imperial College London
South Kensington Campus
SW7 2AZ
UK

Email: a.bismarck@imperial.ac.uk

Chapter 11

M. Misra* and A. K. Mohanty
School of Engineering
Thornbrough Building
University of Guelph
Guelph
Ontario N1G 2W1
Canada

Email: mmisra@uoguelph.ca

S. S. Ahankari
Bioproducts Discovery and
    Development (BDDC)
Department of Plant Agriculture
University of Guelph
Guelph
Ontario N1G 2W1
Canada

A. D. Ngo
Department of Mechanical
    Engineering
University of Quebec
Montreal
Quebec H3C 1K3
Canada

Chapter 12

C. Scarponi
Sapienza University of Rome
Department of Aerospace
    Engineering and Mechanics
Via Eudossiana 18
00183 Rome
Italy

Email: claudio.scarponi@uniroma1.it

Chapter 13

S. Eichhorn
School of Materials
Grosvenor Street
University of Manchester
Manchester
M1 7HS
UK

Email: stephen.j.eichhorn@manchester.
    ac.uk

Part I

Processing and surface treatments to compose the interface in natural fibre composites

Modifying cellulose fiber surfaces in the manufacture of natural fiber composites

A. GANDINI, University of Aveiro, Portugal and
M. N. BELGACEM, Polytechnic Institute of Grenoble, France

Abstract: Several original methods for modifying the surfaces of cellulose fibers are reviewed including: (i) plasma discharge activation solvent-free grafting; (ii) reversible hydrophobic–hydrophilic tailoring of cellulose surface; (iii) 'grafting from' and 'grafting onto' fiber–matrix continuous chemical bonding composites through click-chemistry or living polymerization; (iv) layer-by-layer polyelectrolyte systems combined with the precipitation of metallic or metal oxides nanoparticles, in order to prepare hybrid materials with highly hydrophobic and lipophobic surfaces; and (v) facile gas–solid modifications bearing a green connotation. The variously grafted materials require a thorough characterization and special emphasis is given here to studies in which such techniques as contact angle measurements, cross polarization magic angle spinning (CP-MAS) ^{13}C-nuclear magnetic resonance (NMR), secondary ion mass spectrometry (SIMS) and x-ray photoelectron spectroscopy (XPS), are used to assess the modifications.

Key words: cellulose fibers, chemical grafting, surface modification.

1.1 Introduction

The exploitation of cellulose is a rapidly growing sector, not only for the production of high-volume commodities, such as textiles and paper, but also for novel high-added-value materials, such as functionalized fibers and reinforcing elements in natural fiber-based composite materials (Belgacem and Gandini, 2008a). Such a marked increase is the result of lignocellulosic fibers having several potential advantages, such as low density, a bio-renewable character, ubiquitous availability at low cost and, in a variety of forms, modest abrasivity and easy recyclability in the energy recovery stream. In fact, compared with glass-fiber based composites, cellulose-based composites do not leave any solid residue after their combustion at the end of their life cycle. Unfortunately, natural fibers also have some drawbacks because they possess a strong hydrophilic character, which gives rise to two major limitations when used as reinforcing elements in composite materials. These are: (i) their strong sensitivity to water and even moisture, which results in composites subject to loss of mechanical properties upon

3

moisture adsorption, and (ii) their poor compatibility with the hydrophobic macromolecular matrices generally used in this context, which causes weak interfacial adhesion.

In order to overcome these major drawbacks associated with the possibility of poor performance in the corresponding composites, lignocellulosic fibers are generally submitted to various surface modifications. It follows, therefore, that the main aims of such specific surface modifications are, on the one hand, to provide an efficient hydrophobic barrier and, on the other, to minimize the interfacial energy between the fibers and the non-polar polymer matrix, thus generating optimum adhesion. The chemical processes applied for this purpose are the same as those used to prepare cellulose derivatives, but their impact is limited here to the macromolecular layers constituting the fiber surface, since, in this way, the mechanical properties of the reinforcing elements are preserved, whereas the surface properties are modified to ensure an optimum compatibility with the matrix.

Various modification strategies have been reported in the literature, since the improvement of the interfacial strength and of the hydrophobic character can be attained by different approaches, namely physical treatments such as corona, plasma, laser, vacuum ultraviolet and γ-radiation, and/or, chemical grafting by direct condensation, including (i) fiber–matrix surface compatibilization using hydrophobic moieties, (ii) co-polymerization with the matrix using bifunctional molecules capable of reacting with the OH groups of the cellulose surface and leaving their second function available for further exploitation, and (iii) grafting long chains appended to the fiber surface (by calling upon 'grafting from' and/or 'grafting onto' procedures) (Belgacem and Gandini, 2005, Belgacem and Gandini, 2008b, Belgacem and Gandini, 2009). The two latter strategies ideally generate a continuity of covalent bonds between the surface cellulose chains and the matrix, thus providing the best mechanical performance for a given system, because no irreversible fiber slippage is possible for these situations.

Various cellulose substrates have been employed in these studies, notably lignocellulosic fibers from numerous wood and annual plants (bleached, unbleached, mechanically treated, etc.), solution-regenerated continuous fibers having a regular diameter, cellophane films, microcrystalline powders (Avicel, Technocel, etc.), Whatman filter paper made from high-purity cellulose and laboratory-made paper sheets from different substrates.

The main characterization techniques usually applied to assess the occurrence and the extent of the modification include Fourier transform infra-red (FTIR) and X-ray photoelectron spectroscopy (XPS), elemental analysis, contact angle measurements, inverse gas chromatography (IGC) and scanning electron microscopy (SEM). Recently, new emerging techniques, such as angle take-off photoelectron spectroscopy, cross-polarization magic-angle-spinning nuclear magnetic resonance (CP-MAS NMR), confocal fluorescence microscopy, secondary ion mass spectrometry (SIMS) and atomic force microscopy (AFM) have started to be used in this field.

Although we have published extensively on the state-of-the-art associated with this realm in recent monographs (Belgacem and Gandini, 2005, Belgacem and Gandini, 2008b, Belgacem and Gandini, 2009), the coverage of the two most recent contributions only went up to 2006, and hence the present update deals mostly with the advances, which are numerous, since then. In writing this chapter, we focused our attention on investigations in which the modification of the surface of the fibers for the purpose of preparing optimized composite materials was accompanied by convincing evidence of the actual occurrence and extent of that transformation. Studies dealing empirically with the incorporation of cellulose fibers into polymer matrices, and with the mechanical properties of the ensuing composites, without any proof of surface modification, were therefore left out, considering, moreover, that the present book contains chapters devoted to the mechanical properties of these composites. In addition, interesting reports are included on the surface modification of cellulose fibers aimed at generating specific functionalities, such as omniphobic properties, electrical conductivity, antimicrobial properties and absorption capacity.

1.2   Physical treatments

1.2.1   Plasma grafting

An interesting and environmentally sound surface treatment concerns the use of atmospheric air pressure plasma (AAPP), which was recently applied to various lignocellulosic fibers (abaca, flax, hemp and sisal) using treatments limited to a few minutes (Baltazar-Y-Jimenez & Bismarck, 2007). The wettability of the treated surface was determined using the capillary rise technique, whereas the changes in the surface chemistry were characterized by means of zeta-potential measurements. Some bulk properties, such as the fibers' swelling behavior, were also assessed. The surface energy of the lignocellulosic fibers was found to remain practically constant, even for prolonged treatment times, with the exception of the abaca fibers, for which this parameter decreased with increasing AAPP treatment time. The ζ-potential measurements were performed as a function of pH and the ensuing results were found difficult to rationalize or to correlate with the nature of the treatments because, except for the abaca fibers, this parameter increased and became positive, a fact that is hard to associate with the oxidation mechanism induced by the AAPP electrical discharge. One would, instead, have expected an increase in the concentration of anionic sites following the treatment. This apparent anomaly was compounded by the unusual observation that the ζ-potential of the treated fibers was found to vary strongly with time.

Following a careful characterization of a low pressure SF_6 discharge by means of plasma diagnostics, including radiofrequency electrical probes, Langmuir probes and optical emission spectroscopy (Barni *et al.*, 2007), the authors investigated the modification of PET fabrics, cellulose (paper) and PET films induced by this technique. The hydrophilic character of all the surfaces studied totally disappeared after the treatment and, for the cellulose substrate, the water contact angle increased from 30° to more than 120°, for the most fluorinated hydrophobic surfaces. The plasma treatment at low pressure was also found to produce some surface etching. The XPS analysis confirmed the occurrence of the grafting by detecting bound fluorine atoms.

Belgacem's group recently published two papers dealing with the treatment of cellulose samples with cold plasma in the presence of several coupling agents, namely vinyl trimethoxysilane (VTS), γ-methacrylopropyl trimethoxysilane (MPS) (Gaiolas *et al.*, 2008), myrcene (My) and limonene (LM) (Gaiolas *et al.*, 2009). At the end of the treatment, the modified substrate was extensively Soxhlet extracted, in order to remove the physically adsorbed unbound molecular moieties, before being characterized. Contact angle measurements and XPS showed that the surface cellulose chains had indeed been chemically grafted, as indicated by an increase in the water contact angle from 40 to more than 100° and the corresponding decrease of the polar component to the surface energy from about 23 mJ m^{-2}, to almost zero, for all the treated samples, as summarized in Table 1.1.

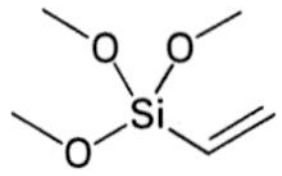

VTS

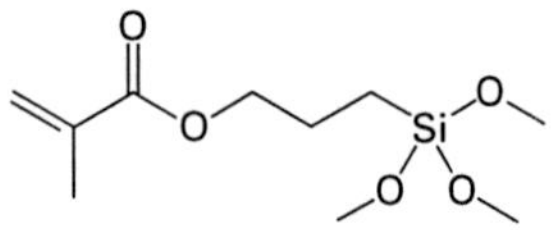

MPS

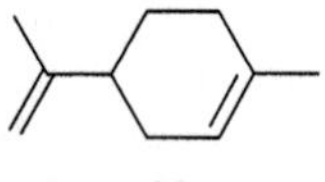

LM

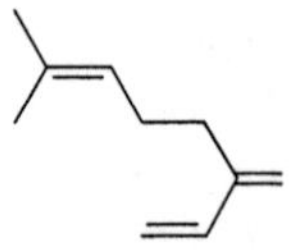

My

The XPS spectra of the silane-bearing coupling agents showed the appearance of two new peaks at 102 and 150 eV, revealing the presence of Si atoms. The C_{1s} spectra of the treated cellulose showed that the intensity of the C1 peak increased from about 17 to around 48, 57, 55 and 92%, for VTS-, MPS-, LM- and My-treated samples, respectively (Table 1.2). The higher reactivity of My, compared with that of LM, was to be expected, given its higher extent of unsaturation and consequently the My-treated surface

Table 1.1 Surface energy and water contact angle (θ_w) of cellulose samples before and after different modifications

Cellulose	θ_w (deg)	Surface energy (mJ m^{-2})			References
		Dispersive	Polar	Total	
Untreated abaca	–	–	–	43.4	Baltazar-Y-Jimenez & Bismarck, 2007
Untreated flax	–	–	–	33.6	Baltazar-Y-Jimenez & Bismarck, 2007
Untreated hemp	–	–	–	31.0	Baltazar-Y-Jimenez & Bismarck, 2007
Untreated sisal	–	–	–	33.1	Baltazar-Y-Jimenez & Bismarck, 2007
AAPP-treated abaca, 3'	–	–	–	33.2	Baltazar-Y-Jimenez & Bismarck, 2007
AAPP-treated flax, 3'	–	–	–	34.2	Baltazar-Y-Jimenez & Bismarck, 2007
AAPP-treated hemp, 3'	–	–	–	34.8	Baltazar-Y-Jimenez & Bismarck, 2007
AAPP-treated sisal, 3'	–	–	–	34.8	Baltazar-Y-Jimenez & Bismarck, 2007
Virgin	40–50	25–30	25–30	50–60	Gaiolas et al., 2008 and 2009, Ly et al., 2010a, 2010b
VTS-treated	98	27.0	0.6	27.6	Gaiolas et al., 2008
MPS-treated	100	32.0	0.2	32.2	Gaiolas et al., 2008
LM-treated	98	32.8	0.3	33.1	Gaiolas et al., 2009
My-treated	112	27.4	0.8	28.2	Gaiolas et al., 2009
Bleached eucalyptus (PC)	56	28	20	48	Cunha et al., 2007a, 2007b and 2007c
Bacterial cellulose (BC)	46	–	–	–	Cunha et al., 2007c
PFB-treated PC	120	–	–	–	Cunha et al., 2007c
PFB-treated BC	120	–	–	–	Cunha et al., 2007c
TFA-treated PC	>120	–	–	–	Cunha et al., 2007b
TFP-treated PC	>113	–	–	–	Cunha et al., 2007a
TFPS-treated	115	15	1	16	Ly et al., 2010b
PFOS-treated	125	26	1	27	Ly et al., 2010b
Avicell (AV)	47	20	0.5	20.5	Ly et al., 2008
PPDI-treated AV	86	29.8	22.9	52.7	Ly et al., 2008
Kraft softwood pulps (KSP)	65	30.7	5.1	35.8	Ly et al., 2008
MDI-treated KSP	90	34.9	10.0	44.9	Ly et al., 2008
AV	–	26.9	5.8	32.7	De Menezes et al., 2009a
Filter paper (FP)	–	28.1	31.3	59.4	De Menezes et al., 2009a
Kraft pulp fibers (KPF)	–	26.2	29.6	55.8	De Menezes et al., 2009a
Regenerated rayon fibers (RRF)	–	33.6	20.4	54.0	De Menezes et al., 2009a
Oxypropylated AV	–	30.8	24.1	54.9	De Menezes et al., 2009a
Oxypropylated FP	–	48.1	4.5	52.6	De Menezes et al., 2009a
		46.5	4.4	50.9	De Menezes et al., 2009a

Continued

Table 1.1 Continued

Cellulose	θ_w (deg)	Surface energy (mJ m^{-2})			References
		Dispersive	Polar	Total	
Oxypropylated KPF	–	46.6	5.7	52.3	De Menezes et al., 2009a
Oxypropylated RRF	–	46.5	4.4	50.9	De Menezes et al., 2009a
PCL$_{10000}$-grafted AV	90	39.5	1.0	40.5	Paquet et al., 2010
PCL$_{42500}$-grafted AV	95	40.7	0.2	40.9	Paquet et al., 2010
PPG$_{2000}$-treated	95	31.9	3.9	35.8	Ly et al., 2010a
PPG$_{2500}$-treated	99	30.1	0.8	30.9	Ly et al., 2010a
PPG$_{3300}$-treated	105	26.1	0.2	26.3	Ly et al., 2010a
PPG$_{4000}$-treated	113	32.4	0.3	32.7	Ly et al., 2010a
PPGMBE-grafted	98	29.9	0.3	30.2	Ly et al., 2010a
TFPS-SiO$_2$-treated	136.3	4.5	0.1	4.6	Gonçalves et al., 2009
PFOS-SiO$_2$-treated	146.8	1.77	0.10	1.87	Gonçalves et al., 2009
Bleached eucalyptus (BE)	–	40.0	4.0	44.0	Tonoli et al., 2009
MPS-treated BE	–	37.5	6.5	44.0	Tonoli et al., 2009
APS-treated BE	–	39.0	20.0	59.0	Tonoli et al., 2009
Untreated cotton fabrics (CF)	73.3	17.21	13.15	30.36	Simončič et al., 2008
Imidazolidinone-treated CF	79.0	16.03	10.67	26.71	Simončič et al., 2008

was mainly composed of aliphatic sequences (92% of C–H moieties). For the silane coupling agents, MPS was found to be more reactive than VTS.

A superhydrophobic cellulose surface was reported (Balu *et al.*, 2008) as a result of a double plasma treatment, viz. first, a nano-etched morphology was produced by the ablation of the amorphous regions of the fibers' surface and then a thin fluorocarbon film was generated by plasma-enhanced chemical vapor deposition using pentafluoroethane. Contact angles as high as 167° were attained under optimized conditions.

Vesel *et al.* (2010) treated viscose textiles with oxygen, nitrogen or hydrogen plasmas for 5 s. High-resolution XPS showed that the use of oxygen and nitrogen atmospheres induced a strong oxidation of the surface, whereas hydrogen, as expected, caused a substantial decrease in oxidized moieties (Table 1.2). Moreover, plasma treatment under a stream of nitrogen caused the fixation of N atoms, as detected by XPS. SEM images showed an increase in the fiber surface roughness after treatment with hydrogen or oxygen plasma. Figure 1.1 shows the XPS spectra and SEM micrographs obtained in this study.

1.2.2　Other techniques

An original method for imparting superhydrophobic properties onto paper surfaces using alkyl ketene dimer in a crystalline form, was recently reported (Werner *et al.*, 2010). Three different techniques were adopted, viz. (i) air blasting with cryo-ground microparticles; (ii) crystallizing from organic solvents; and (iii) spraying with rapid expansion of supercritical solutions. All gave significant results, but the latter method proved particularly effective in terms of the surface morphology. Much work was devoted to the measurements of wetting, but no comment was provided about the long-time stability of these physical coatings. The most significant aspect of this study appears to be its potential to be applied in a continuous fashion without the use of a solvent, i.e. the possibility of industrial implementation.

1.3　Chemical grafting

1.3.1　Surface hydrophobization

In 2007, Cunha *et al.* published three papers dealing with the surface esterification of cellulose fibers with different perfluorinated reagents, viz. trifluoroacetic anhydride (TFA) (Cunha *et al.*, 2007b), pentafluorobenzoyl chloride (PFB) (Cunha *et al.*, 2007c), and 3,3,3-trifluoropropanoyl chloride (TFP) (Cunha *et al.*, 2007a), under controlled heterogeneous conditions and characterization of the ensuing materials by elemental analysis, FTIR spectroscopy, x-ray diffraction (XRD), thermogravimetry, and surface analyses

Table 1.2 XPS analysis of cellulose samples before and after different modifications

Substrate	O/C	Si/C	F/C	N/C	Binding energy (eV)						References
					C1 H-C 284.9	C2 C-O 286.4	C3 O-C-O 287.7	C4 O-C=O 289	CF$_2$ 287.7	CF$_3$ 289	
Virgin viscose	–	–	–	–	30.7	60.0	9.3	–	–	–	Vesel et al., 2010
O$_2$-treated viscose	–	–	–	–	12.2	57.7	17.7	12.4	–	–	Vesel et al., 2010
N$_2$-treated viscose	–	–	–	–	25.7	52.0	15.5	6.8	–	–	Vesel et al., 2010
H$_2$-treated viscose	–	–	–	–	36.5	48.7	12.8	2.0	–	–	Vesel et al., 2010
Virgin pulps	0.74	–	–	–	21.5	57.6	17.5	3.4	–	–	Gaiolas et al., 2008 & 2009
Solvent extracted	0.75	–	–	–	16.9	64.5	18.6	0.0	–	–	Gaiolas et al., 2008 & 2009
VTS-treated	1.7	1.3	–	–	48.5	26.7	17.8	7	–	–	Gaiolas et al., 2008
MPS-treated	0.8	1.0	–	–	57.9	25.2	7.5	9.0	–	–	Gaiolas et al., 2008
LM-treated	0.5	–	–	–	54.6	28.4	11.9	5.1	–	–	Gaiolas et al., 2009
My-treated	0.06	–	–	–	92.4	7.6	0.0	0.0	–	–	Gaiolas et al., 2009
Bleached eucalyptus (PC)	0.72	–	–	–	22.3	58.5	16.0	3.4	–	–	Cunha et al., 2007b & 2007c
Bacterial cellulose (BC)	0.80	–	–	–	19.9	61.3	15.9	2.8	–	–	Cunha et al., 2007c
PFB-treated PC (5 h/65 °C)	0.4	–	0.41	–	7.8	45.4	24.3	8.9	–	13.6 (C–F)	Cunha et al., 2007c
PFB-treated BC (5 h/65 °C)	0.7	–	0.26	–	10.1	54.7	18.5	4.6	–	12.1 (C–F)	Cunha et al., 2007c
TFP-treated PC (2 h/room T)	0.71	–	0.28	–	6.9	57.0	12.7	7.4	–	4.8	Cunha et al., 2007a
TFP-treated PC (2 h/65 °C)	0.60	–	0.42	–	4.2	44.1	12.6	13.1	–	10.4	Cunha et al., 2007a
TFP-treated PC (2 h/100 °C)	0.58	–	0.44	–	2.1	35.3	12.1	15.4	–	11.2	Cunha et al., 2007a
Virgin Avicell (AV)	–	–	–	–	26.0	57.0	15.0	2.0	–	–	Ly et al., 2010b
Extracted Avicel	–	–	–	–	15.0	72.7	12.4	0.0	–	–	Ly et al., 2010b
FPTS-treated WP	0.31	0.06	0.27	–	21.2	54.0	13.7	2.1	–	5.7	Ly et al., 2010b
PFOS-treated WP	0.63	0.03	0.19	–	20.3	58.4	14.9	2.2	4.2	0.9	Ly et al., 2010b
FPTS-treated AV	0.55	0.21	0.43	–	19.9	44.7	15.7	2.4	–	15.7	Ly et al., 2010b
PFOS-treated AV	0.48	0.08	0.53	–	20.1	40.4	13.2	2.5	19.2	4.2	Ly et al., 2010b
PMDA-treated AV	0.74	–	–	–	11.4	69.0	17.1	2.4	–	–	Ly et al., 2008

Sample									Reference
BPDA-treated AV	0.66	–	–	23.5	50.5	9.6	16.4	–	Ly et al., 2008
PPDI-treated AV	0.28	–	0.22	46.4	31.2	12.0	1.4	–	Ly et al., 2008
Kraft softwood pulps (KSP)	0.69	–	–	20.3	61.5	16.4	1.7	–	Ly et al., 2008
PMDA-treated KSP	0.51	–	–	50.4	27.2	11.4	12.5	–	Ly et al., 2008
BPDA-treated KSP	0.52	–	–	58.4	22.6	1.4	9.6	–	Ly et al., 2008
MDI-treated KSP	0.16	–	0.09	75.1	22.4	0.4	3.4	–	Ly et al., 2008
MPS-treated KSP	0.70	0.19	–	42.2	35.4	13.3	9.2	–	Ly et al., 2008
MRPS-treated KSP	0.68	0.10	–	26.5	53.6	17.0	2.8	–	Ly et al., 2008
Undecynoate-grafted AV	–	–	–	29.0	48.0	16.0	6.0	–	Krouit et al., 2008
PCL-grafted AV	–	–	–	64.0	21.0	6.0	8.0	–	Krouit et al., 2008
PCL$_{1000}$-treated AV	0.63	–	0.038	28.7	54.9	12.1	4.2	–	Paquet et al., 2010
PCL$_{42500}$-treated AV	0.59	–	0.036	31.2	50.0	13.0	5.7	–	Paquet et al., 2010
PCL$_{1000}$-treated KSP	0.53	–	0.022	52.9	33.2	4.3	9.5	–	Paquet et al., 2010
PCL$_{42500}$-treated KSP	0.59	–	0.035	37.1	35.0	12.3	8.0	–	Paquet et al., 2010
Virgin Whatman paper (WP)	0.77	–	–	13.8	67.7	16.6	2.0	16.6	Ly et al., 2010b
Extracted WP	0.80	–	–	11.60	69.01	17.30	2.09	2.0	Ly et al., 2010a & 2010b
PEGME$_{2000}$-treated WP	0.34	–	0.17	38.91	48.98	1.35	9.81	–	Ly et al., 2010a
PPGMBE$_{2500}$-treated WP	0.37	–	0.14	40.66	46.91	3.22	9.22	–	Ly et al., 2010a
PTHF$_{2900}$-treated WP	0.28	–	0.19	53.46	32.77	2.93	10.84	–	Ly et al., 2010a
PPG$_{3300}$-treated WP	0.29	–	0.18	46.00	42.57	2.01	9.43	–	Ly et al., 2010a
Cel–CDI–C$_{12}$–NH$_2$	0.61	–	0.05	29.4	57.8	11.0	1.9	–	Alila & Boufi, 2009, Alila et al., 2009
Cel–CDI–C$_{12}$–NH$_2$–CDI–MM	0.53	–	0.15	41.5	45.5	8.7	4.3	–	Alila & Boufi, 2009, Alila et al., 2009
Cel–CDI–C$_{12}$–NH$_2$–CDI–MM– NH–C$_{12}$–NH$_2$	0.27	–	0.14	59.3	31.0	4.0	5.6	–	Alila et al., 2009

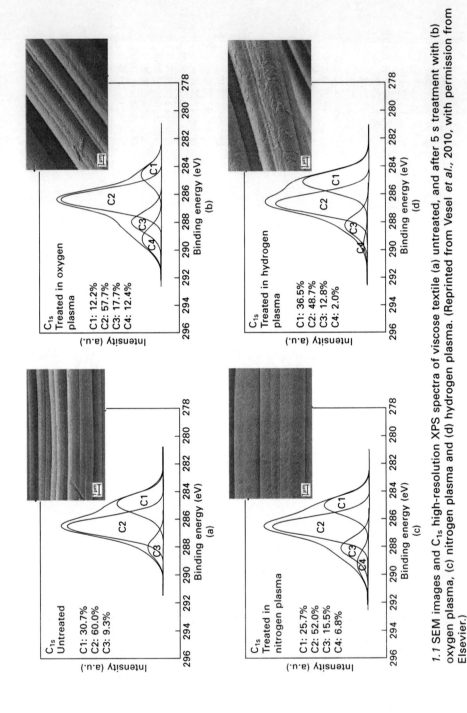

1.1 SEM images and C_{1s} high-resolution XPS spectra of viscose textile (a) untreated, and after 5 s treatment with (b) oxygen plasma, (c) nitrogen plasma and (d) hydrogen plasma. (Reprinted from Vesel et al., 2010, with permission from Elsevier.)

(XPS, time-of-flight (ToF)-SIMS, and contact angles measurements). The occurrence of the grafting was demonstrated by direct techniques, such as FTIR, XPS and ToF-SIMS, whereas the hydrophobic and lipophobic character of the fluorine-containing modified surfaces were found to have increased significantly compared with those of the pristine fibers. The degree of substitution of the pentafluorobenzoylated substrate ranged from 0.014 to 0.39, whereas that of the trifluoropropanoylated counterpart ranged from less than 0.006 to 0.30. The thermal stability decreased only slightly following this treatment, whereas the degree of crystallinity decreased significantly under the most severe experimental conditions. The hydrolytic stability of these perfluorinated cellulose fibers was evaluated both by the effect of water vapor and that of liquid water at different pHs. Whereas the trifluoroacetate displayed a high sensitivity to hydrolysis, making these modified fibers readily revertible to their original cellulose surface, the other two treatments generated more stable esters which were only hydrolized in contact with basic water.

Figure 1.2 gives the FTIR spectra of cellulose fibers (PC), before and after grafting with TFA in different reaction conditions and Fig. 1.3 shows the ToF-SIMS spectra of TFP-grafted cellulose substrates. Tables 1.1 and 1.2 summarize some relevant data related to surface energy and XPS, associated with this series of papers.

The sizing of paper is generally carried out at the wet-end part of the papermaking process by the introduction of a sizing agent into the fiber suspension in the form of an emulsion. Zhang *et al.* (2007) studied a new method of sizing consisting of the possibility of delivering the sizing molecules in a vapor state. Alkyl ketene dimer (AKD) and alkenyl succinic acid anhydride (ASA) are the two sizing chemicals used in the papermaking industry. This original study focused on analyzing the chemical composition of AKD and ASA vapors at typical papermaking temperatures, using FTIR spectroscopy. The ASA vapors were composed of the same molecules as those of this compound in the liquid state, whereas the AKD vapors were rich in fatty acids, indicating that the thermal decomposition of AKD had yielded the molecules from which it had been initially synthesized. The main conclusion of this investigation is that cellulose can be sized by ASA vapor, but not by AKD.

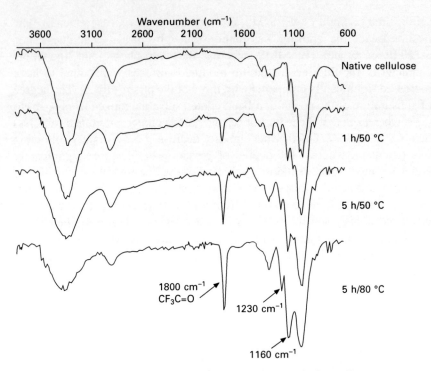

1.2 FTIR spectra of cellulose fibers before and after trifluoro-acetylation at various reaction temperatures and times. (Reprinted from Cunha *et al.*, 2007b, with permission from Elsevier.)

Cellulose fibers from sugar cane bagasse obtained from an organosolvent/supercritical carbon dioxide pulping process were modified with octadecanoyl and dodecanoyl acid chloride by their incorporation as reinforcing elements into amorphous polyethylene (Pasquini *et al.*, 2008). The occurrence of the chemical modification was checked by XPS. The degree of polymerization of the fibers' cellulose and their intrinsic properties (zero tensile strength) were determined and showed that this surface treatment had resulted in a decrease in these parameters, because of the acidic conditions associated with the modification.

Ly *et al.* (2010b) recently reported the grafting of the surface of two model cellulose fibers, Avicell (AV), and Whatman paper (WP), with the two fluorine-bearing alkoxysilanes previously used by Gonçalves *et al.* (2008)

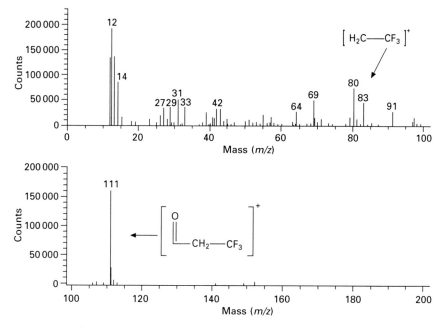

1.3 Partial positive ToF-SIMS spectrum of trifluoropropanoylated cellulose fibers (2 h at 100 °C). (Reprinted from Cunha *et al.*, 2007a, with permission from the American Chemical Society.)

(see below). The ensuing modified fibers were Soxhlet extracted before being characterized by elemental analysis, contact angle measurement, XPS and SEM, which showed that the grafting had indeed occurred.

FPTS   FOTS

The fluorinated moieties appended onto the surface cellulose macromolecules produced a strong hydrophobic effect, as revealed by the increase in water contact angles on both substrates, which went from about 50° before modification, to 115 and 125° for FPTS- and TFOS-modified samples, respectively. Moreover, the determination of the polar contribution to the surface energy using Owens-Wendt's approach, gave a dramatic decrease from about 20 mJ m^{-2} for the pristine celluloses to almost 0 mJ m^{-2} after modification (see Table 1.1). In the full XPS spectra related to the grafted samples, Si peaks appeared at 102 and 150 eV and a F signal arose at 688 eV. The first Si peak was associated with the photoelectrons ejected from Si$_{2s}$ orbitals, whereas the second originated from Si$_{2p}$. The corresponding deconvoluted C_{1s} spectra displayed significant changes (Table 1.2), namely, (i) C–Si and CF_3 peaks at 283 and 292 eV, respectively, appeared for all the modified substrates; (ii) a CF_2 signal at 290 eV appeared in the FPTS-grafted substrate; and (iii) the C1 signal increased in all the spectra.

A different approach for appending long aliphatic chains, perfluoro homologues, or PDMS onto the surface of cellulose fibers was recently proposed (Nyström *et al.*, 2009), based on atom transfer radical polymerization (ATRP), but the processes necessary to produce highly hydrophobic surfaces appear to be rather laborious and cumbersome, compared with the more straightforward systems discussed here and in the hybrid materials section.

All the above approaches called upon the well-known properties of perfluorinated moieties or of long aliphatic chains to impart non-polar low-surface energy to the modified cellulose surfaces. In the search for superhydrophobicity (Feng *et al.*, 2002), a second surface feature is often sought to enhance the role of the chemical grafting, viz. the creation of micro- and nano-asperities capable of generating imprisoned air pockets which constitute obstacles to water spreading and hence minimize wetting. Approaches based on the combination of these two features are discussed below within the context of hybrid cellulose materials.

1.3.2   Matrix–fiber linked systems

This section covers, on the one hand, those systems in which a continuity of covalent bonds extends from the fiber surface to the matrix macromolecules and, on the other, systems in which the links are of a more physical nature, such as chain entanglements and/or strong secondary bonds such as hydrogen bonds. Studies of the grafting of cellulose fibers with polymer chains without the subsequent step of their incorporation into a matrix are also briefly discussed.

Ly *et al.* (2008) prepared composite materials based on cellulose acetobutyrate (CAB) and natural rubber (NR) as matrices, reinforced with long cellulose fibers from softwood and with avicell powder (Ly *et al.*, 2008). The surface of the two cellulose substrates was chemically modified with various difunctional coupling agents, viz. pyromellitic dianhydride (PMDA), benzophenone-3,3′,4,4′-tetracarboxylic dianhydride (BPDA), 1,4-phenylene di-isocyanate (PPDI), methylene-bis(phenylisocyanate) (MDI), γ-mercaptopropyltriethoxysilane (MRPS), and γ-methacrylopropyltriethoxysilane (MPS). The strategy here followed a study by Gandini *et al.* (2001) based on the use of stiff bifunctional reagents in order to ensure that their reaction with the cellulose surface would only involve one of their functionalities, leaving the other available for coupling with the polymer matrix during the processing of the composite. This approach is particularly efficient, because it gives rise to a continuity of covalent linkages between the matrix and the reinforcing elements and enables a perfect stress transfer to occur between them. After the first part of the study, the modified fibers were submitted to Soxhlet extraction in order to remove all the unreacted and physically adsorbed molecules. The occurrence of the grafting was confirmed by elemental analysis, FTIR, XPS, SEM and contact angle measurements. The data related to surface energy and XPS analysis are collected in Tables 1.1 and 1.2, respectively. The second step of this investigation involved the incorporation of the modified fibers into CAB and NR matrices. PMDA- and BPDA-treated Avicell fibers were found to reinforce efficiently the mechanical properties of CAB, whereas MPS-treated cellulose enhanced those of NR.

PMDA   BPDA

PPDI   MDI

MRPS   MPS

The partial oxypropylation of different types of cellulose fibers, namely Avicel, rayon, kraft, and filter paper was carried out and the ensuing materials used to prepare new biphasic mono-component materials (de Menezes *et al.*, 2009a & 2009b). The reaction with filter paper was carried out using a basic catalyst in order to activate the cellulose hydroxyl groups, thus providing the anionic initiation sites for the 'grafting-from' polymerization of propylene oxide. By optimizing the reaction parameters of this gas–solid reaction under pressure, a sleeve of thermoplastic grafted material could be generated around the fibers, while preserving the crystallinity and hence the mechanical properties of their inner core. The ensuing modified fibers were hot-pressed to form films of cellulose fibers dispersed into a thermoplastic poly(propylene oxide) matrix. FTIR spectroscopy, XRD, SEM, differential scanning calorimetry, thermogravimetric analysis and contact angle measurements were used to assess the occurrence and the extent of this chemical modification. The optimal molar ratio between the cellulose OH groups and propylene oxide was established for each substrate. The data related to contact angle measurements are given in Table 1.1.

Very recently, three papers have appeared from Belgacem's laboratory, in which the main idea is based on attaching the matrix to the surface of the cellulose fibers *via* chemical linkages, in such a way that the long chains appended to the fiber surface form macromolecular entanglements with those of the polymer matrix (Krouit *et al.*, 2008, Paquet *et al.*, 2010, Ly *et al.*, 2010a). This new approach was called 'continuous fiber-reinforced composites' and involved polycaprolactone (PCL), because of its biodegradable character, and poly(ethylene oxide) (PEO), poly(propylene oxide) (PPO) and poly(tetrahydrofuran) (PTHF) grafts.

The first paper called upon the use of 'click-chemistry' (Krouit *et al.*, 2008) using a two-step reaction pathway. The first step consisted in preparing cellulose esters by reacting the fiber surface with undecynoic acid. The success of this modification (presence of C≡C terminal moieties) was confirmed by FTIR spectroscopy, XPS and elemental analysis. Concurrently, azido-terminated PCL was prepared and characterized by FTIR and ^{13}C-NMR spectroscopy. The second step consisted of reacting the azido-PCL chains with the C≡C-grafted cellulose substrates in heterogeneous conditions, through click chemistry. The XPS gave clear-cut evidence for the occurrence of this 'grafting onto' reaction. The weight gain achieved reached about 20%. Tables 1.1 and 1.2 provide some relevant data of surface energy and XPS of these modified fibers.

The second paper reported that the grafting of cellulose fibers by PCL could also be achieved through another reaction pathway (Paquet *et al.*, 2010) based on the use of phenyl isocyanate as an OH-blocking agent, and 2,4-toluene di-isocyanate (TDI) as the PCL-cellulose coupling mediator. After this modification, the polar component of the surface energy of cellulose decreased from about 30 mJ m^{-2}, for the virgin substrate, to practically zero with PCL (Table 1.1). The appearance of nitrogen atoms (peak at around 398 eV) at the surface of the grafted cellulose was revealed by XPS. The deconvolution of the C_{1s} spectra showed that the intensity of the cellulose C1 and C4 peaks changed significantly after grafting, as indicated by the increase in the intensity of C1 from about 15% to more than double for the less grafted samples, and up to more that 50%, for the most modified ones, whereas, the intensity of the C4 peak increased from less than 2 to more than 9% for the efficiently modified substrates (Table 1.2). Moreover, the surface was substantially enriched in C atoms, because the O/C ratios moved from around 0.8 to less than 0.3. These modified cellulose materials were found to maintain their biodegradable character, although with slower biodegradation kinetics, as shown in Fig. 1.4.

The third paper of this series reported the coupling of cellulose with PEO, PPO and PTHF oligoether chains (Ly *et al.*, 2010b) following the same reaction pathways described above for the PCL-cellulose preparation (Paquet *et al.*, 2010). The ensuing grafted fibers were characterized using the same techniques and the results indicated similar grafting efficiencies. The water contact angles of PPO-grafted cellulose substrates were found to increase with increasing PPO length, as summarized in Table 1.1. The XPS spectra revealed the appearance of a signal around 398 eV, associated with the presence of N atoms at the surface of the modified cellulose. The deconvolution of the C_{1s} spectra showed that the intensities of the C1 and C4 peaks increased from about 12 to 53% for the PTHF-grafted substrate, whereas that of the C4 peak also increased from around 2 to about 10% (Table 1.2). The biodegradability of these oligoether-grafted fibers was also maintained, as illustrated by Fig. 1.4.

An alternative grafting strategy in the area of fiber–matrix linking by appending macromolecular grafts at the fiber surface (physical entanglement between the grafts and the matrix polymer) is based on the two recent techniques of controlled radical polymerization, viz. free-radical reversible addition-fragmentation chain transfer (RAFT) polymerization and ATRP, which are both 'grafting-from' processes. A recent review (Roy *et al.*, 2009) summarizes the various approaches to cellulose grafting and covers these two more recent mechanisms. Here, we limit our comments to a few examples and advise the interested reader to consult that more comprehensive monograph.

Polystyrene (PS) was grafted from cellulose (Barsbay *et al.*, 2007) in the presence of cumyl phenyldithioacetate as chain transfer agent (CTA). The

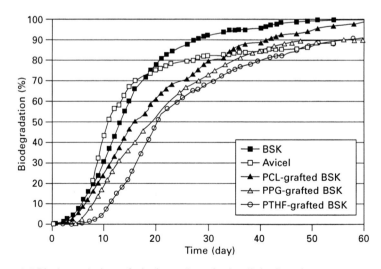

1.4 Biodegradation of virgin and grafted cellulosic substrates (Reprinted from Ly *et al.*, 2010b, and from Paquet *et al.*, 2010, with permission from Elsevier.)

originality of this study consists in having carried out the controlled grafting without prior functionalization of the surface, by fixing the CTA at the surface of the cellulose fibers. The occurrence of the reaction was confirmed by Raman spectroscopy and XPS, DSC, TGA, SEM and contact angle measurements. Figure 1.5 shows the SEM and water contact angle images corresponding to different stages of the fiber modification. Subsequently, the grafted polystyrene chains were cleaved by acid hydrolysis and the ensuing macromolecules characterized by size-exclusion chromatography (SEC).

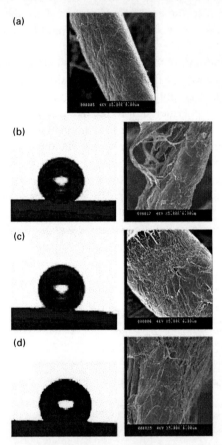

1.5 SEM photomicrographs of: (a) pristine cellulose fibers; (b) water contact angle 138° (left) and corresponding SEM photomicrograph of cellulose-grafted polystyrene, 39% grafting (right); (c) water contact angle 134° (left) and corresponding SEM photomicrograph of cellulose-grafted polystyrene, 30% grafting (right); (d) water contact angle 123° (left) and corresponding SEM photomicrograph of cellulose-grafted polystyrene, 20% grafting (right). (Reprinted from Barsbay *et al.*, 2007, with permission from the American Chemical Society.)

Concurrently, a blank experiment was carried out in which styrene was polymerized under identical conditions, but in the absence of cellulose fibers. The characterization of this PS by SEC gave a number-average molecular weight and a polydispersity index which corresponded to the calculated values and, more importantly, which were essentially the same as those of the cleaved grafts.

In another vein, the RAFT polymerization of 2-(dimethylamino)ethyl methacrylate (DMAEMA) was carried out from activated filter paper cellulose and the ensuing grafts subsequently quaternized with alkyl bromides of different chain lengths (C8–C16) (Roy *et al.*, 2008), as shown in the following mechanism.

These modified fiber surfaces possessed a large concentration of quaternary ammonium groups and they were tested in the field of antibacterial substrates against *Escherichia coli*. Their antibacterial activity was found to depend on the alkyl chain length and on the amount of quaternary amino groups present on the grafts, the highest activity being displayed by fibers bearing grafts with the highest degree of quaternization and the shortest alkyl chains.

Ramie fibers were grafted with acrylic polymers by RAFT polymerization (Chen *et al.*, 2009) and the ensuing materials thoroughly characterized to prove the occurrence of the desired modification, which reached high degrees of substitution. This work opened the way to the preparation of composites using a very abundant cellulose source.

ATRP grafting from cellulose has been the subject of several investigations aimed at preparing liquid crystalline grafts (Westlund *et al.*, 2007) and thermo- and pH-sensitive materials (Ifuku and Kadla, 2008, Lindqvist *et al.*, 2008). More recently, cellulose was grafted with various monomers using activators regenerated by electron transfer (ARGET), a version of ATRP (Hansson *et al.*, 2009).

Cellulose membranes (CM) were modified using surface-initiated ATPR for blood compatibility improvement (Liu *et al.*, 2009) calling upon a two-step process consisting of grafting first 2-bromoisobutyl bromide and then polymerizing *p*-vinylbenzyl sulfobetaine (DMVSA) from it. The modified CM substrates were characterized by attenuated-total-reflectance Fourier transform infra-red spectroscopy (ATR-FTIR), XPS, water contact angle measurements, AFM and TGA, which showed that the zwitterionic grafted brushes had been successfully appended onto the CM surfaces, and that their density had increased gradually with increasing polymerization time. The XPS tracings pointed to the occurrence of grafting by detecting the different new atoms borne by the initiating moiety and the monomer i.e. Br, S and N. These materials were then tested for blood compatibility by their capacity of adsorbing proteins and by *in vitro* platelet adhesion tests and showed improved resistance to these effects.

DMVSA

Other recent ATRP studies applied to cellulose grafting (Meng *et al.*, 2009 and Zampano *et al.*, 2009) did not provide any original feature with respect to the previously established state of the art.

A novel approach to 'grafting-from' polymerization was recently described, which made use of photochemical initiation. Cotton fabrics were coupled with a benzophenone photoactive moiety using butyl tetracarboxylic acid (Hong *et al.*, 2009), as shown in the following mechanism.

These UV-sensitive substrates were then grafted with polyacrylamide by exposing the fabrics to UV irradiation in the presence of the monomer, and the ensuing materials characterized by SEM, FTIR, XRD and TGA, which clearly corroborated the successful occurrence of the attachment of polyacrylamide grafts onto the cotton fibers.

All-cellulose composites are a new family of materials which have attracted some attention in recent years. Although, strictly speaking, their morphology does not involve direct linkages between the fiber and the matrix, the principle of their preparation implies, in general, the regeneration of cellulose macromolecules (the matrix) onto highly crystalline cellulose fibers (the reinforcing elements) and, hence, a strong interaction at the interface through hydrogen bonds. A selective dissolution method was used to prepare one such composite based on three different types of fibers, viz. two kinds of Lyocell and Bocell fibers (Soykeabkaew *et al.*, 2009). The fibers were first activated by subsequent immersion in distilled water, acetone, and DMAc, each solvent treatment being carried out at room temperature for 40 min. They were then immersed (from 2 to 18 min) in a mixture of LiCl/DMAc

(8% wt/v) before coagulation in water. The morphology of the ensuing regenerated fibers thus consisted of a highly crystalline core (undissolved cellulose) surrounded by the reprecipitated more amorphous cellulose matrix. Although this study focused essentially on the mechanical properties of these high modulus/strength all-cellulose composites, we considered it relevant to quote it here as an example of the originality of the approach and the thoroughness with which the materials were characterized (SEM, XRD, dynamic mechanical analysis and tensile testing).

1.3.3   Hybrid inorganic–organic cellulose materials

The association of inorganic nanoparticles with cellulose fibers is a rather recent field of research that began in earnest with a pioneering study on the simple *in situ* synthesis of noble metal nanoparticles on cellulose papers (He *et al.*, 2003). Particles of Ag, Au, Pt and Pd having a narrow size distribution below 10 nm were deposited and visualized by SEM and TEM. This seminal investigation set the stage for further approaches and, as it announced, for application of the novel hybrid materials.

Further work on this topic developed progressively, including recent contributions, such as the electrostatic deposition of Au nanoparticles on wood pulp and bacterial cellulose (Pinto *et al.*, 2007), the preparation of cellulose/SiO_2 nanocomposites by an original *in situ* synthesis utilizing polyelectrolyte layer-by-layer assembly (Pinto *et al.*, 2008) and the surface modification of cellulose fibers with TiO_2 (Gonçalves *et al.*, 2009) to give sunlight photostable nanocomposites. Other recent investigations have focused on increasing the fiber surface coverage (Dong and Hinestroza, 2009) and exploring alternative deposition processes (Ferraria *et al.*, 2009). All these studies concentrated on the choice and optimization of the generation of the nanoparticles and the characterization of the ensuing composites, as further discussed below for some of them. An interesting example of promising applications deals with the antibacterial activity of cellulose/Ag nanocomposites (Pinto *et al.*, 2009) in a publication in which it was shown that Ag concentrations as low as 5×10^{-4} wt% provided an effective action in this context.

Superhydrophobic surfaces were prepared using fluoroalkylsilane (FAS)-modified layer-by-layer (LBL) structured film-coated electrospun nanofibrous membranes (Ogawa *et al.*, 2007). Whereas the electrostatic LBL coating of TiO_2 nanoparticles and poly(acrylic acid) (PAA) yielded a rough surface consisting of nanosized grooves, the FAS modification gave rise to an increase in the surface hydrophobicity thanks to the perfluorinated moieties, although, unfortunately, the chemical structure of the fluoroalkylsilane used was not provided by the authors. Moreover, the modified substrates were not submitted to solvent extraction (at least this information is not available), which made any surface characterization meaningless, because the detection of the chemical moieties or atoms, as well as the water contact angle, could not be used as a proof of successful grafting. This situation emphasizes the need of rigorous procedures in order to provide unquestionable evidence about the surface characterization. In the present context, the fact that XPS indicated the presence of Si, F and Ti atoms, in addition to the C and O present in the initial substrate, together with water contact angles as high as 162°, suggest the elaboration of a superhydrophobic surface, but, regrettably, there is no clear-cut evidence.

A fluoroalkyl functional water-borne siloxane, nanosized silver particles and a reactive organic–inorganic binder were used to prepare water and oil repellent and antimicrobial cotton fibers using various procedures (Tomsic *et al.*, 2008). Contact angle measurements using different liquid probes (water, di-iodomethane and *n*-hexadecane) were used to demonstrate the hydrophobic and oleophobic properties of the treated cotton fabrics, which were maintained even after 10 washings, indicating that such a treatment was quite durable, at least when the optimized treatment was applied. FTIR spectroscopic and XPS measurements revealed the presence of fluorine atoms and C–F vibrational bands. The antibacterial activity of the treated cotton samples was demonstrated by the reduction of the *E. coli* and *Staphylococcus aureus* bacteria. Their physical properties were assessed and showed that the fabric softness and flexibility increased, whereas its breaking strength and air permeability decreased slightly.

An innovative multi-step nanoengineering process was recently developed in order to prepare superhydrophobic cellulose nanocomposites (Gonçalves *et al.*, 2008), by combining three techniques, as shown in Fig. 1.6, namely: (i) a layer-by-layer polyelectrolyte adsorption; (ii) a sol–gel silica-precipitation and (iii) a chemical grafting consisting in appending perfluorinated structures using the FPTS and FOTS coupling agents which readily reacted with the surface Si–OH of the silica nanoparticles. The polyelectrolyte used was poly(diallyldimethylammonium chloride) and the silica deposition precursor was tetraethoxysilane (TEOS). As already discussed, such an approach had the merit of providing a highly efficient hydrophobic character by combining an increased surface roughness induced by the amorphous silica nanoparticles with a reduction in its surface energy guaranteed by the perfluoro-moieties. Each modification step was followed by SEM and AFM. The final surfaces bore strong hydrophobic features that were associated with water contact angles approaching 150°, also suggesting a self-cleaning character. The most relevant surface energy and XPS data of these surfaces is collected in Tables 1.1 and 1.2.

The surface of cellulose fibers was modified using the hydrolysis of TEOS, octyltrimethoxysilane (OTMS) or phenyltrimethoxysilane (PTMS), followed by the layer-by-layer deposition of previously synthesized TiO_2 nanoparticles

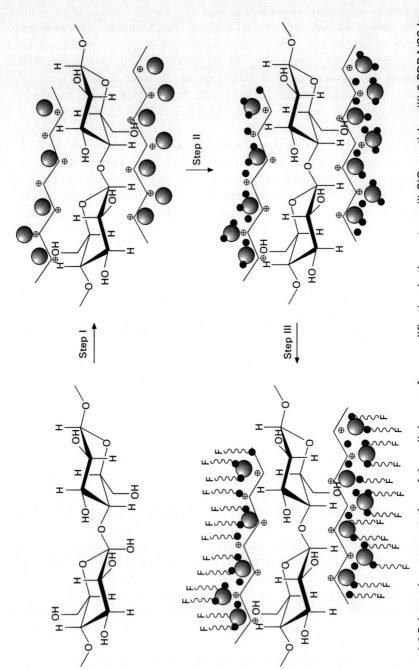

1.6 Schematic representation of the cellulose surface modification in three steps: (I) SiO₂ particles, (II) 5 PDDA/SS layer by layer, and (III) fluorosiloxane. (Reprinted from Gonçalves *et al.*, 2008, with permission from Elsevier.)

(Gonçalves *et al.*, 2009). The ensuing nanocomposites were characterized by Raman, FTIR and ^{29}Si solid state NMR, SEM and water contact angle measurements. The spectroscopic and microscopic analyses confirmed that the grafting had indeed occurred. After grafting with alkoxysilanes, the water contact angle increased from about 50° to more than 90°. These cellulose hybrid materials have a potential to be used as self-cleaning surfaces and as reinforcing agents in polymer matrices. Furthermore, the presence of TiO$_2$ nanoparticles provided the additional property of solar photoactivity. Yet another investigation of omniphobic cellulose fibers organic–inorganic hybrids was recently conducted using (3-isocyanatopropyl)triethoxysilane as the source of cellulose OH coupling (through the NCO moiety) and of sol–gel processing (Cunha *et al.*, 2010b). The following overall scheme illustrates the various pathways adopted in this study.

The hydrophobic micro/nano surface asperity thus generated on the fibers was complemented by incorporation of perfluoromoieties as a source of chemical phobic character. These materials displayed strong biphobic surfaces with contact angles as high as 140° for water and 134° for the non-polar di-iodomethane.

Love *et al.* (2008) reported a new strategy for natural fiber modification consisting of imparting them with glass surface properties. This process is based on filling the internal voids and micropores of the fibers with silica. Before treating the fibers, they were dried by azeotropic distillation, in order to substitute the water molecules entrapped in the cell walls and in the lumen by the silane solution. Tetraethoxy- and triethoxychlorosilane were used as silica precursors. Different techniques were applied to characterize the silane sol–gel system and the modified fibers, namely, silica content, ^{13}C- and ^{29}Si-NMR, FTIR and energy-dispersive spectroscopy, fiber cross-sectional dimensions analysis, SEM, and transmission electron microscopy (TEM). The microscopy images and the energy-dispersion spectroscopy (EDS) showed that the filling had indeed occurred. Figure 1.7 illustrates the SEM micrographs of the fibers after the treatment, clearly showing the filling of the lumen with silica.

Surface-initiated ATRP was applied to graft poly(*tert*-butyl acrylate) (P*t*BA) brushes onto cellulose filter papers (Tang *et al.*, 2009), before hydrolyzing them into poly(acrylic acid) (PAA) in the presence of trifluoroacetic acid. The PAA-grafted paper was used to form chelate complexes with Ag^+, which was then reduced *in situ* to generate silver nanoparticles on the surface of the cellulose fibers. FTIR, XRD and field emission (FE) SEM were used to characterize the chemical structure of the ensuing materials. These cellulose substrates bearing silver nanoparticles were found to possess a good antibacterial action against *E. coli*, compared with that of both the original filter paper and the PAA-modified one.

Ceric ammonium nitrate was used to initiate the graft polymerization of acrylamide onto filter paper fibers in the presence of *N,N'*-methylene bisacrylamide as a cross-linker (Tankhiwale & Bajpai, 2009) before depositing silver nanoparticles onto the ensuing substrate. The silver nanoparticles were precipitated onto the grafted fibers by the citrate reduction of a silver nitrate solution and their formation was corroborated by TEM and SAED analyses. This process gave rise to filter papers with strong antibacterial properties

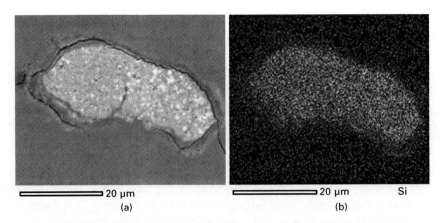

■━━━━━ 20 µm	■━━━━━ 20 µm Si
(a)	(b)

1.7 Scanning electron micrograph (a) and silicon EDS micrograph (b) of the cross-section of a kraft fiber modified using ethyl acetate azeotrope containing 'Silbond 40' at 30% w/v concentration. (Reprinted from Love *et al.*, 2008, with permission from Elsevier.)

against *E. coli*, viz. interesting potential candidates for the preparation of antibacterial food-packaging materials.

The effect of surface modification of cellulose pulp fibers on the mechanical properties and microstructure of fiber–cement composites was recently reported (Tonoli *et al.*, 2009). The surface modification was carried out using γ-methacryloxypropyltrimethoxysilane (MPTS) and γ-aminopropyltriethoxysilane (APTS) and was found to occur efficiently and to influence the fiber–matrix interface and the mineralization of the fiber lumen, as shown by SEM images obtained with a backscattered electron detector. The application of ageing cycles indicated that the MPTS-modified fibers were well protected because they were free from cement hydration products. Instead, the APTS-modified fibers underwent rapid mineralization. These observations helped in the understanding of the mechanisms of degradation of fiber–cement composites.

Ultrathin cellulose films containing immobilized silver nanoparticles were recently prepared using a mild wet chemistry spin-coating method (Ferraria *et al.*, 2009). The substrates were first modified by grafting diaminoalkanes activated by *N,N'*-carbonyldi-imidazole, which acted as anchoring centers for silver nanoparticle generation and immobilization. The final product, as well as some intermediate materials, were characterized by FTIR spectroscopy in the attenuated-total-reflection multiple-internal-reflection mode (ATR/MIR), XPS and AFM. The presence of Ag nanoparticles (7 to 30 nm) was detected at the surface and in the bulk, even with the untreated cellulose substrates, although the amine-modified films displayed a higher density of Ag nanoparticles located at their surface, and this density could be further increased by generating carboxylic acid salts, as shown in Fig. 1.8.

Another study of a hybrid cellulose material displaying superhydrophobic properties (Li *et al.*, 2010) called upon coating the fibers with TiO_2 and thereafter introducing an alkyl-chain silica layer produced by sol–gel chemistry. The materials were only characterized by water contact angle measurements (150–160°) and SEM, without any proof of the suggested surface chemical structures resulting from these treatments.

A very recent investigation (Cunha *et al.*, 2010b) called upon a remarkably simple and straightforward method to attain a high hydrophobicity through the gas-phase coupling of trichloromethylsilane (TCMS) onto the surface of cellulose fibers, in the presence of small amounts of moisture. TCMS reacts with both the surface OH groups of cellulose and with the water present and generates a surface coverage whose structure composition, shown schematically below for the different possible constructs, depends on the actual experimental conditions.

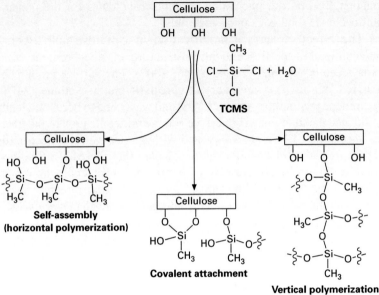

After extraction of the cellulose unbound species, water contact angles higher than 130° were measured together with correspondingly high contact angles with non-polar liquids, showing that highly biphobic cellulose surfaces had been generated by the conjunct action of nano-roughness and chemical inertness. This novel approach, which possesses a clear green chemistry connotation, is being extended to other OH-bearing natural polymers.

1.3.4   Conductive cellulose fibers

Electrically conductive cellulose fibers have attracted considerable attention in recent years with the aim of exploring new possible applications for fiber-based material. A parallel study of the deposition of polyelectrolyte multilayers (PEMs) on the surface of bleached softwood fibers and on that of SiO_2 (used as a model surface) was carried out (Wistrand *et al.*, 2007) using poly(3,4-ethylenedioxythiophene) (PEDOT)/poly(styrene sulfonate) (PSS) and poly(allyl amine) (PAH). The ensuing cellulose-based films were then evaluated in terms of tensile strength, adsorbed amount of polymer, and electrical conductivity. The adsorption of PEM was easily followed because the fiber grew increasingly darker blue throughout the modification sequence. As expected, the conductivity of the fiber network was found to depend directly on the deposited charge density, with variations that spanned 2–3 orders of magnitude. The conductivity also depended on the chemical structure of the adsorbed polyelectrolyte. Thus, PAH-based systems gave lower conductivities compared with those of the PEDOT/PSS counterparts.

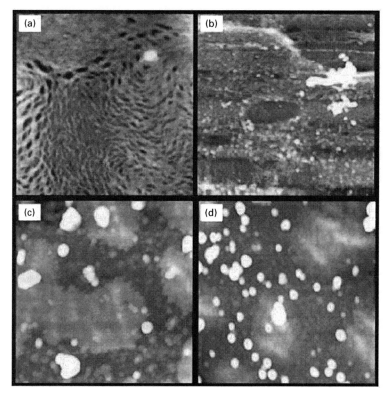

1.8 AFM images (1 μm × 1 μm) of: (a) cellulose (z-scale: 7 nm); (b) cellulose immersed in a AgNO₃ solution for 1 h (z-scale: 15 nm); (c) cellulose film treated with various solutions: *N,N*'-carbonyldi-imidazole (CDI), 1,4-diaminobutane (DAB) and AgNO₃ (z-scale: 20 nm); (d) cellulose film activated by various solutions: CDI, DAB, AgNO₃ containing sodium acetate in a 50/50 v/v water/DMSO medium for 1 h (z-scale: 20 nm). (Reprinted from Ferraria *et al.*, 2009, with permission from the American Chemical Society.)

An important part of the data obtained in this work concerned the wet end chemistry of the papermaking process and tended to establish a relationship between the added PEMs and the mechanical or physical properties of the ensuing papers. This falls outside of the scope of the present chapter and is therefore not discussed further. Nevertheless, the electrical conductivity of these functionalized fibers is an interesting feature worthy to report. Figure 1.9 displays the conductivity as a function of frequency for various sequentially-modified samples.

Polypyrrole and polyaniline conducting polymer composites were successfully generated at the surface of cellulose fibers (Kelly *et al.*, 2007) by the direct polymerization of the respective monomers using ferric chloride and ammonium persulfate as initiators in the presence of individual pulps.

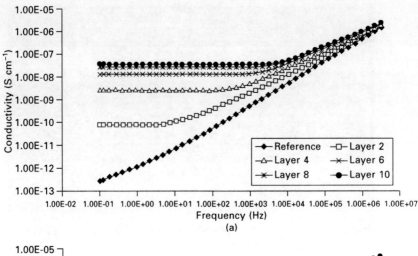

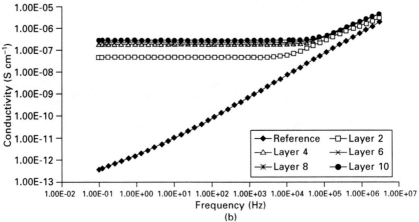

1.9 Conductivity as a function of frequency for the reference sample and for samples sequentially modified with PAH and PEDOT/PSS polyelectrolytes for (a) untreated pulp and (b) carboxymethylated pulp. (Reprinted from Wistrand *et al.*, 2007, with permission from Elsevier.)

The ensuing novel conducting polymer/cellulose fiber composite materials exhibited the inherent properties of both components, viz. the electrical and chemical properties of polypyrrole and polyaniline and the strength, flexibility and available surface areas of the cellulose fibers. Additionally, silver nanoparticles were deposited on the modified fibers to generate hybrid cellulose/conducting polymer/Ag composites. The characterization of these interesting materials was carried out by SEM, cyclic voltammetry, XPS, EDS, electrical conductivity measurements and antimicrobial activity, when silver was present. The highest conductivities of about 3×10^{-1} S cm^{-1} were measured for the pulp/polypyrrole materials, but the presence of silver induced a strong decrease in the composite conducting capacities. The samples based on silver-coated composites actively inhibited the growth of *Staphylococcus aureus* microbes, thus confirming the very active role of silver in the field of antimicrobial activity.

Mini-emulsified poly(3-octylthiophene) nanoparticles were prepared and characterized in terms of their stability and their adsorption onto the surface of cellulose fibers (Sarrazin *et al.*, 2009). Dilution in water induced the depletion of the surfactant (tetradecyltrimethylammonium bromide) from the surface of the particles and yielded a progressive aggregation of the nanoparticles. After the addition of the fiber suspension, the aggregated nanoparticles were adsorbed on the surface of the fibers. This phenomenon was explained by the neutralization of a certain amount of the nanoparticle surface charge and could be rationalized by the ionic interactions between the anionic moieties present at the fiber surface (carboxylic groups) and the remaining cationic sites at the nanoparticles surface, arising from the residual adsorbed surfactant. The adsorption was improved by increasing the amount of negative charge borne by the fibers' surface, achieved through a pre-treatment with carboxymethyl cellulose. The ensuing fibers were highly conductive after doping with iodine ($2–6 \times 10^{-4}\,S\,cm^{-1}$), but their ageing was found to be accompanied by a drastic decrease in conductivity, indicating a progressive de-doping of the conducting polymer and/or its 'over-oxidation' by atmospheric O_2.

1.3.5 Modification to trap organic pollutants

N,N'-carbonyldiimidazole (CDI) was used as an activator to graft the surface of cellulose fibers with different amines under mild conditions (Alila *et al.*, 2009). The preactivated fibers were coupled with a diamine or a triamine to give a carbamate derivative and primary amino end groups, as shown in following most representative structures:

Cel–CDI–NH–C$_{12}$–NH$_2$

Cel–CDI–NH–C$_{12}$–NH$_2$–CDI–MM

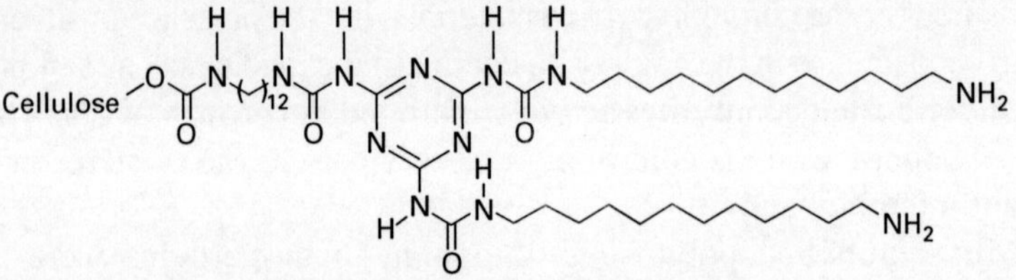

Cel–CDI–NH–C$_{12}$–NH$_2$–CDI–NH–C$_{12}$–NH$_2$

FTIR spectroscopy, CP-MAS NMR and XPS were used to prove the occurrence of the reaction at each modification stage. Figure 1.10 shows the ^{13}C-CP-MAS NMR spectra of the ensuing substrates and Table 1.2 gives the data deduced from the XPS C_{1s} deconvolution spectra. The water contact angle measurements showed that the surface wettability changed after modification. As expected, the values of the contact angle of a droplet of water deposited at the surface of modified cellulose increased with increasing the number of methylene groups borne by the grafts.

These differently modified cellulose fibers were used as adsorbents to remove several aromatic organic compounds and three commercial herbicides, viz. Alachlor, Linuron and Atrazine (Alila & Boufi, 2009). The modified substrates were introduced into a column and their adsorption capacity investigated in a continuous regime as a function of both the solute structure and the nature of the surface modification. It was shown that the chemical modification of the fibers' surface greatly enhanced the adsorption capacity toward water-soluble organic compounds, whereas the adsorption capacity of the pristine cellulose ranged between 20 and 50 mol g^{-1}, the modified substrates increased substantially this aptitude to reach 400 to 1000 mol g^{-1}, depending on the solute structure and the nature of the grafts appended at the cellulose surface. After the saturation of the column, the substrate was washed with ethanol, which completely extracted the trapped compounds. The regenerated column was then used in several adsorption–desorption cycles without any significant loss of activity.

1.3.6   Miscellaneous

The cellulose fiber modification with an imidazolidinone derivative (1,3-dimethyl-4,5-dihydroxyethylene urea, DMeDHEU) was characterized in terms of surface free energy and electrochemical potential (Simončič et al., 2008). The surface energy of both pristine and modified substrates was deduced from liquid penetration tests according to the van Oss–Chaudhury–Good formalism or that based on Shibowski's approach. These tests showed that the incorporation of DMeDHEU into the cellulose structure induced a decrease in the value of the basic component of the fiber surface energy, whereas the dispersive and the acidic contributions remained virtually unchanged.

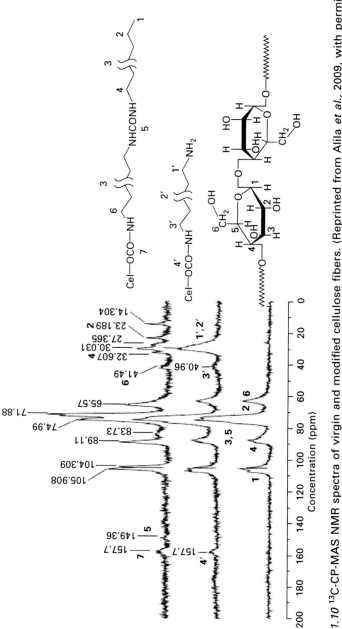

1.10 ^{13}C-CP-MAS NMR spectra of virgin and modified cellulose fibers. (Reprinted from Alila *et al.*, 2009, with permission from Elsevier.)

The absolute values given for the dispersive and total surface energy deduced from the first and second approaches were extremely different, varying by more than a factor of two, which casts doubts about the actual relevance and reliability of this part of the study. The electrokinetic potential measurements showed that the consumption of cellulose hydroxyl groups in the cross-linking reaction with DMeDHEU did not induce a decrease in the electrokinetic properties of the cellulose surface. Such an argument is not based on theoretical considerations, because the electrokinetic properties of cellulose are not conditioned by the amount of hydroxyl groups, since these moieties are not ionized under the neutral pH used in this study. The ζ-potential of cellulose fibers is instead associated with the amounts of carboxylic functions originating from glucoronic acids borne by hemicellulose macromolecules.

Several phenol derivatives, namely caffeic (CA), p-hydroxybenzoic (HB) and gallic acid (GA), eugenol (EG), isoeugenol (iEG), thymol (TH) and dopamine (DA), were subjected to laccase-catalyzed enzymatic polymerization on an unbleached kraft liner solid substrate in order to create covalent linkage with lignin macromolecules present in the fibers. The ensuing polymers were found to be efficient in terms of antimicrobial activity versus *S. aureus* and *E. coli* in liquid media (Elegir *et al.*, 2008). Handsheet papers were prepared from these suspensions and showed higher efficiency against Gram positive and Gram negative bacteria, compared with papers treated only with the phenol derivatives. The antimicrobial activity was found to vary as a function of the structure of the grafted phenol polymer, its concentration, and the reaction time. The substrates were characterized by ^{13}C- and ^{31}P- CP MAS NMR and FTIR spectroscopy, SEC, and titration of acid groups.

Isogai's group recently published a series of papers dealing with the oxidation of various cellulose substrates (bleached sulfite wood pulp, cotton, tunicin, bacterial cellulose and never-dried and once-dried hardwood celluloses) with the 2,2,6,6-tetramethylpiperidine-l-oxyl (TEMPO) radical (Saito *et al.*, 2006, 2007, 2009; Fukuzumi *et al.*, 2009). The ensuing COO^- groups appended on the cellulose chains provided the source of electrostatic repulsion necessary to the following step, where the oxidized fiber suspensions were disintegrated and homogenized using a mechanical treatment (Waring blender), which yielded transparent and highly viscous suspensions composed of individual nanofibrils with a regular width of 3–5 nm and a few micrometres in length. The films prepared from these dispersions were highly crystalline, transparent and flexible, and exhibited a high tensile strength (>312 MPa), even at a low density of 1.47 g cm^{-3} (Saito *et al.*, 2009). The AFM images of the sheets showed that the surface consisted of randomly assembled cellulose nanofibers. When a thin PLA film was coated on these nanofibers, the ensuing sandwich film was found to be a very efficient barrier toward oxygen gas with a permeability dropping from 0.75 L m^{-2} day^{-1} Pa^{-1} for a PLA film down to almost zero, viz. 0.001 L m^{-2} day^{-1} Pa^{-1}. Finally, after the immersion of these double films in an AKD emulsion and a standard drying process, a good hydrophobization was achieved, with water contact angles of 95°.

Regenerated cellulose beads were oxidized with $NaClO_2$, used as a primary oxidant and catalytic amounts of NaClO and 4-acetamide-2,2,6,6-tetramethylpiperidinyl-1-oxyl radicals (Hirota *et al.*, 2009). The amount of carboxylate groups increased to about 2 mmol g^{-1} and the oxidation was limited to C6 primary hydroxyls. These mild conditions allowed the occurrence of such modification without significant molecular weight losses or morphological changes. The ensuing highly porous spheres were then used in cation-exchange applications and exhibited a capacity of fixing small cations and cationic polymers higher than that of carboxymethylcellulose counterparts.

In order to chemically link it to cellulose fabrics, a triazine derivative, 2,4,6-tri-[(2-hydroxy-3-trimethylammonium)propyl]-1,3,5-triazine chloride (Tri-HTAC), was first converted to its triepoxy counterpart (Xie *et al.*, 2009), viz.:

Tri-HTAC

NaOH

Tri-HTAE

The ensuing modified fibers were characterized by FTIR spectroscopy and nitrogen elemental analysis before conducting printability tests on these substrates by measuring the apparent color strength of the printed samples using three reactive dyes. The results clearly indicated an increase in printability with respect to the untreated fabric in terms of both color yield and fixation rate. The wet rubbing and washing fastnesses were also improved thanks to the modification.

1.4   Conclusions

In recent years, substantial progress has been achieved in the field of the surface modification of cellulose substrates, both in widening the quantity and depth of the established approaches and, more significantly, in terms of novel procedures, whose originality has been matched by very promising materials. The latter include solvent-free grafting routes working at room temperature, the use of click chemistry, the preparation of single-component cellulose composites, particularly by the green process of partial oxypropylation, the preparation of superhydrophobic surfaces, and the elaboration of hybrid materials with inorganic components comprising metal and oxide nanoparticles.

This update attempted to highlight these remarkable advances with the aim of stimulating further investigation to give cellulose, in all its natural and processed manifestations, a more prominent place in the realm of composites, bioactive, surface intelligent and, more generally, razor-edge exponents of modern materials science.

1.5   References

Alila S and Boufi S (2009). Removal of organic pollutants from water by modified cellulose fibers, *Ind Crops Prod* **30**, 93–104.

Alila S, Ferraria AM, Botelho do Rego AM and Boufi S (2009). Controlled surface modification of cellulose fibers by amino derivatives using *N,N'*-carbonyldiimidazole as activator, *Carbohydr Polym* **77**, 553–562.

Baltazar-Y-Jimenez A and Bismarck A (2007). Surface modification of lignocellulosic fibers in atmospheric air pressure plasma, *Green Chem* **9**, 1057–1066.

Balu B, Breedveld V and Hess DW (2008). Fabrication of 'roll-off' and 'sticky' superhydrophobic cellulose surfaces via plasma processing, *Langmuir* **24**, 2785–2790.

Barni R, Zanini S, Beretta D and Riccardi C (2007). Experimental study of hydrophobic/hydrophilic transition in SF_6 plasma interaction with polymer surfaces, *Eur Phys J Appl Physics* **38**, 263–268.

Barsbay M, Güven O, Stenzel MH, Davis TP, Barner-Kowollik C and Barner L (2007). Verification of controlled grafting of styrene from cellulose *via* radiation-induced RAFT polymerization, *Macromolecules* **40**, 7140–7147.

Belgacem MN and Gandini A (2005). Physical, chemical and physico-chemical modification of cellulose fibers. *Composite Interface* **12**, 41–75.

Belgacem MN and Gandini A (2008a). *Monomers, polymers and composites from renewable resources*, Elsevier, Amsterdam.

Belgacem MN and Gandini A (2008b). Surface modification of cellulose fibers. In: Belgacem MN and Gandini A. *Monomers, polymers and composites from renewable resources*, Elsevier, Amsterdam, Chapter 18, 385–400.

Belgacem MN and Gandini A (2009). Natural fiber-surface modification and characterization. In: Sabu T and Pothan L, *Cellulose fiber reinforced polymer composites*. Old City Publishing, Philadelphia, Chapter 2, 14–46.

Chen J, Yi J, Sun P and Liu Z-W (2009). Grafting from ramie fiber with poly(MMA) or poly(MA) via reversible addition–fragmentation chain transfer polymerization, *Cellulose* 16, 1133–1145.

Cunha AG, Freire CSR, Silvestre AJD, Pascoal Neto C, Gandini A, Orblin E and Fardim P (2007a). Bi-phobic cellulose fibers derivatives via surface trifluoropropanoylation, *Langmuir* 23, 10801–10806.

Cunha AG, Freire CSR, Silvestre AJD, Pascoal Neto C, Gandini A, Orblin E and Fardim P (2007b). Characterization and evaluation of the hydrolytic stability of trifluoroacetylated cellulose fibers, *J Colloid Interface Sci* 316, 360–366.

Cunha AG, Freire CSR, Silvestre AJD, Pascoal Neto C, Gandini A, Orblin E and Fardim P (2007c). Highly hydrophobic biopolymers prepared by the surface pentafluorobenzoylation of cellulose substrates, *Biomacromolecules* 8, 1347–1352.

Cunha AG, Freire CSR, Silvestre AJD, Pascoal Neto C and Gandini A (2010a). Preparation and characterization of novel highly omniphobic cellulose fibers organic–inorganic hybrid materials, *Carbohydrate Polymers* 80, 1048–1056.

Cunha AG, Freire CSR, Silvestre AJD, Pascoal Neto C, Gandini A, Belgacem MN, Chaussy D and Beneventi D (2010b). Preparation of highly hydrophobic and lipophobic cellulose fibers by a straightforward gas–solid reaction, *J Colloid Interface Sci* 344, 588–595.

de Menezes AJ, Pasquini D, da Silva Curvelo AA and Gandini A, (2009a). Self-reinforced composites obtained by the partial oxypropylation of cellulose fibers. 1. Characterization of the materials obtained with different types of fibers, *Carbohydr Polym* 76, 437–442.

de Menezes AJ, Pasquini D, da Silva Curvelo AA and Gandini A (2009b). Self–reinforced composites obtained by the partial oxypropylation of cellulose fibers. 2. Effect of catalyst on the mechanical and dynamic mechanical properties, *Cellulose* 16, 239–246.

Dong H and Hinestroza JP (2009). Metal nanoparticles on natural cellulose fibers: electrostatic assembly and in situ synthesis, *ACS Appl Mater Interfaces* 1, 797–803.

Elegir G, Kindl A, Sadocco P and Orlandi M (2008). Development of antimicrobial cellulose packaging through laccase-mediated grafting of phenolic compounds. *Enzyme Microbial Technol* 43, 84–92.

Feng L, Li S, Li Y, Li H, Zhang L, Zhai J, Song Y, Liu B, Jiang L and Zhu D (2002). Super-hydrophobic surfaces: from natural to artificial. *Adv Mater* 14, 1857–1860.

Ferraria AM, Boufi S, Battaglini N, Botelho do Rego AM, Rei Vilar M (2009). Hybrid systems of silver nanoparticles generated on cellulose surfaces, *Langmuir* 26, 1996–2001.

Fukuzumi H, Saito T, Iwata T, Kumamoto Y and Isogai A (2009). Transparent and high gas barrier films of cellulose nanofibers prepared by TEMPO-mediated oxidation *Biomacromolecules* 10, 162–165.

Gaiolas C, Costa AP, Nunes M, Santos Siva MJ and Belgacem MN (2008). Grafting of paper by silane coupling agents using cold-plasma discharge, *Plasma Process Polym* **5**, 444–452.

Gaiolas C, Belgacem MN, Silva L, Thielemans W, Costa AP, Nunes M and Santos Silva MJ (2009). Green chemicals and process to graft cellulose fibers, *J Colloid Interface Sci* **330**, 298–302.

Gandini A, Botaro VR, Zeno E and Bach S (2001). Activation of solid polymer surfaces with bifunctional reagents, *Polym Int* **50**, 7–11.

Gonçalves G, Marques PAAP, Pinto RJB, Trindade T and Pascoal Neto C (2009). Surface modification of cellulosic fibers for multi-purpose TiO$_2$ based nanocomposites, *Compos Sci Technol* **69**, 1051–1056.

Gonçalves G, Marques PAAP, Trindade T, Pascoal Neto C and Gandini A (2008). Superhydrophobic cellulose nanocomposites, *J Colloid Interface Sci* **324**, 42–46.

Hansson S, Östmark E, Crlmark A and Malmström E (2009). ARGET ATRP for versatile grafting of cellulose using various monomers, *ACS Appl Mater Interfaces* **1**, 2651–2659.

He J, Kunitake T and Nakao A (2003). Facile in situ synthesis of noble metal nanoparticles in porous cellulose fibers, *Chem Mater* **15**, 4401–4406.

Hirota M, Tamura N, Saito T and Isogai A (2009). Surface carboxylation of porous regenerated cellulose beads by 4-acetamide-TEMPO/NaClO/NaClO$_2$ system, *Cellulose* **16**, 841–851.

Hong KH, Liu N and Sun G (2009). UV-induced graft polymerization of acrylamide on cellulose by using immobilized benzophenone as a photo-initiator, *Eur Polym J* **45**, 2443–2449.

Ifuku S and Kadla JF (2008). Preparation of a therosensitive highly regioselective cellulose/*N*-isopropylacrylamide copolymer through atom transfer radical polymerization, *Biomacromolecules* **9**, 3308–3313.

Kelly FM, Johnston JH, Borrmann T and Richardson MJ (2007). Functionalised hybrid materials of conducting polymers with individual fibers of cellulose. *Eur J Inorg Chem* 5571–5577.

Krouit M, Bras J and Naceur M (2008). Cellulose surface grafting with polycaprolactone by heterogeneous click-chemistry, *Eur Polym J* **44**, 4074–4081.

Li S, Wei Y and Huang J (2010). Facile fabrication of superhydrophobic cellulose materials by a nanocoating approach, *Chem Lett* **39**, 20–21.

Linqvist J, Nyström D, Östmark E, Antoni P, Carlmark A, Johansson M, Hult A and Malström E (2008). Intelligent dual-responsive cellulose surfaces via surface-initiated ATRP, *Biomacromolecules* **9**, 2139–2145.

Liu PS, Chen Q, Liu X, Yuan B, Wu SS, Shen J and Lin SC (2009). Grafting of zwitterion from cellulose membranes via ATRP for improving blood compatibility, *Biomacromolecules* **10**, 2809–2816.

Love KT, Nicholson BK, Lloyd JA, Franich RA, Kibblewhite RP and Mansfield SD (2008). Modification of Kraft wood pulp fiber with silica for surface functionalisation, *Composites: Part A* **39**, 1815–1821.

Ly B, Thielemans W, Dufresne A, Chaussy D, Belgacem MN (2008). Surface functionalization of cellulose fibers and their incorporation in renewable polymeric matrices, *Compos Sci Technol* **68**, 3193–3201.

Ly EB, Belgacem MN, Bras J and Salon-Brochier MC (2010a). Grafting of cellulose by fluorine-bearing silane coupling agents, *Mater Sci Eng, Part C* **30**, 343–347.

Ly EB, Bras J, Sadocco P, Belgacem MN, Dufresne A and Thielemans W (2010b). Surface functionalization of cellulose by grafting oligoether chains, *Mater Chem Phys* **120**, 438–445.

Meng T, Gao X, Zhang J, Yuan J, Zhang Y, He J (2009). Graft copolymers prepared by atom transfer radical polymerization (ATRP) from cellulose, *Polymer* **50**, 447–454.

Nyström D, Lindqvist J, Östmak E, Antoni P, Carlmark A, Hult A and Malmström E (2009). Superhydrophobic and self-cleaning bio-fiber surfaces via ATRP and subsequent postfunctionalization, *ACS Appl Mater Interfaces* **1**, 816–823.

Ogawa T, Ding B, Sone Y and Shiratori S (2007). Super-hydrophobic surfaces of layer-by-layer structured film-coated electrospun nanofibrous membranes, *Nanotechnology* **18**, 165607–165614.

Paquet O, Krouit M, Bras J, Thielemans W and Belgacem MN (2010). Surface modification of cellulose by PCL grafts, *Acta Mater* **58**, 792–801.

Pasquini D, Teixeira EM, Curvelo AAS, Belgacem MN and Dufresne A (2008). Surface esterification of cellulose fibers: processing and characterisation of low-density polyethylene/cellulose fibers composites, *Compos Sci Technol* **68**, 193–201.

Pinto RJB, Marques PAAP, Barros-Timmons A, Trinidade T and Pascoal Neto C (2008). Novel SiO_2/cellulose nanocomposites obtained by in situ synthesis and via polyelectrolyte assembly, *Composite Sci Technol* **68**, 1088–1093.

Pinto RJB, Marques PAAP, Martins MA, Pascoal Neto C and Trinidade T (2007). Electrostatic assembly and growth of gold nanoparticles in cellulosic fibres, *J Colloid Interface Sci* **312**, 506–512.

Pinto RJB, Marques PAAP, Pascoal Neto C, Trinidade T, Daina S and Sadocco P (2009). Antibacterial activity of nanocomposites of silver and bacterial or vegetable cellulosic fibers, *Acta Biomater* **5**, 2279–2289.

Roy D, Knapp JS, Guthrie JT and Perrier S (2008). Antibacterial cellulose fiber *via* RAFT surface graft polymerization, *Biomacromolecules* **9**, 91–99.

Roy D, Semsarilar M, Guthrie JT, Perrier S (2009). Cellulose modification by polymer grafting: a review, *Chem Soc Rev* **38**, 2046–2064.

Saito T, Hirota M, Tamura N, Kimura S, Fukuzumi H, Heux L and Isogai A (2009). Individualization of nano-sized plant cellulose fibrils by direct surface carboxylation using TEMPO catalyst under neutral conditions, *Biomacromolecules* **10**, 1992–1996.

Saito T, Kimura S, Nishiyama Y and Isogai A (2007). Cellulose nanofibers prepared by TEMPO-mediated oxidation of native cellulose. *Biomacromolecules* **8**, 2485–2491.

Saito T, Nishiyama Y, Putaux J-L, Vignon M and Isogai A (2006). Homogeneous suspensions of individualized microfibrils from TEMPO-catalyzed oxidation of native cellulose. *Biomacromolecules* **7**, 1687–1691.

Sarrazin P, Chaussy D, Stephan O, Vurth L and Beneventi D (2009). Adsorption of poly(3-octylthiophene) nanoparticles on cellulose fibers: effect of dispersion stability and fiber pre-treatment with carboxymethyl cellulose, *Colloids Surf A: Physicochem Eng Aspects* **349**, 83–89.

Simončič B, Černe L, Tomšič B and Orel B (2008). Surface properties of cellulose modified by imidazolidinone, *Cellulose* **15**, 47–58.

Soykeabkaew N, Nishino T and Peijs T (2009). All-cellulose composites of regenerated cellulose fibers by surface selective dissolution, *Composites, Part A: Appl Sci Manuf* **40**, 321–328.

Tang F, Zhang L, Zhang Z, Cheng Z and Zhu X. (2009). Cellulose filter paper with antibacterial activity from surface-initiated ATRP, *J Macromol Sci, Part A: Pure Appl Chem* **46**, 989–996.

Tankhiwale R and Bajpai SK (2009). Graft copolymerization onto cellulose-based filter paper and its further development as silver nanoparticles loaded antibacterial food-packaging material. *Colloids Surf B: Biointerfaces* **69**, 164–168.

Tomsic B, Simoncic B, Orel B, Cerne L, Tavcer PF, Zorko M, Jerman I, Vilcnik A and Kovac J (2008). Sol–gel coating of cellulose fibers with antimicrobial and repellent properties, *J Sol–Gel Sci Technol* **47**, 44–57.

Tonoli GHD, Rodrigues FUP, Savastano Jr H, Bras J, Belgacem MN, Rocco Lahr FA (2009). Cellulose modified fibers in cement based composites, *Composites Part A: Appl Sci Manuf* **40**, 2046–2053.

Vesel A, Mozetic M, Strnad S, Persin Z, Stana-Kleinschek K and Hauptman N (2010). Plasma modification of viscose textile, *Vacuum* **84**, 79–82.

Werner O, Quan, C, Turner C, Pettersson B and Wågberg L (2010). Properties of superhydrophobic paper treated with rapid expansion of supercritical CO_2 containing a crystallizing wax, *Cellulose* **17**, 187–198.

Westlund R, Crlamark A, Hult A, Malmström E and Saez IM (2007). Grafting liquid crystalline polymers from cellulose substrates using atom transfer radical polymerization, *Soft Mater* **3**, 866–871.

Wistrand I, Lingstrom R and Wagberg L (2007). Preparation of electrically conducting cellulose fibers utilizing polyelectrolyte multilayers of poly(3,4-thylenedioxythiophene): poly(styrene sulphonate) and poly(allyl amine), *Eur Polym J* **43**, 4075–4091.

Xie K, Liu H and Wang X (2009). Surface modification of cellulose with triazine derivative to improve printability with reactive dyes, *Carbohydr Polym* **78**, 538–542.

Zampano G, Bertoldo M and Bronco S (2009). Poly(ethyl acrylate) surface-initiated ATRP grafting from wood pulp cellulose fibers, *Carbohydr Polym* **75**, 22–31.

Zhang H, Kannangara D, Hilder M, Ettl R and Shen W (2007). The role of vapour deposition in the hydrophobization treatment of cellulose fibers using alkyl ketene dimers and alkenyl succinic acid anhydrides, *Colloids Surf A: Physicochem Eng Aspects* **297**, 203–210.

Interface engineering through matrix modification in natural fibre composites

N.-M. BARKOULA and T. PEIJS, Queen Mary University of London, UK

Abstract: The dependence of the mechanical performances and physical properties of natural fibre reinforced plastics on their interface properties is reviewed, with particular emphasis on low adhesion and the potential ways to overcome this problem via matrix modification. The focus is on polypropylene-based composites and maleic anhydride grafting. The way in which matrix modification improves the adhesion, as well as the parameters that influence the efficiency of bonding are considered. Finally, the effect of matrix modification on interfacial properties and on macroscopic properties of natural fibre composites is discussed. Two main composite categories are considered, long and short natural fibre reinforced polymer composites.

Key words: natural fibre composites, matrix modification, interfaces, maleic anhydride, grafting.

2.1 Introduction

As indicated by the title of this book and chapter, the scope of this chapter is to highlight the research performed in the area of matrix modification in natural fibre composites in order to engineer the interface of these systems and, in turn, to maximize their performance. It is outside the scope of this chapter to compile a full list of papers published regarding natural fibre composites, where matrix modification is performed. Rather, the focus is on systems that have been successfully used, from a scientific as well as from an industrial point of view, by modifying the most common matrices in natural fibre composites. In order to do so, we discuss briefly the motivation behind using natural fibre composites and the trends in their use over the years to allow the reader to understand which systems are the most important to be discussed next. The challenges in using natural fibre composites and the problem of low adhesion is also considered, together with a brief review of the potential ways to overcome this problem.

Next, we discuss the most successful ways to modify the matrix in natural fibre composites. In this section, the focus is on polypropylene (PP) based composites and maleic anhydride (MAH) grafting. Matrix modification of natural fibres composites with polyethylene (PE), poly(ethylene terephthalate)

43

(PET), poly(butylene terephthalate (PBT) and biodegradable matrices is also briefly reviewed. The way matrix modification improves adhesion, as well as the parameters that influence the efficiency of bonding are considered, by employing the different theories of coupling mechanism, adhesion and chemical bonding. Finally, the effect of matrix modification on the interfacial properties and on the macroscopic properties of natural fibre composites is discussed. The synergy between the modification of the matrix and the fibres is discussed for two main categories of composites, i.e. long-fibre and short-fibre composites.

As aforementioned, the focus of this chapter is to review various ways of modifying the polymer matrix in order to enhance adhesion. It is important to keep in mind the difficulty in defining a clear boundary between fibre, matrix modification and coupling agents. In most cases materials used as coupling agents are at the same time modifiers of the fibre and/or the matrix.

2.2   Motivation behind using natural fibre composites and trends

The replacement of conventional glass fibre reinforced plastics (GFRP) by natural fibre ones has been boosted in recent years by ecological concern, environmental awareness and new rules and regulations, which require sustainability and eco-efficiency in technical applications. Lignocellulosic natural fibres are renewable and nonabrasive, and can be incinerated for energy recovery (Bledzki and Gassan, 1999). Furthermore, natural fibre reinforced composites, upon correct design and manufacturing, can be competitive in both mechanical performance and price compared with glass fibre reinforced ones (Berglund and Peijs, 2010; Bledzki and Gassan, 1999; Garkhail *et al.*, 2000; Heijenrath and Peijs, 1996; Joshi *et al.*, 2004; Peijs, 2000; Peijs, 2002; Peijs *et al.*, 1998; Singleton *et al.*, 2003; Stamboulis *et al.*, 2000; Wambua *et al.*, 2003).

The focus until very recently was on composites with polyolefin matrices and, more specifically, on PP, where the effect of natural fibre addition on the physical and mechanical properties has been discussed comprehensively (Barkoula, *et al.*, 2010b; Bledzki and Faruk, 2006; Bledzki and Gassan, 1999, Bos *et al.*, 2006; Cantero *et al.*, 2003; Doan *et al.*, 2006; Espert *et al.*, 2004; Garkhail *et al.*, 2009; 2000; Gassan, 2002; George *et al.*, 2001a; 2001b; Heijenrath and Peijs, 1996; Hornsby *et al.*, 1997; Jayaraman 2003; Keener *et al.*, 2004; Peijs, 2000; Peijs *et al.*, 1998; Singleton *et al.*, 2003; Stamboulis *et al.*, 2000; 2001; Van Den Oever *et al.*, 2000; Zafeiropoulos *et al.*, 2002a; 2002b). Natural fibre reinforced PP is relatively cheap, is thermally stable, and is extensively used in technical applications, e.g. in the automotive industry (Bledzki *et al.*, 2006; Brouwer, 2001; Eisele, 1994; Karus *et al.*, 2005; Mieck *et al.*, 1996; Schlößer and Knothe, 1997; Schüssler, 1998).

Although polyolefin-based natural fibre composites are the most researched ones, during the last decade, biopolymers which originate from renewable raw materials, have been also proposed as matrices in natural fibre composites (Averous and Boquillon, 2004; Barkoula *et al.*, 2010a; Bax and Muessig, 2008; Berglund and Peijs, 2010; Bodros *et al.*, 2007; Mohanty *et al.*, 2002; Nishino *et al.*, 2003; Oksman *et al.*, 2003; 2005; Vila *et al.*, 2008). Generally, biopolymers are thermoplastic materials offering advantages such as low processing time, recyclability along with a feature of biodegradability, and additional recovery options, such as composting. Being fully integrated into natural cycles or carbon dioxide (CO_2) neutral combustion, biocomposites also meet the steadily increasing environmental demands of legislative authorities. Recent reviews have been published on the potential of these materials in various applications (Chiellini *et al.*, 2004; Gatenholm and Mathiasson, 1994; Gatenholm *et al.*, 1992; Mohanty *et al.*, 2000; 2002; Philip *et al.*, 2007; Satyanarayana *et al.*, 2009; Shanks *et al.*, 2004; Van De Velde and Kiekens, 2002; Yu *et al.*, 2006). Among other biodegradable polymers, polyhydroxybutyrate (PHB) and its copolymers are of special interest, because PHB is a highly crystalline polymer and has a melting point, strength and modulus comparable to those of isotactic PP (Jiang *et al.*, 2008).

2.3 Challenges in using natural fibre composites: the problem of low adhesion

Some of the difficulties in using natural fibres as reinforcement in polymer matrix composites are related to: (a) the relatively low thermal stability of natural fibres (up to 230 °C) (Bledzki and Gassan, 1999; George *et al.*, 2001a; 2001b; Myers *et al.*, 1990), and potential fibre degradation when used with matrices requiring processing at high temperatures, (b) their susceptibility to moisture (Bledzki and Faruk, 2004; Stamboulis *et al.*, 2000; 2001), (c) the fibre decomposition during compounding with the polymer matrix (Barkoula *et al.*, 2010b; Keller, 2003; Van Den Oever *et al.*, 2008), which can lead to significant fibre breakage, influencing the morphology and final properties of the composite. Next to fibre breakage, compounding may also be used in an advantageous way to separate the technical cellulose fibres into elementary fibres (Berglund and Peijs, 2010; Snijder, *et al.* 1999).

However, one of the biggest problems in natural fibre reinforced plastics (NFRP), as in case of polymer matrix composites reinforced with inorganic fibres/fillers, is the low adhesion between the reinforcement and the matrix (Bledzki and Gassan, 1999; Bos, 2004; John and Thomas, 2008; Mieck *et al.*, 1995a, 1995b; Mohanty *et al.*, 2001; Nechwatal *et al.*, 2005; Peijs *et al.*, 1998; Saheb and Jog, 1999; Snijder and Bos, 2000). Bonding between the matrix and the fibre is dependent on the atomic arrangement and chemical properties of the fibre and on the molecular conformations and chemical constitution of the polymer matrix. Therefore, the interface is specific to each fibre/matrix system. Natural fibres are hydrophilic and polar in nature, whereas common thermoplastic matrices are hydrophobic and non-polar. The weak bonding between the non-polar matrix and the polar fibres is caused by the large difference in the respective surface energies (Belgacem *et al.*, 1994; Myers *et al.*, 1990) of the two materials linked to the existence of the hydroxyl groups and C–O–C links on the surface of the cellulose fibre. The only bonding mechanisms in this case are relative weak dispersion (chemical) and morphological (mechanical) bonding (Felix *et al.*, 1993). This fact results in poor bonding between the fibres and the matrix, which, in turn, prevents the necessary wet-out of the fibres by molten polymer leading to poor dispersion of the fibres, insufficient reinforcement, and poor mechanical properties. More importantly, a weak interface means insufficient stress transfer from the matrix to the fibres through shear stresses at the interface.

The significance of the fibre/matrix interface for the macroscopic response of fibre reinforced composites is well known. Felix and Gatenholm, (1991) showed that a natural fibre composite with poor adhesion responded at small strain as a material with holes the shape of the filler. A strong interface allows effective stress transfer between the matrix and fibres. When the interface is too strong then brittle scission of fibres occur, whereas a too weak interface leads to pull-out of fibres. It is therefore not straightforward to find an ideal interface to optimize all the macroscopic properties simultaneously (Gamstedt and Almgren, 2007). Besides affecting the tensile strength of a composite, many other properties including fracture toughness, impact toughness, resistance to creep, fatigue and dimensional stability, environmental degradation, are also affected by the characteristics of the interface. In these instances, the relationship between properties and interface characteristics are generally complex (Gamstedt and Almgren, 2007). Based on the above, it can be concluded that in any composite system, interfacial properties need to be optimized rather than maximized, and this is particularly true for natural fibre composites, which often suffer from a lack of toughness, attributed to too strong interfacial interactions relative to the fibre's strength.

Physical and chemical methods can be used to optimize the interface in natural fibre composites (Araújo *et al.*, 2008; Arbelaiz *et al.*, 2005a; Arbelaiz *et al.*, 2005b; Bledzki and Gassan, 1999; Bledzki *et al.*, 1996; 2004; Bos, 2004; Cantero *et al.*, 2003; Dalväg *et al.*, 1985; Doan *et al.*, 2006; Erasmus and Anandjiwala, 2009: Felix and Gatenholm, 1991; Gamstedt and Almgren, 2007; Garkhail *et al.*, 2000; Gassan, 2002; George *et al.*, 2001b; Keener *et al.*, 2004; Kim *et al.*, 2007; Li *et al.*, 2007; Manchado *et al.*, 2003; Mieck *et al.*, 1995a, 1995b; Nechwatal *et al.*, 2005; Nyström *et al.*, 2007; Pickering *et al.*, 2007; Roberts and Constable, 2003; Snijder and Bos, 2000; Van de Velde and Kiekens, 2001; Zafeiropoulos *et al.*, 2002a; Zafeiropoulos *et al.*, 2002b). These modification methods have varying efficiencies for the adhesion between matrix and fibre. The nature of adhesion in composites of modified cellulose fibres and PP has been discussed extensively over the past two decades, and numerous publications can be found on the subject (Arbelaiz *et al.*, 2005b; Cantero *et al.*, 2003; Doan *et al.*, 2006; Erasmus and Anandjiwala, 2009; Felix and Gatenholm, 1991; Felix *et al.*, 1993; George *et al.*, 2001b; Roberts and Constable, 2003; Van de Velde and Kiekens, 2001). A detailed analysis is given by Felix and Gatenholm (1991) and George *et al.* (2001b), where the various physical and chemical methods for improved adhesion have been discussed. More recently a review article was published on the chemical treatments of natural fibre for use in polymer composites (Li *et al.*, 2007). In summary adhesion can be promoted by:

(a) modification of the fibre surface,
(b) use of coupling/compatibilizing agents,
(c) modification of the matrix.

Natural fibres have surfaces covered with accessible hydroxyl and also carboxyl end groups. The surfaces of carbon and glass fibres are relatively inert. Graphite and silica surfaces contain relatively few functional groups (Gamstedt and Almgren, 2007). Natural fibres therefore show more versatility because the interface may be engineered to achieve an optimal interface with regard to the desired property (Gamstedt and Almgren, 2007). The various surface chemical modifications of natural fibres such as treatment by alkali, silane, isocyanate or permanganate, latex coating, acetylation, or monomer grafting under UV radiation have achieved various levels of success in improving fibre strength and fibre/matrix adhesion in natural fibre composites. Methods for natural fibre treatment have been reviewed extensively and are the subject of the chapter 1 of this book.

The substance which promotes or establishes a stronger bond at the matrix/reinforcement interface is known as the coupling agent. Another term that describes the third interphase component, used to improve adhesion, is compatibilizer. The role of the coupling agent/compatibilizer is to interact with both the fibre and matrix, in order to bridge the properties of the two different systems (Felix and Gatenholm, 1991). In general, to enhance the compatibility between hydrophobic and hydrophilic components of a system, the compatibilizing agents have a functional group able to react with the hydroxyl groups of cellulose, an alkyl chain, which decreases the hydrophilicity of the fibre and, at the same time, makes its surface more compatible for good adhesion to the matrix. Lu *et al.* (2000) provided a review of the various coupling agents used to modify the fibre/matrix adhesion in natural fibre reinforced plastics.

The adhesion between the fibre and matrix in natural fibre composites can also be adjusted by modifying the matrix. For polyolefin matrices, one of the most appropriate ways to modify the matrix is by changing their chemistry by attaching polar groups onto the polymer backbone. Biopolyesters, like PHB, on the other hand are polar in nature and are expected to show better adhesion with natural fibres than polyolefins (Shanks *et al.*, 2004). The problem of adhesion seems also to be less relevant when all-cellulose composites are considered (Duchemin *et al.*, 2007; Eichhorn *et al.*, 2010; Gea *et al.*, 2007; 2010; Gindl and Keckes 2005; Gindl *et al.*, 2006; Nishino and Arimoto, 2007; Nishino *et al.*, 2004; Qin *et al.*, 2008; Soykeabkaew *et al.*, 2008; 2009a; 2009b). As well as modifying the matrix with additives, it has been suggested that the morphology of the matrix and more specifically the crystallization process could be controlled so as to create a transcrystalline layer between the polymer matrix and the cellulose fibre. Existence of a transcrystalline layer has been reported to have a significant effect on the fibre/matrix adhesion of natural fibre composites (Felix and Gatenholm 1994; Garkhail *et al.*, 2009; Gray, 1974; Quillin *et al.*, 1993; Son *et al.*, 2000; Zafeiropoulos *et al.*, 2001a; 2001b; 2001c).

2.4   Matrix modification, coupling mechanism and efficiency of bonding

The most important and efficient group in improving interfacial adhesion between a natural fibre and a polyolefin matrix is based on maleic anhydride-grafted PP (MAH-PP) (Arbelaiz *et al.*, 2005a; 2005b; Bledzki and Gassan, 1999; Bledzki *et al.*, 1996; Cantero *et al.*, 2003; Dalväg *et al.*, 1985; Felix and Gatenholm, 1991; Feng *et al.*, 2001; Fung *et al.*, 2002; Garkhail *et al.*, 2000; Gauthier *et al.*, 1998; George *et al.*, 2001b; Joly *et al.*, 1996a; Joseph *et al.*, 2003; Karmaker, 1997; Karmaker and Youngquist, 1996; Karmaker *et al.*, 1994; Manchado *et al.*, 2003; Mieck *et al.*, 1995b; Mishra *et al.*, 2000; Myers *et al.*, 1991a; 1991b; Oksman and Clemons, 1998; Qiu *et al.*, 2003; Rana *et al.*, 1998; 2003; Sanadi *et al.*, 1994a; Sanadi and Caulfield, 2000; Sanadi *et al.*, 1995; Snijder and Bos, 2000; Van Den Oever *et al.*, 2000).

MAH is used to modify the polymer matrix in the presence of a free radical initiator. It is then grafted on to cellulose fibres. The MAH groups of the polymer chain either react directly with the hydroxyl groups of the cellulose or interact by hydrogen bonding with the cellulose. The molecular chain of MAH is much shorter than that of the polymer matrix and cellulose fibres. Long chains of high molecular weight are obtained usually by grafting MAH with PE, PP, and polystyrene (Rowell *et al.*, 1996). Some other systems reported for modifying polyolefins are maleated styrene–ethylene/butylene-styrene triblock copolymers and ethylene–propylene random copolymers (Long and Shanks, 1996), polyethylene copolymer grafted with MAH (Sgriccia and Hawley, 2007), and treatment with acrylic acid, 4-pentanoic acid, 2,4-pentadienoic acid and 2-methyl-4-pentanoic acid (Erasmus and

Anandjiwala, 2009). The formed copolymers (MAH-PE, MAH-PP, S-MAH), styrene–ethylene–butylene–styrene/MAH (SEBS-MAH) are used as coupling agents (Raj and Kokta, 1991) to improve interfacial properties of the interface. Besides the grafting of MAH with polyolefin matrices, it has been found to modify also PS matrix (Maldas and Kokta, 1990; 1991), while ethylene/*n*-butyl acrylate/glycidyl methacrylate (EBGMA) and ethylene methylacrylate (EMA) are copolymers that have been used to modify the adhesion between natural fibres and PET (Corradini *et al.*, 2009).

The effectiveness of MAH is the result of a better compatibility (Arbelaiz *et al.*, 2005a; 2005b; Erasmus and Anandjiwala, 2009) and the ability of MAH to decrease the amount of hydrogen bonding between the fibres (Kazayawoko *et al.*, 1997), and to form covalent bonding between hydroxyl groups of the cellulosic fibres and the anhydride groups of the MAH (Rana *et al.*, 1998).

Other important coupling agents for PPs are silanes. According to Mieck *et al.* (1994; 1995a; 1995b) and Mieck and Reußmann (1995) the application of alkyl-silanes does not lead to chemical bonds between the cellulose fibres and the PP matrix. It seems, however, that the hydrocarbon chains do affect the wettability of the fibres and the chemical affinity of the PP with the natural fibres.

Although biopolyesters, such as poly(lactic acid) (PLA) and poly(hydroxybutyrate) (PHB) are more hydrophilic than polyolefins, research suggests that the interaction between PLA and PHB with inorganic fillers can be improved. For this reason, PLA matrices modified with MAH have been investigated (Zhang and Sun, 2004; Huneault and Li, 2007). Mitomo *et al.*, (1995) reported the radiation graft behaviour of methyl methacrylate (MMA), 2-hydroxyethyl methacrylate (HEMA), acrylic acid (AA) and styrene (ST) onto PHB and its copolymers. Grondahl *et al.* (2005) grafted AA onto PHB for tissue engineering applications. Chen *et al.* (2003), reported the synthesis and characterization of MAH-PHB beyond surface grafting. For the composites based on thermoset matrices, such as epoxy, polyester, and phenol formaldehyde, very little is done to modify the thermoset matrices to increase adhesion. Here, the most common treatments to enhance adhesion are simple alkali treatments of the natural fibres, which cleans and dewaxes the fibre (Gassan and Bledzki, 1999b; Mishra *et al.*, 2001; Mwaikambo and Ansell, 2003: Ray *et al.*, 2001; 2002a; 2002b, 2002c).

In order to understand how the modification of the matrix improves the fibre/matrix adhesion in natural fibre composites it is necessary to explain why materials adhere to each other. For that purpose a large number of adhesion mechanisms have been proposed and recognised (Felix and Gatenholm, 1991; Oksman and Clemons, 1998; Voyutskii, 1963). These include (a) adsorption and chemical bonding, (b) diffusion, (c) electrostatic attraction and (d) mechanical interlocking. The main chemical bonding theory alone is not sufficient. So, the consideration of other concepts appears to be necessary, including the morphology of the interface, the acid–base reactions at the interface, surface energy and the wetting phenomena (Felix and Gatenholm, 1991; Maldas *et al.*, 1989; Westerlind and Berg, 1988). Bledzki and Gassan (1999) proposed several mechanisms of coupling in natural fibre composites, viz (a) elimination of weak boundary layers, (b) production of a tough and flexible layer, (c) development of a highly cross-linked interphase region, (d) improvement of the wetting between polymer and substrate, (e) formation of covalent bonds with both materials and (f) alteration of acidity of substrate surface.

The exact mechanisms of the grafting reactions have been and continue to be debated in the literature (Gaylord and Mishra, 1983; Heinen *et al.*, 1996). A schematic representation of the most accepted interaction between the MAH group and the cellulosic fibre is presented in Fig. 2.1. The MAH group or the AA groups grafted on the polymer matrix react with functional groups present on the surface of the cellulose fibre to form chemical bonds as primary interaction mechanism. A second type of interaction consists of co-crystallization of the high molecular weight tail with the molecular chains of the polymer matrix, giving physical entanglements. As aforementioned, the MAH group, present in the MAH-PP, not only provides polar interactions but can covalently link to the hydroxyl groups on the cellulosic fibre (Qingxiu and Matuana, 2003). The formation of ester linkages and hydrogen bonds between the MAH– and the –OH of cellulose has been indicated through Fourier transform infrared spectroscopy (FTIR) and electron spectroscopy for chemical analysis (ESCA) by Felix and Gatenholm, (1991). Whereas, the

2.1 Schematic representation of the adhesion process between MAH-PP and the cellulose fibre (Garkhail, 2002).

alkyl group molecules chain ends promote adhesion with the bulk polymer matrix through chain entanglements (Ranganathan *et al.*, 1999). Also, the similarity of the additive and matrix structure can permit the occurrence of segmental crystallization, which is desirable for cohesive coupling between the copolymer and the PP matrix (Felix and Gatenholm, 1991). After the MAH-PP treatment, the surface energy of the fibres is increased to a level much closer to the surface energy of the matrix (Garkhail, 2002).

Lu *et al.* (2001) suggested a broader interfacial bonding consisting of covalent bonds, secondary bonding (such as hydrogen bonding and Van der Waals forces) and mechanical interlocking. The mechanical interlocking may occur between the fibre and MAH and the polymer and MAH. All of these bonding mechanisms may concurrently exist across the interface at varying degrees (Keener *et al.*, 2004).

The proposed mechanisms for PP grafted with AA have been discussed in detail in Erasmus and Anandjiwala (2009). Briefly, grafting of AA onto PP is initiated by peroxide radicals. The peroxide grafting of the AA occurs at the tertiary carbons of the polymer chain or at the terminal unsaturated part of the chain. The radicals extract hydrogen atoms, preferably from the tertiary carbon of the polymer chain, leading to the creation of new reactive sites, which are expected to react with other monomers or, as in this instance, with AA. A drawback of this process is that, as the polymer is grafted with AA, the molecular weight is lowered owing to chain degradation which results in a reduction in viscosity. In the same study, chemicals that contain the two functional groups (AA, 4-pentanoic acid, 2,4-pentadienoic acid and 2-methyl-4-pentanoic acid) were used as coupling agent in the treatment of the PP. The double bond (of the coupling agent) acts as the anchoring point for the PP and the carboxylic acid (of the coupling agent) ultimately forms the ester linkage to the cellulose.

There are several theories on how silanes work as coupling agents (Plueddeman, 1982). One is the chemical reaction theory. It simply states that a covalent bond is formed between the filler and polymer. The silano group reacts with the surface of the filler, whereas the organofunctional group reacts with the polymer. Another mechanism is formation of an interpenetrating network, in which the silane molecules diffuse into the polymer matrix, forming an interphase network of polymer and silane. These two explanations are the most widely accepted, although neither one alone completely explains chemical coupling (Plueddeman, 1982).

The two most discussed properties of MAH-grafted polymers that could influence their effectiveness as coupling agents for cellulose fibre composites are molecular weight, which affects entanglement with the matrix chains, and acid number, which determines the functionality present in the coupling agent (Canche-Escamilla *et al.*, 1999; Felix *et al.*, 1993; Gassan and Bledzki, 2000; Olsen, 1991; Panthapulakkal *et al.*, 2005; Rana *et al.*, 1998). In addition to molecular weight and acid number, the choice of peroxide and reaction temperature can be controlled to create a polyolefin with the desired level of grafted MAH (Keener *et al.*, 2004).

2.4.1   Molecular weight

The molecular weight of the MAH-polymer determines how well the grafted polymer molecules will diffuse into the polymer matrix, create entanglements and co-crystallize (Rana *et al.*, 1998; Olsen, 1991). When polymer chains are very short, there is little chance of entanglement between chains and they can easily slide past one another (Neilsen, 1977). In addition, no co-crystallization is expected. When the polymer chains are longer, entanglement between chains can occur and chains are able to diffuse deeper into the matrix. Thus, MAH-polymer chains become more involved in inter-chain entanglements (Felix *et al.*, 1993). A minimum chain length is necessary to develop these entanglements (Neilsen, 1977). If the molecular weight of the coupling agent is too high, it may entangle with the polymer molecules so that the polar groups on the coupling agent have difficulty 'finding' the –OH groups on the fibre surface (Rowell, 1996).

2.4.2   Acid number

Acid number measures the functionality present in the coupling agent, which is dependent upon MAH units grafted per polymeric chain. If the acid number is too high, then the MAH may be too close to the polar surface and thus, the interaction with the non-polar phase might not be sufficient (Keener *et al.*, 2004).

The acid or anhydride functional group reacts with the functional groups present on the surface. Then grafting of acid or anhydride occurs on a single polymer molecule leading to adhesion which is localized and not sufficient. For enhanced bonding, numerous acid or anhydride groups are required to react at multiple sites. Although this is beneficial in terms of uniform interactions, it limits the co-crystallization ability with the matrix (Roberts and Constable, 2003).

There are a few studies that investigated the effects of MAH-PP or AA-PP on the interphase thickness of PP composites. In a study by Felix and Gatenholm (1991), the interphase thickness was determined for a group of cellulose-PP composites containing MAH with a range of molecular weights. It was found that the increase in the molecular weight led to an increase in the thickness of the interphase layer along with the mechanical properties. Other studies evaluated several different coupling agents consisting of MAH-PP with various molecular weights and acid contents (Olsen, 1991; Sanadi *et al.*, 1994b). All MAH agents improved the mechanical properties. However, the coupling agents with both high molecular weight and high acid content produced the highest mechanical properties. Low molecular weight on the other hand is expected to act more as a dispersing agent instead of a true coupling agent (Lu *et al.*, 2000).

As mentioned above, another way to control the fibre/matrix stress transfer is through the control of the morphology of the matrix material and, more specifically, of the crystallization process near the interface. Nucleation of crystals is the first stage of the crystallization process, and is the result of homogenous as well as heterogeneous processes. Heterogeneous nucleation occurs in the presence of foreign surfaces e.g. particles, fibres, impurities, dust or additives. For composites, when heterogeneous nucleation occurs with a sufficiently high density along a reinforcing fibre surface, the resulting crystal growth is restricted to the lateral direction so that a columnar layer develops around the fibre. This phenomenon is known as transcrystallization (Garkhail *et al.*, 2009). First Gray, (1974) reported that cellulose fibres induced transcrystallinity in isotactic PP (iPP). He documented the growth of a transcrystalline layer under isothermal crystallization at 136 °C with purified wood fibres. Less purified fibres developed nucleation, but this required a longer time than purified fibres. Felix and Gatenholm (1994) conducted a more systematic study for highly purified cotton fibre/iPP. They suggested that under isothermal crystallization the iPP chains have sufficient time to adopt the most favourable conformation and, because the crystal structures of cellulose and iPP are matching, there is an increased Van der Waals interaction between the iPP and the cellulose. The effect of various conditions such as the crystallization temperature, time and cooling rates on the formation of transcrystallinity was investigated on dew retted and green flax/iPP systems by Zafeiropoulos *et al.* (2001a; 2001b; 2001c). The transcrystallinity phenomenon in the flax/PP system was affected not only by the different types of flax fibres but also by the different types of PP. The isothermal crystallization behaviour and mechanical properties of ionomer-treated sisal fibres reinforced high-density polyethylene (HDPE) composites was investigated by Choudhury (2008). It was concluded that the crystallization behaviour of HDPE was influenced by the presence and concentration of short sisal fibres. Also here the surfaces of sisal fibres acted as nucleating sites for the crystallization of the matrix polymer, promoting the growth and formation of a transcrystalline layer in the composites. The concentration of nucleating sites in the HDPE/sisal composites increased as the fibre content increased, which was reflected in the increase of the overall crystallinity of the composites. Finally, Garkhail *et al.* (2009) investigated the effect of flax fibre reinforcement on the crystallization behaviour of PP and found that flax fibres are good nucleating agents for PP matrix leading to the development of a well-defined transcrystalline interphase zone. After isothermal crystallization from the melt a transcrystalline layer was found having lamellar crystals grown perpendicular to the fibre axis. Figure 2.2 shows the transcrystalline zone grown around the flax fibre in a PP matrix at a constant crystallization temperature of 138 °C with increasing crystallization time. It is clear that the thickness of the transcrystalline layer or interphase increased with time and reached a maximum value after a certain time. The nucleation density on the fibre surface was higher than in the bulk matrix; hence, there was less competition from spherulitic nucleation and growth in the polymer melt. This led to further growth of the transcrystalline layers, which became thicker and more uniform, before they impinged with spherulites grown in the bulk matrix. The effect of MAH on the crystallization process has been also reported in the literature (Yin *et al.*, 1999; Manchado *et al.*, 2003). It was found that the addition of MAH-PP increased the

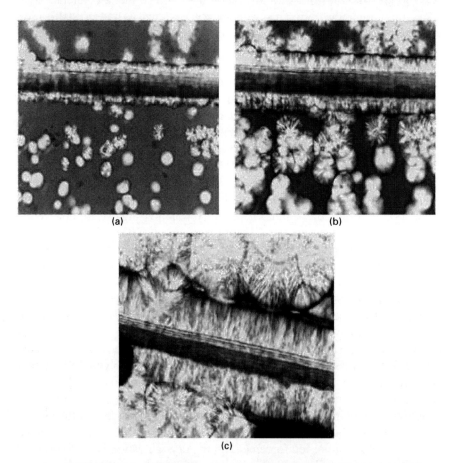

(a)   (b)

(c)

2.2 Effect of crystallization time on transcrystalline thickness in PP/flax system at a crystallization temperature of 138 °C. Cooling rate 10 °C min^{-1}; (a) 2 min, (b) 10 min, (c) 34 min (Garkhail *et al.*, 2009).

nucleation capacity, and accelerated the crystallization process, leading to a transcrystalline region. PE modified with MAH and processed with wood flour showed an increase in overall crystallinity compared with the pure and unmodified PE. The hypothesis that explains this phenomenon is again the nucleation of crystallites (Villar and Marcovich, 2003). Similar results were reported by Mildner and Bledzki, (1999) who studied the systems jute fibre (untreated and alkali treated)/iPP and jute fibre/MAH-PP. They found that the thickness of the transcrystalline layer varied with the cooling rate, with the system jute/MA-PP having the thickest layer.

2.5   Effect of matrix modification on interfacial properties

It is beyond the scope of this chapter to review all potential methods to evaluate the interfacial adhesion after modification. One direct measure of fibre/matrix adhesion is the interfacial shear strength (IFSS). Although much has been reported on the effect of modification on the mechanical properties, much less is published on actual values of the interfacial shear strength after modification.

Previous studies showed that the modification with MAH led to a significant improvement in the apparent IFSS (Doan *et al.*, 2006; Nechwatal *et al.*, 2005; Manchado *et al.*, 2003; Stamboulis *et al.*, 1999; Van Den Oever *et al.*, 2000). Stamboulis *et al.* (1999) used the pull out test to determine the IFSS of flax/PP and flax/PP/MAH-PP. For the same system, Garkhail (2002) used the micro-debond test. Van Den Oever and Bos (1998) calculated the IFSS of elementary and technical flax fibre with PP and MAH-PP. They calculated IFSS values of 13 MPa for the elementary fibres, which increased to 28 MPa with the addition of 1 wt% MAH in PP matrix. Similarly, in the case of a technical fibre, the addition of the MAH led to an increase from 8 to 12 MPa. Other studies showed that the addition of 2 wt% MAH-PP in jute/PP composites (with two different PP grades) led to IFSS values of 16.3 and 19.8 MPa, which were very close to the shear yield strength values of the pure PP matrices as calculated from the Von Mises criterion $(= \sigma_y/\sqrt{3})$. These results indicate a good interaction between the jute fibres and the PP matrix with MAH-PP (Doan *et al.*, 2006). In a system of wood fibre with PP, the IFSS increased from 14 to 23 MPa, with the addition of MAH-PP, which means that here the addition of coupling agent improved adhesion between fibre and matrix by almost 65% (Nyström *et al.*, 2007). These results are similar to the results of Van de Velde and Kiekens (2001) who report an increase in the interlaminar shear strength in a unidirectional PP/flax system from circa 10 to 24 MPa upon the addition of MAH-PP.

Garkhail (2002) comprehensively studied the effect of both fibre and matrix modification on the IFSS when MAH-PP was added to a PP/flax composite. Using the micro-debond test, the effect of the MAH-content on the IFSS, when the matrix is modified was documented. A photomicrograph of a PP/MAH-PP blend on a flax fibre is shown in Fig. 2.3. The effect of the MAH-content on the IFSS, when the matrix is modified is presented in Fig. 2.4.

As can be seen from this figure, the introduction of 5 wt% MAH-PP

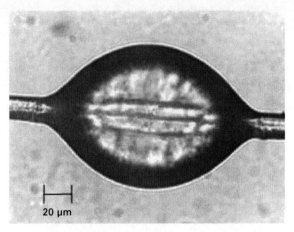

2.3 Droplet of PP/MAH-PP blend (80:20) on flax fibre for the micro-debond test (Garkhail, 2002).

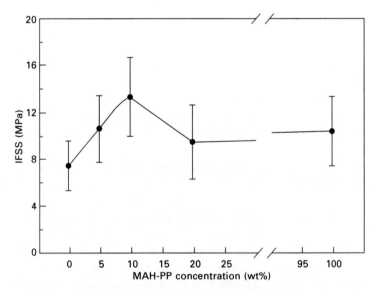

2.4 Interfacial shear strength (IFSS) values of flax fibre with varying amounts of MAH-PP concentration in the PP matrix (Garkhail, 2002).

in the PP matrix resulted in a substantial improvement in the IFSS of the untreated fibres. With the amount of MAH-PP increasing to 10 wt%, the average IFSS reached nearly 14 MPa, which is an increase of 90% compared with the matrix without MAH-PP and is close to the shear yield stress of the matrix (~16 MPa). However, the IFSS declined with the further increase in the concentration of MAH-PP in the matrix. So the optimal (or critical) MAH-PP concentration for the system used, was 10 wt%. Other studies reported an optimum in adhesion, which existed at a specific concentration found through examining the influence of the composition of copolymers on bonding strength (Garkhail, 2002; Van Den Oever and Bos, 1998). Van Den Oever and Peijs (1998) stated that the critical concentration results from the variation of the secondary structure and molecular connectivity (i.e. the orientation and conformation) of the polymers within the interface. The IFSS depends on the mode of fracture i.e. adhesive type or cohesive type. When the amount of compatibilizing agent is low (less than the critical concentration, ϕ_c), the limiting factor is the adhesive strength. With increasing compatibilizing agent concentration, the adhesive strength is improved through an increase in polar interactions. However, at concentrations greater than ϕ_c, the cohesive strength (i.e. within the matrix) may be reduced owing to the dense attachment of the polar molecular chains to the solid substrate, leading to a reduction in chain entanglements of polar molecules with the bulk matrix (Fig. 2.5). As a result, the cohesive strength can be lower than the adhesive strength and the interface may fail cohesively instead of adhesively i.e. through the 'interphase' or bulk matrix, leading to a poor stress transfer (Garkhail, 2002).

The effect of a transcrystalline region on composite interfacial and macromechanical properties has been also studied in various fibre reinforced thermoplastic systems with contradictory results (Folkes and Hardwick, 1987; 1990; Gati and Wagner, 1997; Pompe and Mader, 2000; Son et al., 2000; Wagner et al., 1993; Wood and Marom, 1997; Zafeiropoulos et al., 2001a; 2001b; 2001c; Zhang et al., 1996). In some instances a positive effect of the transcrystalline layer on the interfacial properties has been reported, owing

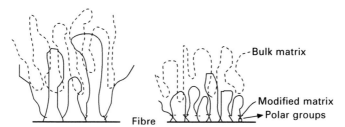

2.5 Schematic representation of the transformation in the fibre/matrix interface/interphase on increasing the polar groups in the matrix (Garkhail, 2002).

to some kind of physical coupling between the fibre and polymer matrix (Wagner *et al.*, 1993; Zafeiropoulos *et al.*, 2001a; 2001b; 2001c, Choudhury, 2008), whereas other studies report negative effects on both interfacial and macroscopic properties (Garkhail *et al.*, 2009). By applying the single fibre fragmentation test (Felix and Gatenholm, 1994; Zafeiropoulos *et al.*, 2001a; 2001b; 2001c) showed that there was an increase of the interfacial shear strength (IFSS) when transcrystallinity was present. They attributed this increase to improved interactions between the fibre and the matrix because of the transcrystalline layer. However, a number of studies reported no or even negative effects of transcrystalline interfaces on the interfacial properties of composites, especially for inorganic fibre reinforced composites (Folkes and Hardwick, 1987; 1990; Garkhail *et al.*, 2009; Gati and Wagner, 1997; Pompe and Mader, 2000; Son *et al.*, 2000; Wagner *et al.*, 1993; Wood and Marom, 1997; Zhang *et al.*, 1996).

Garkhail *et al.* (2009) investigated two different cooling conditions (quench and slow cooling), for studying the effect of transcrystallization on the IFSS in micro PP/flax composites by the fibre pull-out test method. For the PP/flax average IFSS values of 7.8 and 9.8 MPa were measured for samples with and without a transcrystalline layer, respectively, without adding any coupling agent. Both values were relatively high IFSS compared with PP/glass, probably owing to the rougher surface of the natural fibres, which can result in additional physical coupling between fibre and matrix. As can be seen, the presence of the transcrystalline layer here had a negative effect on the IFSS and led to a reduction in the stress transfer capability from the matrix to the fibre. As aforementioned these findings were in contrast to those reported for a dew-retted flax/PP system by Zafeiropoulos *et al.* (2001a). They reported IFSS values of 12 and 23 MPa for samples without and with a transcrystalline layer, respectively, with a positive effect of the transcrystalline layer on the interface properties. Although this difference could be partly caused by differences in the type of flax, PP, sample processing and data reduction, the main difference is most likely the result of the very different micromechanical tests used.

2.6   Effect of matrix modification on macroscopic properties

It is well known that fibre/matrix adhesion is very significant for the macroscopic response of fibre-reinforced composites. The most important properties when a matrix modifier is used are the strength and modulus in tension and bending. The effect of matrix modification on properties, such as fracture toughness, impact toughness, resistance to creep and fatigue has been less studied. The dimensional stability and environmental degradation of modified natural fibre composites has received considerable attention and data are mostly available on MAH-PP modified natural fibre reinforced PP composites. In general, it can be stated that compatibilization can lead to significant improvements in strength and moisture resistance. However, maximizing the interfacial strength also hinders fibre pull-out which in turn leads to a reduction in toughness. A typical natural fibre used in most of today's composites has a tensile strength of only one fourth that of glass (Garkhail *et al.*, 2000). At similar fibre/matrix adhesion levels, for example, as a result of similar compatibilization schemes, fracture processes in natural fibre composites involve fewer energy-absorption mechanisms, such as debonding and fibre pullout, than in GFRPs, leading to a more brittle fracture. Hence, interfaces in natural fibre composites need to be weaker than in GFRPs in order to obtain similar levels of toughness. Moreover, unlike isotropic glass fibres, natural fibres are highly anisotropic, (Garkhail *et al.*, 2000, Bos *et al.*, 2002; Bos, 2004) and after optimization of the interface, internal adhesion within the fibre bundle structure often becomes the weakest link to initiate failure. As is shown by Bos *et al.* (2002), the lateral strength and thus the shear strength within the fibre cell wall is lower than the strength of the fibre in the length direction. Once fibre matrix adhesion becomes stronger than the lateral or shear strength of the secondary cell wall, the lateral strength of the cell wall becomes the limiting factor, and composite strength cannot be further increased by optimization of the compatibilizer (Van Den Oever *et al.*, 2000). This has been well illustrated on the example of a flax stem by Van Den Oever *et al.* (2000) (see Fig. 2.6).

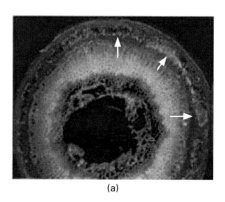
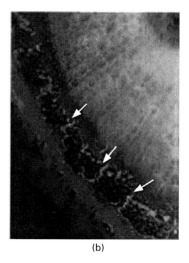

(a)   (b)

2.6 Photographs of a cross section of a flax stem (a) with the bast fibres marked and (b) with spots marked where weak technical fibre bonding is expected (Van Den Oever *et al.*, 2000).

Most of the studies on the effect of matrix modification in natural fibre composites with approximately 2–5 wt% MAH-PP report a substantial increase in the strength of the modified systems, both in tension and flexure, whereas the stiffness is less influenced (Arbelaiz *et al.*, 2005a; Arbelaiz *et al.*, 2005b; Barkoula *et al.*, 2010b; Bledzki *et al.*, 2004; Bos, 2004; Bos *et al.*, 2006; Dalväg *et al.*, 1985; Doan *et al.*, 2006; Garkhail *et al.*, 2000; Gassan and Bledzki, 1997; 1999a; Karmaker, 1997; Karmaker and Youngquist, 1996; Karmaker *et al.*, 1994; Keener *et al.*, 2004; Kim *et al.*, 2007; Manchado *et al.*, 2003; Pickering *et al.*, 2007; Snijder *et al.*, 1998; Van De Velde and Kiekens, 2001; 2003). Depending on the system studied, the amount, molecular weight and MAH functionality, improvements in composite strength of 20–100% were reported. In terms of stiffness both in tension and flexure, an increase in the range of 20% was documented. Young's modulus reflects the capability of both fibre and matrix material to transfer the elastic deformation in the case of small strains without interface fracture. Therefore, it is not surprising that the Young's modulus is less sensitive to variations in interfacial adhesion than strength which is strongly associated with debonding and interfacial failure (Doan *et al.*, 2006).

Gassan and Bledzki (1997; 1999a) showed that the impact energy decreases owing to the lower energy absorption in the interface of PP/jute composite. Gassan and Bledzki, (1999a and 1999b) investigated the influence of MAH-PP on the impact behaviour of PP/jute composites and found that damage initiation is shifted to higher loads in the case of a strong fibre/matrix adhesion. The dissipation factor was also higher for MAH-PP modified systems. It was explained that the absence of a coupling agent results in impact energy being absorbed by debonding and frictional effects at the fibre/matrix interface.

Kim *et al.* (2007) on the other hand showed increased Izod impact values with the addition of various types of MAH-PP modifiers. Dalväg *et al.* (1985) also presented the effect of different MAH-PP content on the impact properties of cellulosic thermoplastic composites, and found a remarkable improvement. Oksman and Clemons (1998) studied the optimization of the mechanical properties of PP/wood flour composites by means of the addition of various compatibilizers, in particular the impact behaviour of the composites, and they succeeded in doubling the unnotched impact strength (from 86 to 167 J m^{-1}) and notched Izod impact strength (from 26 to 54 J m^{-1}) while retaining a tensile strength of 30 MPa and a modulus of 1.9 GPa, by adding a maleated styrene–ethylene/butylene–styrene triblock copolymer (SEBS-MA). These seemingly inconsistent results on the effect of matrix modification on the impact performance can often be explained on the basis of the role of fibre length on the energy absorption mechanisms during impact. As mentioned earlier, too strong an interface leads to brittle failure, with no energy absorption, whereas too weak an interface leads to fibre pull-out without significant effort and energy dissipation. For short fibre composites with fibre lengths well below the critical fibre length, such as wood fibre composites, improved adhesion may lead to improved energy absorption through an increase in pull-out or debonding stress. Composites on the other hand based on longer fibres such as jute fibres may show embrittlement because improved adhesion leads to a reduction in pull-out length.

Sain *et al.* (2000) showed only a marginal improvement in creep properties by maleic and maleimide interfacial modification in wood fibre PP composites. A stronger effect of the interface is expected in cyclic loading under fatigue conditions, where an imperfect interface can lead to repeated frictional sliding, which induces temperature increase and energy dissipation. Gassan and Bledzki (1997) measured a significant decrease in energy dissipation during cyclic loading for jute fibre reinforced PP with improved interface from MAH-PP. Doan *et al.* (2006) reported improved thermomechanical behaviour of MAH-PP modified PP/jute, with an increase in the storage modulus values. Gassan and Bledzki (1998) created a graph (see Fig. 2.7) to provide an overview of the influence of MAH-PP coupling agent on the most discussed mechanical properties of natural fibre reinforced composites on the example of jute/PP composites (Gassan and Bledzki, 1998).

A comprehensive study on the effect of AA as a matrix modifier on the response of natural fibre composites is performed by Erasmus and Anandjiwala (2009). Similarly to the addition of MAH, AA addition leads to enhancement of the strength when used as a matrix modifier. Both

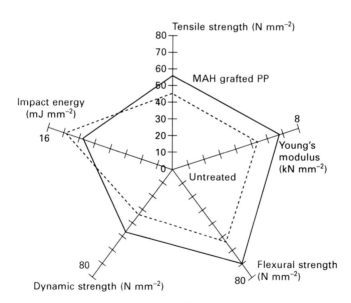

2.7 Influence of coupling agent (MAH-grafted PP) on the mechanical properties of jute PP composites (fibre content = 37 vol. %) (Gassan and Bledzki, 1998).

tensile strength and modulus increase with increasing AA content up to 2 wt% where a maximum is achieved, and then decreases with a further increase in AA content. Addition of 1 wt% AA causes a decrease in flexural strength, but as the amount of AA increases, there is an increase in flexural strength. However, as the amount of AA increases the benefits of fibre/ matrix coupling come into play and the stress is transferred from the matrix to the fibre more effectively. The drop in modulus and tensile strength at the higher concentrations could possibly be attributed to damage caused to the fibre, instead of being caused by coupling. Another possible explanation is the increased use of peroxide during the 4 wt% AA modification, which causes an increase in β-scission. A decrease in molecular weight of the PP could lead to a reduction in composite strength (Erasmus and Anandjiwala, 2009). Again, the addition of AA to the matrix has a negative effect on the Charpy impact strength. As the amount of AA increased, the composites showed a reduction in impact strength. This result is consistent with the other findings reported, namely that a too strong interaction between fibre and matrix leads to a poor impact strength (Sain *et al.*, 2005). This might indeed suggest that AA is effective in improving interfacial interaction (Erasmus and Anandjiwala, 2009). Son *et al.* (2000) found that the tensile strength of macrocomposites decreased with the presence of a transcrystalline layer and also with increasing isothermal crystallization time. To investigate the effect of transcrystallinity on the macromechanical properties of the PP/ flax composites, two different cooling conditions (quench and slow) were used and reported by Garkhail *et al.* (2009). PP/flax composites had a lower modulus and strength with the presence of a transcrystalline interphase. This was attributed to two causes: first, the weaker fibre/matrix interface, as also measured by the IFSS test; second, the different morphology of the PP matrix itself in composites processed using different cooling conditions (fast quench cooled versus slowly cooled).

Several studies looked at the effect of matrix modification on the environmental response of natural fibre reinforced plastics. As mentioned before, one of the major restrictions in the applicability of these composites comes from the fact that these materials are very susceptible to moisture. This results in water uptake and dimensional instability (swelling), and the fibre/ matrix interface is one of the weakest links in this process, in addition to the hydrophilic nature of the natural fibres. Typically, the cellulosic natural fibres swell more than the polymer matrix. Improved interfacial bonding through matrix modification resulted in reduction in the water uptake and diffusion coefficient in natural fibre composites as stated in various publications (Arbelaiz *et al.*, 2005c; Avella *et al.*, 1995; Bailie *et al.*, 2000; Espert *et al.*, 2004; Gauthier *et al.*, 1998; George *et al.*, 1998; Joly *et al.*, 1996b; Joseph *et al.*, 2002; Marcovich *et al.*, 2005; Naik and Mishra, 2006; Peijs *et al.*, 1998; Rana *et al.* 1998; Stamboulis *et al.*, 2000; Thwe and Liao, 2003). From the above studies, the reasons for this improvement can be summarized as: (a) better wetting of the fibre, (b) fewer gaps at the fibre/matrix interface, and (c) ester linkages from the reaction of MAH with the –OH groups of the fibre, resulting in a less hydrophilic fibre surface. All these effects are inter-related and lead to less moisture uptake since capillary absorption through narrow channels along the interfaces is suppressed.

2.6.1   Short versus long fibre composites

In most engineering applications natural fibres are found in the form of short or long discontinuous reinforcements in the polymer matrix. Two manufacturing routes are mainly utilized, (i) the mat technology, where non-woven natural-fibre mats are compression moulded with the polymer matrix to produce random natural-fibre mat reinforced thermoplastic composites (NMT) (Garkhail *et al.*, 2000; Heijenrath and Peijs, 1996; Jolly and Jayaraman, 2006; Nechwatal *et al.*, 2005; Peijs *et al.*, 1998; Stamboulis *et al.*, 2000; Van Den Oever *et al.*, 2000; Wielage *et al.*, 2003) and (ii) the granule technology where natural fibre reinforced polymer granules are injection moulded or extrusion compression moulded (ECM) to produce mainly short fibre reinforced composites (Arbelaiz *et al.*, 2005a; 2005b; Aurich and Mennig, 2001; Bos *et al.*, 2006; Cantero *et al.*, 2003; Hull, 1981; Li and Sain, 2003; Nyström *et al.*, 2007; Wielage *et al.*, 2003). In discontinuous fibre reinforced composites, the parameters that influence the macroscopic response are the intrinsic fibre properties, their length and diameter, and the fibre/matrix adhesion (Hull, 1981; Nechwatal *et al.*, 2005; Van Den Oever and Bos, 1998). There is a critical fibre length below which no effective reinforcement can be expected. This critical fibre length can be predicted based on micromechanical models and is different for stiffness, strength and impact properties (Thomason and Vlug, 1996; Thomason *et al.*, 1996; Nechwatal *et al.*, 2005).

The critical fibre length is interrelated with the fibre/matrix adhesion, and more specifically is dependent on the IFSS. The critical fibre length (L_c) can be calculated using the Kelly–Tyson theory (Kelly and Tyson, 1965):

$$L_c = \frac{\sigma_f d}{2\tau}$$

where σ_f is the fibre strength, d is the fibre diameter, and τ is the interfacial shear strength.

It is obvious that the improvement of the fibre/matrix adhesion with matrix modification results in lower critical fibre lengths required for improvement of the composite's macroscopic response. The influence of interface compatibilizer was studied in compression moulded PP/flax, where no significant improvements on the mechanical properties (Garkhail *et al.*, 2000; Nechwatal *et al.*, 2005) were found with the addition of compatibilizers.

On the other hand, for injection-moulded PP/flax composites, positive effects of coupling agents on final mechanical properties were reported in numerous studies (Arbelaiz *et al.*, 2005a; 2005b; Bos *et al.*, 2006; Cantero *et al.*, 2003; Li and Sain, 2003; Nechwatal *et al.*, 2005; Nyström *et al.*, 2007; Wielage *et al.*, 2003). These studies conclude that an adhesion promoter based on MAH-PP behaves as a true coupling agent, i.e. improves the mechanical performance of the PP/flax composites (Arbelaiz *et al.*, 2005a; 2005b; Bos *et al.*, 2006; Cantero *et al.*, 2003; Li and Sain, 2003; Nechwatal *et al.*, 2005; Nyström *et al.*, 2007). This is because NMTs contain fibres that are long enough to provide reinforcement ($L_f > L_c$), which is not the case in injection moulded composites, where normally shorter fibres are found compared with compression moulded composites ($L_f < L_c$). Therefore, fibre/matrix adhesion becomes more critical.

In a recent study by Barkoula *et al.* (2010b) the influence of MAH-PP concentration on flax/PP composites, which were manufactured through injection moulding (compounding through kneading), was investigated. The materials investigated consisted of composites with 28 wt% of short flax fibres. As reference a pure PP matrix was used.

The tensile modulus of the injection moulded MAH-PP/flax composites is given in Fig. 2.8 which shows that the modulus results are approximately constant, 6.2 and 1.7 GPa for the 28 wt% flax series and the reference PP material, respectively. As expected no influence of fibre/matrix adhesion on composite stiffness was observed.

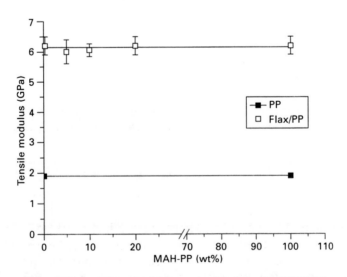

2.8 Tensile modulus of injection moulded flax/PP/MAH-PP composites with varying amount of MAH-PP content in the matrix (28 wt% flax fibres) (Barkoula *et al.*, 2010b).

The tensile strength values of the injection moulded flax/PP/MAH-PP composites are given in Fig. 2.9. The matrix tensile strength shows a constant value of around 30 MPa. On increasing the amount of MAH-PP in the flax/PP/MAH-PP composite a maximum tensile strength was observed for 20 wt% MAH-PP. This behaviour is in line with a previous observation on the effect of MAH-PP on the mechanical properties of GFRPP (Van Den Oever and Peijs, 1998), where first an increase in strength with compatibilizer is found, followed by a subsequent decrease at higher fractions of MAH-PP. An optimum in interfacial shear strength was also observed for MAH-PP/flax systems. However, here an optimum was found at 10 wt% MAH-PP content in microcomposites, measured through micro-debond tests (Mieck *et al.*, 1995b). The observed optimum in micro-debond tests is not directly reflected in the macro-composite properties, which may be because of effects related to fibre volume fractions (single fibre versus bulk composite). However, such high weight fractions of MAH-PP are not economical when a balance in price and property enhancement is desired. Such balance can, however, be obtained for 3–5 wt% addition of MAH-PP in PP.

A comparison was also made with results on long fibre NMTs published previously (Garkhail *et al.*, 2000). Figures 2.10 and 2.11 show the stiffness and strength of compression moulded NMT (Garkhail *et al.*, 2000) and injection moulded (compounding done through kneading) flax/3 wt% MAH-PP composites, respectively. In spite of a reduction in fibre length during injection moulding it can be seen that injection moulded samples showed tensile properties similar to those of NMT (fibre length ~25 mm) compression moulded composites. This could be because of reasons like: (a) improved fibre efficiency because of dimensional changes, i.e. a reduction in fibre

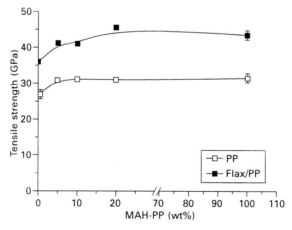

2.9 Tensile strength of injection moulded flax/PP/MAH-PP composites with varying amount of MAH-PP content in the matrix (28 wt% flax fibres) (Barkoula *et al.*, 2010b).

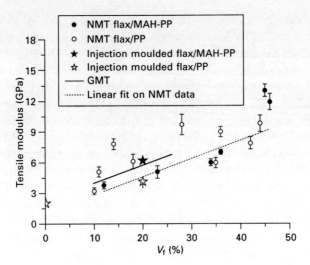

2.10 Tensile modulus of PP/flax composites manufactured through compression moulding of NMT (Garkhail *et al.,* 2000) and injection moulding as a function of fibre volume fraction. Also shown is glass fibre based GMT (Garkhail *et al.,* 2000).

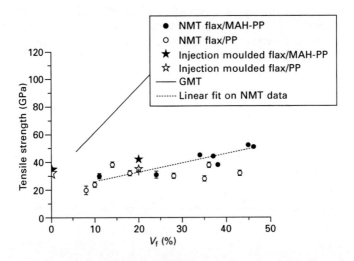

2.11 Tensile strength of PP/flax composites manufactured through compression moulding of NMT (Garkhail *et al.,* 2000) and injection moulding as a function of fibre volume fraction. Also shown is glass fibre based GMT (Garkhail *et al.,* 2000; Barkoula *et al.,* 2010b).

diameter through fibre opening during compounding and (b) changes in fibre orientation along the direction of polymer flow in the case of injection moulded composites. Dimensional changes can play an important role since the flax fibres used for composite reinforcement are often actually fibre bundles (so-called technical fibres) consisting of fibre cells (so-called elementary fibres) (Garkhail *et al.*, 2000; 2009). Also, it has been shown that the fibre tensile strength is strongly dependent on the fibre length (Thomason *et al.*, 1996) and therefore a reduction in fibre length could have led to a separation of these fibre bundles into fibre cells, hence leading to improved fibre efficiency through enhanced fibre tensile properties (of fibre cells) or improved fibre aspect ratio.

The above claims can be better understood via the calculation of the critical fibre length for technical and elementary flax fibres. For this calculation, strength values are taken from the literature (technical fibre: 800 MPa (Heijenrath and Peijs, 1996), elementary fibre: 1500 MPa (Heijenrath and Peijs, 1996), for fibres with 80 and 12.5 μm diameter, respectively). The IFSS are those stated above, i.e. 8 MPa for PP/flax (Garkhail, 2002; Garkhail *et al.*, 2000; 2009; Mieck *et al.*, 1995b) and 16 MPa (Garkhail *et al.*, 2000) for MAH-PP/flax. Using these values the critical fibre length was calculated as shown in Table 2.1 (Barkoula *et al.*, 2010b). Based on this data a critical fibre length of approximately 2 mm was found in the case of MAH-PP/PP with technical flax fibres. To attain short fibre composite strengths which are equivalent to 90% of the tensile strength of a continuous fibre composite, fibre lengths $>5L_c$ are required (Thomason and Vlug, 1996). This means that, for effective strength improvement, fibre lengths of approximately 10 mm are required for MAH-PP/flax and of approximately 15 mm for PP/flax. These results are in agreement with Nechwatal *et al.* (2005) where optimum strength for MAH-PP/flax is obtained for approximately 8 mm fibre lengths. If elementary fibres are assumed in combination with MAH-PP matrix then $5L_c$ is equal to 3 mm. A considerable fraction of the fibres can be above 3 mm after injection moulding, and this could be the reason why injection moulded composites can have similar strength to NMTs.

In another publication (Stamboulis *et al.*, 2000) the environmental response with regard to MAH-PP addition was evaluated on three different types of NMTs: (i) treated Duralin™ flax/PP, (ii) treated Duralin™ flax/MAH-PP and (iii) untreated green flax/PP as a reference. The Duralin™ treatment,

Table 2.1 Critical fibre lengths for PP/flax and MAH-PP/flax systems with elementary and technical fibres (Barkoula *et al.*, 2010b)

	Critical fibre length	
	PP/flax (mm)	MAH-PP/flax (mm)
Elementary fibre	1.2	0.6
Technical fibre	4	2

which consisted of a steam or water-heating step of the flax fibre above 160 °C for 30 minutes and a drying/curing step above 150 °C for 2 hours, led to improved moisture resistance of cellulose fibres. The results of the moisture content as a function of the exposure time are presented in Fig. 2.12. As shown in Fig. 2.12, green flax fibre based composites are more sensitive to moisture than the other two types of composites. Also, the diffusivity of Duralin™ flax fibre composites (Table 2.2), as calculated from the initial slope of the moisture uptake curves, is much lower than that for NMTs based on green flax fibre mats. It is interesting to note that the use of MAH-PP as a compatibilizer lowers the diffusivity even further. Clearly, the initial moisture uptake in this composite system is taking place at a lower rate than for the PP system without compatibilizer. The maximum moisture content level is, however, similar for both PP and MAH-PP based Duralin™ flax systems. The higher diffusivity for the PP system without compatibilizer indicates that initially a fair amount of moisture uptake takes place along the fibre/matrix interface.

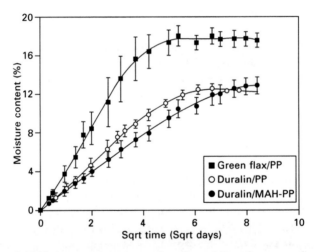

2.12 Moisture content as a function of time for green flax/PP, Duralin/PP and Duralin/MAH-PP NMT composites (Stamboulis *et al.*, 2000).

Table 2.2 Maximum moisture content and diffusivity of PP/flax composites (Stamboulis *et al.*, 2000)

Composite	Max. moisture content, $M_{m,c}$ (%)	Diffusivity, D_c (cm^2 s^{-1})
PP/green flax	18.0	1.3×10^{-2}
PP/Duralin™ flax	12.8	7.8×10^{-3}
MAH-PP/Duralin™ flax	13.5	5.0×10^{-3}

2.7   Future trends

The most recent trends in natural fibre composites comprise the introduction of all-cellulose composites and cellulose nanofibres for nanocomposites. Although research in the area of cellulose nanofibres and nanocomposites has shown that these materials have an exciting potential as reinforcement in nanocomposites (Berglund and Peijs, 2010; Eichhorn *et al.*, 2010), they do not offer any particular solution to the problem of low fibre/matrix adhesion. Moreover, it can be said that since the surface area increases substantially when fibres are of nanosize, the importance of fibre/matrix adhesion and particularly dispersion is even greater in these materials. Moreover, so far most cellulose nanofibres that have been produced exhibit fairly low aspect ratios, implying that interface engineering in these materials will become even more important in order to achieve high reinforcement efficiencies. On the other hand all-cellulose composites might offer alternative solutions towards better fibre/matrix adhesion.

Recently, a completely new route to cellulose-based composites with good interfacial adhesion was proposed by the groups of Nishino and Peijs. (Nishino *et al.*, 2004; Nishino and Arimato, 2007; Qin *et al.*, 2008; Soykeabkaew *et al.*, 2008; 2009a; 2009b). They focused on approaches following self-reinforced polymer concepts for thermoplastic fibres (Alcock *et al.*, 2006a; 2006b; 2007). Such an approach was followed to create high-strength all-cellulose composites, having the essential advantages of being fully biobased and biodegradable materials. These composites exhibited excellent mechanical properties, e.g. a tensile strength of approximately 500 MPa which compared very favourably to more traditional unidirectional natural fibre based composites (Heijenrath and Peijs, 1996; Van Den Oever *et al.*, 2000; Oksman *et al.*, 2002; Liu *et al.*, 2005). Moreover, because both fibre and matrix are composed of the same material they show excellent interfacial compatibility. During composite preparation, the surface layer of the cellulose fibres is partially dissolved to form the matrix phase of 'all-cellulose' composites. The developed surface dissolution method results not only in very high fibre volume fractions, but also in a gradual interphase or interfacial region, which minimizes voids and stress concentrations, and can even lead to transparent composites.

2.8   Sources of further information and advice

Dalväg *et al.* (1985) carried out one of the first studies on the efficiency of cellulosic fillers in common thermoplastics and the effect of processing aids and coupling agents. Felix and Gatenholm (1991) published a seminal study on the nature of adhesion in composites of modified cellulose fibres and PP. Mieck *et al.*, (1995a, 1995b) studied fibre/matrix adhesion in flax fibre reinforced thermoplastics; one study was on the effect of silane modification and the second on the use of functionalized PP. Bledzki and Gassan (1999) provided a review on natural fibre reinforced plastics and their potential application. Lu *et al.* (2001) published a review of coupling agents and treatments for natural fibre composites. George *et al.* (2001b) and Mohanty *et al.* (2001) have compiled reviews on interfacial modifications in natural-fibre composites and their effects on the composite properties. Another important study is the one by Nechwatal *et al.* (2005) on the processing technologies used for the addition of the MAH in natural fibre reinforced composites. Important studies conducted by Bos and coworkers are summarized by Bos (2004). Current international research into cellulose composites is well documented by Berglund and Peijs (2010) and Eichhorn *et al.* (2010).

Interesting books and book chapters include the following:

(a) *Handbook of polypropylene and polypropylene composites*, ed. Karian HG, second edition, Chapter 3: Chemical coupling agents for filled and grafted polypropylene composites (Roberts and Constable, 2003), which provides a good review of coupling agents for PP and the chemistry behind grafting.

(b) *Properties and performance of natural-fibre composites*, ed. K Pickering, Woodhead Publishing Limited, which provides an overview of the types of natural fibres used in composites, a discussion of the fibre-matrix interface and how it can be engineered to improve performance, an examination of the increasing use of natural-fibre composites in automotive and structural engineering and the packaging and energy sector and, finally, a consideration of the methods used to assess the general mechanical performance of natural fibre composites.

More useful information can be obtained at the website
http://www.nova-institut.de/.

2.9   References

Alcock B, Cabrera NO, Barkoula N-M, Loos J, Peijs T (2006a), 'The mechanical properties of unidirectional all-polypropylene composites', *Compos Pt A-Appl Sci Manuf*, **37**(5), 716–726.

Alcock B, Cabrera NO, Barkoula N-M, Peijs T (2006b), 'Low velocity impact performance of recyclable all-polypropylene composites', *Compos Sci Technol*, **66**(11–12), 1724–1737.

Alcock B, Cabrera NO, Barkoula N-M, Spoelstra AB, Loos, J, Peijs T (2007), 'The mechanical properties of woven tape all-polypropylene composites', *Compos Pt A: Appl Sci Manuf*, **38**(1), 147–161.

Araújo JR, Waldman WR, De Paoli MA (2008), 'Thermal properties of high density polyethylene composites with natural fibres: coupling agent effect', *Polym Degrad Stabil*, **93**(10), 1770–1775.

Arbelaiz A, Cantero G, Fernández B, Gañán P, Kenny JM, Mondragon I (2005c), 'Flax fibre surface modifications, effect on fibre physico mechanical and flax/polypropylene interface properties', *Polym Compos*, **26**, 324–332.

Arbelaiz A, Fernández B, Cantero G, Llano-Ponte R, Valea A, Mondragon I (2005a), 'Mechanical properties of flax fibre/polypropylene composites. Influence of fibre/ matrix modification and glass fibre hybridization', *Compos Pt A: Appl Sci Manuf*, **36**(12), 1637–1644.

Arbelaiz A, Fernández B, Ramos JA, Retegi A, Llano-Ponte R, Mondragon I (2005b), 'Mechanical properties of short flax fibre bundle/polypropylene composites: Influence of matrix/fibre modification, fibre content, water uptake and recycling', *Compos Sci Technol*, **65**(10), 1582–1592.

Aurich T, Mennig G (2001), 'Flow-induced fiber orientation in injection molded flax fiber reinforce polypropylene', *Polym Compos*, **22**(5), 680–689.

Avella M, Bozzi C, dell' Erba R, Focher B, Marzetti A, Martuscelli E (1995), 'Steam-exploded wheat straw fibres as reinforcing material for polypropylene-based composites', *Angew Makromol Chem*, **233**, 149–166.

Averous L, Boquillon N (2004), 'Biocomposites based on plasticized starch: thermal and mechanical behaviours', *Carbohydr Polym*, **56**, 111–122.

Baillie C, Tual D, Terraillon JC (2000), 'Interfacial pathways in wood', *Adv Compos Lett*, **9**, 45–57.

Barkoula NM, Garkhail SK, Peijs T (2010a), 'Biodegradable composites based on flax/ polyhydroxybutyrate and its copolymer with hydroxyvalerate', *Ind Crop Prod*, **31**, 34–42.

Barkoula NM, Garkhail SK, Peijs T (2010b), 'Effect of compounding and injection moulding on the mechanical properties of flax fibre polypropylene composites', *J Reinf Plast Comp*, **29**(9), 1366–1385.

Bax B, Muessig J (2008), 'Review: Impact and tensile properties of PLA/Cordenka and PLA/flax composites', *Compos Sci Technol*, **68**, 1601–1607.

Belgacem MN, Bataille P, Sapieha S (1994), 'Effect of corona modification on the mechanical properties of polypropylene/cellulose composites', *J Appl Polym Sci*, **53**, 379–385.

Berglund LA, Peijs T (2010), 'Cellulose biocomposites: From bulk moldings to nanostructured systems', *MRS Bull*, **35**(3), 201–207.

Bledzki AK, Faruk O (2004), 'Creep and impact properties of wood fibre–polypropylene composites: Influence of temperature and moisture content', *Compos Sci Technol*, **64**, 693–700.

Bledzki AK, Faruk O (2006), 'Injection moulded microcellular wood fibre-polypropylene composites', *Compos Pt A: Appl Sci Manuf*, **37**(9), 1358–1367.

Bledzki AK, Faruk O, Sperber VE (2006), 'Cars from bio-fibres', *Angew Makromol Chem*, **291**(5), 449–457.

Bledzki AK, Fink H-P, Specht K (2004), 'Unidirectional hemp and flax EP- and PP-composites: influence of defined fiber treatments', *J Appl Polym Sci*, **93**(5), 2150–2156.

Bledzki AK, Gassan J (1999), 'Composites reinforced with cellulose based fibers', *Prog Polym Sci*, **24**(2), 221–274.

Bledzki AK, Reihmane S, Gassan J (1996), 'Properties and modification methods for vegetable fibers for natural fiber composites', *J Appl Polym Sci* **59**, 1329–1336.

Bodros E, Pillin I, Montrelay N, Baley C (2007), 'Could biopolymers reinforced by randomly scattered flax fibre be used in structural applications?', *Compos Sci Technol*, **67**(3–4), 462–470.

Bos HL (2004), 'PhD Thesis: The potential of flax fibres as reinforcement for composite materials', Eindhoven: Technische Universiteit Eindhoven.

Bos HL, Müssig J, Van Den Oever MJA (2006), 'Mechanical properties of short-flax-fibre reinforced compounds', *Compos Pt A: Appl Sci Manuf*, **37**(10), 1591–1604.

Bos HL, Van Den Oever MJA, Peters OCJJ (2002), 'Tensile and compressive properties of flax fibres for natural fibre reinforced composites', *J Mater Sci*, **37**, 1683–1692.

Brouwer W D (2001), 'Natural fibre composites, saving weight and cost with renewable materials', BD-1414. In: *Thirteenth international conference on composite materials*, Beijing, China.

Canche-Escamilla G, Cauich-Cupul JI, Mendizabal E, Puig JE, Vazquez-Torres H, Herrera-Franco PJ (1999), 'Mechanical properties of acrylate-grafted henequen cellulose fibres and their application in composites', *Compos Pt A: Appl Sci Manuf*, **30**, 349–359.

Cantero G, Arbelaiz A, Llano-Ponte R, Mondragon I (2003), 'Effects of fibre treatment on wettability and mechanical behaviour of flax/polypropylene composites', *Compos Sci Technol*, **63**(9), 1247–1254.

Chiellini E, Cinelli P, Chiellini F, Iman SH (2004), 'Environmentally degradable bio-based polymeric blends and composites', *Macromol Biosci*, **4**, 218–231.

Chen C, Peng S, Fei B, Zhuang Y, Dong L, Feng Z, Chen S, Xia H (2003), 'Synthesis and characterization of maleated poly(3-hydroxybutyrate)', *J Appl Polym Sci*, **88**(3), 659–668.

Choudhury A (2008), 'Isothermal crystallization and mechanical behavior of ionomer treated sisal/HDPE composites', *Mat Sci Eng A: Struct*, **491**(1–2), 492–500.

Corradini E, Ito EN, Marconcini JM, Rios CT, Agnelli JAM, Mattoso LHC (2009), 'Interfacial behaviour of composites of recycled poly(ethylene terephthalate) and sugarcane bagasse fiber', *Polym Test*, **28**(2), 183–187.

Dalväg H, Klason C, Strömvall H-E (1985), 'The efficiency of cellulosic fillers in common thermoplastics. Part II. Filling with processing aids and coupling agents', *Int J Polym Mater*, **11**(I), 9–38.

Doan T-T-L, Gao S-L, Mäder E (2006), 'Jute/polypropylene composites I. Effect of matrix modification', *Compos Sci Technol*, **66**(7–8), 952–963.

Duchemin B, Newman R, Staiger M (2007), 'Phase transformations in microcrystalline cellulose due to partial dissolution', *Cellulose*, **14**(4), 311–320.

Eichhorn SJ, Dufresne A, Aranguren M, Marcovich NE, Capadona JR, Rowan SJ, Weder C, Thielemans W, Roman M, Renneckar S, Gindl W, Veigel S, Keckes J, Yano H, Abe K, Nogi M, Nakagaito AN, Mangalam A, Simonsen J, Benight AS, Bismarck A, Berglund LA, Peijs T (2010), 'Review: current international research into cellulose nanofibres and nanocomposites', *J Mater Sci*, **45**(1), 1–33.

Eisele D (1994), 'Faserhaltige Bauteile für die Automobilausstattung Zur Leistungsfähigkeit von Naturfasern', *Textil Praxis Int*, **49**(1–2), 68–76.

Erasmus E, Anandjiwala R (2009), 'Studies on enhancement of mechanical properties and interfacial adhesion of flax reinforced polypropylene composites', *J Thermoplast Compos*, **22**(5), 485–502.

Espert A, Vilaplana F, Karlsson S (2004), 'Comparison of water absorption in natural cellulosic fibres from wood and one-year crops in polypropylene composites and its influence on their mechanical properties', *Compos Pt A: Appl Sci Manuf*, **35**(11), 1267–1276.

Felix JM., Gatenholm P (1991), 'The nature of adhesion in composites of modified cellulose fibers and polypropylene', *J Appl Polym Sci*, **42**(3), 609–620.

Felix JM, Gatenholm P (1994), 'Effect of transcrystalline morphology on interfacial adhesion in cellulose/polypropylene composites', *J Materials Sci*, **29**(11), 3043–3049.

Felix JM, Gatenholm P, Schreiber HP (1993), 'Controlled interactions in cellulose–polymer composites. I. Effect on mechanical properties', *Polym. Composites*, **14**(6), 449–457.

Feng D, Caulfield DF, Sanadi AR (2001), 'Effect of compatibilizer on the structure–property relationships of kenaf-fiber/polypropylene composites', *Polym Compos*, **22**(4), 506–517.

Folkes MJ, Hardwick ST (1987), 'Direct study of the structure and properties of transcrystalline layers', *J Mater Sci Lett*, **6**, 656–658.

Folkes MJ, Hardwick ST (1990), 'The mechanical properties of glass/polypropylene multilayer laminates', *J Mater Sci*, **25**, 2598–2606.

Fung KL, Li RKY, Tjong SC (2002), 'Interface modification on the properties of sisal fiberreinforced polypropylene composites', *J Appl Polym Sci*, **85**, 169–176.

Gamstedt EK, Almgren KM (2007), 'Natural fibre composites – with special emphasis on effects of the interface between cellulosic fibres and polymers', Proceedings of the 28th Riso International Symposium on Materials Science: *Interface Design of Polymer Matrix Composites – Mechanics, Chemistry, Modelling and Manufacturing*, Editors: B. F. Sorensen, L. P. Mikkelsen, H. Lilholt, S. Goutianos, F. S. Abdul-Mahdi, Riso National Laboratory, Roskilde, Denmark.

Garkhail, S. (2002), PhD Thesis, Queen Mary University of London, Chapter 3.

Garkhail S, Wieland B, George J, Soykeabkaew N, Peijs T (2009), 'Transcrystallisation in PP/flax composites and its effect on interfacial and mechanical properties', *J Mater Sci*, **44**(2), 510–519.

Garkhail SK, Heijenrath RWH, Peijs T (2000), 'Mechanical properties of natural-fibre mat-reinforced thermoplastics based on flax fibres and polypropylene', *Appl Compos Mater*, **7**(5–6), 351–372.

Gassan J (2002), 'A study of fibre and interface parameters affecting the fatigue behaviour of natural fibre composites', *Compos Pt A: Appl Sci Manuf*, **33**(3), 369–374.

Gassan J, Bledzki AK (1997), 'The influence of fiber-surface treatment on the mechanical properties of jute–polypropylene composites', *Compos Pt A: Appl Sci Manuf*, **28**(12), 1001–1005.

Gassan J, Bledzki AK (1998), 'Possibilities to improve the properties of natural fiber reinforced plastics by fiber modification', *In Proceedings: ECCM-8*, Ed. I. Crivelli Visconti, 111–118.

Gassan J, Bledzki AK (1999a), 'Influence of fiber surface treatment on the creep behavior of jute fiber-reinforced polypropylene', *J Thermoplast Compos*, **12**(5), 388–398.

Gassan J, Bledzki AK (1999b), 'Possibilities for improving the mechanical properties of jute/epoxy composites by alkali treatment of fibres', *Compos Sci Technol*, **59**, 1303–1309.

Gassan J, Bledzki AK (2000), 'Possibilities to improve the properties of natural fibre reinforced plastics by fibre modification-jute polypropylene composites', *Appl Compos Mater*, **7**, 373–385.

Gatenholm P, Kubat J, Mathiasson A (1992), 'Biodegradable natural composites. I. Processing and properties', *J Appl Polym Sci*, **45**, 1667–1677.

Gatenholm P, Mathiasson A (1994), 'Biodegradable natural composites. II. Synergistic effects of processing cellulose with PHB', *J Appl Polym Sci*, **51**, 1231–1237.

Gati A, Wagner HD (1997), 'Stress transfer efficiency in semicrystalline-based composites comprising transcrystalline interlayers', *Macromolecules*, **30**, 3933–3938.

Gauthier R, Joly C, Compas A, Gauthier H, Escoubes M (1998), 'Interfaces in polyolefin/cellulosic fibre composites; chemical coupling, morphology, correlation with adhesion and ageing in moisture', *Polym Compos*, **19**, 287–300.

Gaylord NG, Mishra MK (1983), 'Nondegradative reaction of maleic anhydride and molten polypropylene in the presence of peroxides', *J Polym Sci: Polym Lett Ed*, **21**, 23–30.

Gea S, Bilotti E, Reynolds CT, Soykeabkeaw N, Peijs T (2010), 'Bacterial cellulose-poly(vinyl alcohol) nanocomposites prepared by an in situ process', *Mater Lett*, **64**(8), 901–904.

Gea S, Torres FG, Troncoso OP, Reynolds CT, Vilasecca F, Iguchi M, Peijs T (2007), 'Biocomposites based on bacterial cellulose and apple and radish pulp', *Int Polym Proc*, **22**(5), 497–501.

George J, Bhagawan SS, Thomas S (1998), 'Effects of environment on the properties of low-density polyethylene composites reinforced with pineapple-leaf fibre', *Compos Sci Technol*, **58**, 1471–1485.

George J, Klompen ETJ, Peijs T (2001a), 'Thermal degradation of green and upgraded flax fibres', *Adv Compos Lett*, **10**(2), 81–88.

George J, Sreekala MS, Thomas S (2001b), 'A review on interface modification and characterization of natural fiber reinforced plastic composites', *Polym Eng Sci*, **41**(9), 1471–1485.

Gindl W, Keckes J (2005), 'All-cellulose nanocomposite', *Polymer*, **46**(23), 10221–10225.

Gindl W, Martinschitz KJ, Boesecke P, Keckes J (2006), 'Structural changes during tensile testing of an all-cellulose composite by in situ synchrotron x-ray diffraction', *Compos Sci Technol*, **66**(15), 2639–2647.

Grondahl L, Chandler-Temple A, Trau M (2005), 'Polymeric grafting of acrylic acid onto poly(3-hydroxybutyrate-co-valerate): surface functionalization for tissue engineering applications', *Biomacromolecules*, **6**, 2197–2203.

Gray DG (1974), 'Polypropylene transcrystallization at the surface of cellulose fibres', *J Polym Sci Polymer Lett Ed*, **12**, 645–650.

Heijenrath R, Peijs T (1996), 'Natural-fibre-mat-reinforced thermoplastic composites based on flax fibres and polypropylene', *Adv Compos Lett*, **5**(3), 81–85.

Heinen W, Rosenmoler CH, Wenzel CB, de Groot HJM, Lugtenburg J and zan Duin M (1996), '^{13}C NMR study of the grafting of maleic anhydride onto polyethylene, polypropylene, and ethene–propene copolymers', *Macromolecules*, **1151**, 29.

Hornsby PR, Hinrichsen E, Tarverdi K (1997), 'Preparation and properties of polypropylene composites reinforced with wheat and flax straw fibres. Part I. Fibre characterization', *J Mater Sci*, **32**(2), 443–449.

Hull D (1981), *An introduction to composite materials*, Cambridge University Press, Cambridge.

Huneault MA, Li H (2007), 'Morphology and properties of compatibilized polylactide/thermoplastic starch blends', *Polymer*, **48**(1), 270–280.

Jayaraman K (2003), 'Manufacturing sisal-polypropylene composites with minimum fibre degradation', *Compos Sci Technol*, **63**(3–4), 367–374.

Jiang L, Morelius E, Zhang J, Wolcott M, Holbery J (2008), 'Study of the poly(3-hydroxybutyrate-co-3-hydroxyvalerate)/cellulose nanowhisker composites prepared by solution casting and melt processing', *J Compos Mater*, **42**, 2629–2645.

John MJ, Thomas S (2008), 'Biofibres and biocomposites', *Carbohydr Polym*, **71**(3), 343–364.

Jolly M, Jayaraman K (2006), 'Manufacturing flax fibre-reinforced polypropylene composites by hot-pressing', *Int J Mod Phys B*, **20–25**(26–27), 4601–4606.

Joly C, Gauthier R, Chabert B (1996a), 'Physical chemistry of the interface in polypropylene/cellulosic-fibre composites', *Compos Sci Technol*, **56**(7), 761–765.

Joly C, Kofman M, Gauthier R (1996b), 'Polypropylene/cellulosic fiber composites: chemical treatment of the cellulose assuming compatibilization between the two materials', *JMS Pure Appl Chem*, **A33**, 1981–1996.

Joseph PV, Joseph K, Thomas S, Pillai CKS, Prasad VS, Groeninckx G, Sarkissova M (2003), 'The thermal and crystallisation studies of short sisal fibre reinforced polypropylene composites', *Compos Pt A: Appl Sci Manuf*, **34**, 253–266.

Joseph PV, Rabello MS, Mattoso LHC, Joseph K, Thomas S (2002), 'Environmental effects on the degradation behaviour of sisal fibre reinforced polypropylene composites', *Compos Sci Technol*, **62**, 1357–1372.

Joshi S V, Drzal LT, Mohanty AK, Arora S (2004), 'Are natural fiber composites environmentally superior to glass fiber reinforced composites?' *Compos Pt A: Appl Sci Manuf*, **35**, 371–376.

Karmaker AC (1997), 'Effect of water absorption on dimensional stability and impact energy of jute fibre reinforced polypropylene', *J Mater Sci Lett*, **16**, 462–464.

Karmaker AC, Hoffmann A, Hinrichsen G (1994), 'Influence of water uptake on the mechanical properties of jute fiber-reinforced polypropylene', *J Appl Polym Sci*, **54**, 1803–1807.

Karmaker AC, Youngquist JA (1996), 'Injection molding of polypropylene reinforced with short jute fibres', *J Appl Polym Sci*, **62**, 1147–1151.

Karus M, Ortmann S, Vogt D (2005), 'All natural on the inside? Natural fibre in automotive interiors', *Kunststoffe Plast Europe*, **7**, 1–3.

Kazayawoko M, Balatinecz JJ, Woodhams RT, Law S (1997), 'Effect of ester linkages on mechanical properties of wood fibre-polypropylene composites', *Reinf Plast Compos*, **16**, 1383–1392.

Keener TJ, Stuart RK, Brown TK (2004), 'Maleated coupling agents for natural fibre composites', *Compos Pt A: Appl Sci Manuf*, **35**(3), 357–362.

Keller A (2003), 'Compounding and mechanical properties of biodegradable hemp fibre composites', *Compos Sci Technol*, **63**(9), 1307–1316.

Kelly A, Tyson WR (1965), 'Tensile properties of fibre-reinforced metals: copper/tungsten and copper/molybdenum', *J Mech Phys Solids*, **13**(6), 329–350.

Kim H-S, Lee B-H, Choi S-W, Kim S, Kim H-J (2007), 'The effect of types of maleic anhydride-grafted polypropylene (MAPP) on the interfacial adhesion properties of bio-flour-filled polypropylene composites', *Compos Pt A: Appl Sci Manuf*, **38**(6), 1473–1482.

Li X, Tabil LG, Panigrahi S (2007), 'Chemical treatments of natural fiber for use in natural fiber-reinforced composites: a review', *J Polym Environ*, **15**(1), 25–33.

Li H, Sain MM (2003), 'High stiffness natural fiber-reinforced hybrid polypropylene composites', *Polym-Plast Technol Eng*, **42**(5), 853–862.

Liu W, Misra M, Askeland P, Drzal LT, Mohanty AK (2005), 'Green composites from soy based plastic and pineapple leaf fiber: fabrication and properties evaluation', *Polymer*, **46**, 2710–2721.

Long Y, Shanks RA (1996), 'PP/elastomer/filler hybrids. II. Morphologies and fracture', *J Appl Polym Sci*, **62**, 639–646.

Lu JZ, Wu Q, McNabb HS (2000), 'Chemical coupling in wood fiber and polymer composites: A review of coupling agents and treatments', *Wood Fiber Sci*, **32**(1), 88–104.

Lu JZ, Wu Q, Negulescu I (2001), 'The influence of maleation on polymer adsorption and fixation, wood surface wetability, and interfacial bonding strength in wood-PVC composites', *Wood Fibre Sci*, **34**(3), 434–459.

Maldas D, Kokta BV (1990), 'Influence of polar monomers on the performance of wood fiber reinforced polystyrene composites. I. Evaluation of critical conditions', *Int J Polym Mater*, **14**(3–4), 165–189.

Maldas D, Kokta BV (1991), 'Influence of maleic anhydride as a coupling agent on the performance of wood fiber-polystyrene composites', *Polym Eng Sci*, **31**(18), 1351–1357.

Maldas D, Kokta BV, Daneaulf C (1989), 'Influence of coupling agents and treatments on the mechanical properties of cellulose fiber-polystyrene composites', *J Appl Polym Sci*, **37**, 751–775.

Manchado MAL, Arroyo M, Biagiotti J, Kenny JM (2003), 'Enhancement of mechanical properties and interfacial adhesion of PP/EPDM/flax fiber composites using maleic anhydride as a compatibilizer', *J Appl Polym Sci*, **90**(8), 2170–2178.

Marcovich NE, Reboredo MM, Aranguren MI (2005), 'Lignocellulosic materials and unsaturated polyester matrix composites: Interfacial modifications', *Compos Interfaces*, **12**, 3–24.

Mieck KP, Lutzkendorf R, Reussmann T (1996), 'Needle-punched hybrid nonwovens of flax and PP fibers: textile semiproducts for manufacturing of fiber composites', *Polym Compos*, **17**, 873–878.

Mieck KP, Nechwatal A, Knobelsdorf C (1994), 'Anwendungsmöglichkeiten von Naturfasern bei Verbundmaterial', *Melliand Textilberichte*, **75**(11), 892–898.

Mieck KP, Nechwatal A, Knobelsdorf C (1995a), 'Faser-matrix-haftung in kunststoffverbunden aus thermoplastischer matrix und flachs. 1. Die ausrüstung mit silanen', *Angew Makromol Chem*, **224**(1), 73–88.

Mieck KP, Nechwatal A, Knobelsdorf C (1995b), 'Faser-matrix-haftung in kunststoffverbunden aus thermoplastischer matrix und flachs, 2. Die anwendung von funktionalisiertem polypropylen', *Angew Makromol Chem*, **225**(1), 37–49.

Mieck KP, Reußmann T (1995), 'Flachs versus Glas', *Kunststoffe*, **85**(3) 366–370.

Mildner I, Bledzki AK (1999), In: *Proceedings of the second international wood natural fiber composites symposium*, Kassel, Germany, 17.

Mitomo H, Enjôji T, Watanabe Y, Yoshii F, Makuuchi K, Saito T (1995), 'Radiation-induced graft polymerization of poly(3-hydroxybutyrate) and its copolymer', *J Macromol Sci A*, **32**: 3, 429–442.

Mishra S, Misra M, Tripathy SS, Nayak SK, Mohanty AK (2001), 'Graft copolymerization of acrylonitrile on chemically modified sisal fibers', *Macromol Mater Eng*, **286**(2), 107–113.

Mishra S, Naik J, Patil Y (2000), 'The compatibilizing effect of maleic anhydride on swelling and mechanical properties of plant-fibre reinforced novolac composites', *Compos Sci Technol*, **60**, 1729–1735.

Mohanty AK, Misra M, Drzal LT (2001), 'Surface modifications of natural fibers and performance of the resulting biocomposites: An overview', *Compos Interfaces*, **8**, 663–699.

Mohanty AK, Misra M, Drzal LT (2002), 'Sustainable bio-composites from renewable recourses: opportunities and challenges in the green materials world', *J Polym Environ*, **10**(1–2), 19–26.

Mohanty AK, Misra M, Hinrichsen G (2000), 'Biofibres, biodegradable polymers and biocomposites: an overview', *Macromol Mater Eng*, **276/277**, 1–24.

Mwaikambo LY, Ansell MP (2003), 'Hemp fibre reinforced cashew nut shell liquid composites', *Compos Sci Technol*, **63**, 1297–1305.

Myers GE, Chahyadi IS, Coberly CA, Ermer DS (1991a), 'Wood flour polypropylene composites. Influence of maleated polypropylene aand process and composition variables on mechanical properties', *Int J Polym Mater*, **15**(1), 21–44.

Myers GE, Chahyadi IS, Gonzalez C, Coberly CA, Ermer DS (1991b), 'Wood flour and polypropylene or high-density polyethylene composites-influence of maleated polypropylene concentration and extrusion temperature on properties', *Int J Polym Mater*, **15**(3–4), 171–186.

Myers EG, Kolosick PC, Chahyadi IS, Coberly CA, Koutsky JA, Ermer DS (1990), 'Extruded wood-flour polypropylene composites: effect of a maleated polypropylene coupling agent on filler–matrix bonding and properties', *Materials Research Society Symposium Proceedings*, **97**, 67–77.

Naik JB, Mishra S (2006), 'The compatibilizing effect of maleic anhydride on swelling properties of plant-fiber-reinforced polystyrene composites', *Polym Plast Technol Eng*, **45**, 923–927.

Nechwatal A, Reussmann T, Böhm S, Richter E (2005), 'The dependence between the process technologies and the effect of MAH-PP-adhesives in natural fibre reinforce thermoplastic composites', *Adv Eng Mater*, **7**(1–2), 68–73.

Neilsen LE (1977), *Polymer rheology*, Marcel Dekker, Inc, New York.

Nishino T, Arimoto N (2007), 'All-cellulose composite prepared by selective dissolving of fiber surface', *Biomacromolecules*, **8**(9), 2712–2716.

Nishino T, Matsuda I, Hirao K (2004), 'All-cellulose composite', *Macromolecules*, **37**, 7683–7687.

Nishino T, Nakamae K, Hirao K, Kotera M, Inagaki H (2003), 'Kenaf reinforced biodegradable composite', *Compos Sci Technol*, **63**, 1281–1286.

Nyström B, Joffe R, Långström R (2007), 'Microstructure and strength of injection molded natural fiber composites', *Reinf Plast Compos*, **26**(6), 579–599.

Oksman K, Clemons C (1998), 'Mechanical properties and morphology of impact modified polypropylene-wood flour composites', *J Appl Polym Sci*, **67**(9), 1503–1513.

Oksman K, Gatenholm P, Bengtsson M, Dammstrom S (2005), 'Green composites: the latest development from micro to nanoscale' *In: Proceedings of the 229th ACS national meeting*, San Diego, CA, March 13–17, 938.

Oksman K, Skrifvars M, Selin JF (2003), 'Natural fibres as reinforcement in polylactic acid (PLA) composites', *Compos Sci Technol*, **63**, 1317–1324.

Oksman K, Wallstrom L, Berglund LA, Filho RDT (2002), 'Morphology and mechanical properties of unidirectional sisal–epoxy composites', *J Appl Polym Sci*, **84**, 2358–2365.

Olsen DJ (1991), 'Effectiveness of maleated polypropylenes as coupling agents for wood flour/polypropylene composites', In: *ANTEC, Proceeding of the 49th 38 Annual Technical Conference*, Montreal, Canada, May 5–9, 1886–1891.

Panthapulakkal S, Sain M, Law S (2005), 'Effect of coupling agents on rice-husk-filled HDPE extruded profiles', *Polym Int*, **54**, 37–142.

Peijs T (2000), 'Natural fiber based composites', *Mater Technol*, **15**(4), 281–285.

Peijs T (2002), 'Composites turn green' *e-Polymers*, T_002, 1–12.

Peijs T, Garkhail S, Heijenrath R, Van Den Oever M, Bos H (1998), 'Thermoplastic composites based on flax fibres and polypropylene: Influence of fibre length and fibre volume fraction on mechanical properties', *Macromol Symp*, **127**, 193–203.

Peijs T, Van Melick HGH, Garkhail SK, Pott GT, Baillie CA (1998), 'Natural-fibre-mat reinforced thermoplastics based on upgraded flax fibres for improved moisture resistance', *In: ECCM-8 conference,* Naples, 119–126.

Philip S, Keshavarz T, Roy I (2007), 'Review. Polyhydroxyalkanoates: biodegradable polymers with a range of applications', *J Chem Technol Biotechnol,* **82**, 233–247.

Pickering KL, Beckermann GW, Alam SN, Foreman NJ (2007), 'Optimising industrial hemp fibre for composites', *Compos Pt A: Appl Sci Manuf,* **38**(2), 461–468.

Plueddeman ED (1982), *Silane coupling agents, New York: Plenum Press,* 17–20.

Pompe G, Mader E (2000), 'Experimental detection of a transcrystalline interphase in glass-fibre/polypropylene composites', *Compos Sci Technol,* **60**, 2159–2167.

Qin C, Soykeabkaew N, Xiuyuan N, Peijs T (2008), 'The effect of fibre volume fraction and mercerization on the properties of all-cellulose composites', *Carbohydr Polym,* **71**(3), 458–467.

Qingxiu L, Matuana LM (2003), 'Effectiveness of maleated and acrylic acid functionalized polyolefin coupling agents for HDPE/wood–flour composites', *J Thermoplast Compos,* **16**(6), 551–564.

Qiu W, Zhang F, Endo T, Hirotsu T (2003), 'Preparation and characteristics of composites of high-crystalline cellulose with polypropylene: effects of maleated polypropylene and cellulose content', *J Appl Polym Sci,* **87**, 337–345.

Quillin DT, Caulfield DF, Koutski JA (1993), 'Crystallinity in the polypropylene/cellulose system. Nucleation and crystalline morphology', *J Appl Polymer Sci,* **50**(7), 1187–1194.

Raj RG, Kokta BV (1991), 'Improving the mechnical properties of HDPE-wood fibre composites with additives/coupling agents', *49th Annual Technical Conference,* Montreal, Canada, May 5–9. Society of Plastics Engineers, 1883–1885.

Rana AK, Mandal A, Bandyopadhyay S (2003), 'Short jute fibre reinforced polypropylene composites: effect of compatibiliser, impact modifier and fibre loading', *Compos Sci Technol,* **63**, 801–806.

Rana AK, Mandal A, Mitra BC, Jacobson R, Rowell R, Banerjee AN (1998), 'Short jute fiber-reinforced polypropylene composites: effect of compatibilizer', *J Appl Polym Sci,* **69**, 329–338.

Ranganathan S., Baker WE, Russell KE, Whitney RA (1999), 'Peroxide initiated grafting of maleic anhydride onto linear and branched hydrocarbons', *J Polym Sci: Part A,* **37**, 3817–3825.

Ray D, Sarkar BK, Basak RK, Rana AK (2002b), 'Study of thermal behaviour of alkali-treated jute fibres', *J Appl Polym Sci,* **85**, 2594–2599.

Ray D, Sarkar BK, Das S, Rana AK (2002a), 'Dynamic mechanical and thermal analysis of vinylester-resin-matrix composites reinforced with untreated and alkali-treated jute fibres', *Compos Sci Technol,* **62**, 911–917.

Ray D, Sarkar BK, Rana AK (2002c), 'Fracture behavior of vinylester resin matrix composites reinforced with alkali-treated jute fibres', *J Appl Polym Sci,* **85**, 2588–2593.

Ray D, Sarkar BK, Rana AK, Bose NR (2001), 'Effect of alkali treated jute fibres on composite properties, *Bull Mater Sci,* **24**(2), 129–135.

Roberts D, Constable RC (2003), 'Chemical coupling agents for filled and grafted polypropylene composites', In: *Handbook of polypropylene and polypropylene composites,* ed. Karian HG, second edition, New York: Marcel Dekker, 35–80.

Rowell RM, Young RA, Rowell JK (1996), *Paper and composites from agro-based resources.* Lewis Publishers.

Saheb DN, Jog JP (1999), 'Natural fiber polymer composites: A review', *Adv Polym Technol*, **18**(4), 351–363.

Sain MM, Balatinecz J, Law S (2000), 'Creep fatigue in engineered wood fiber and plastic compositions', *J Appl Polym Sci*, **77**, 260–268.

Sain M, Suhara P, Law S, Bouilloux A (2005), 'Interface modification and mechanical properties of natural fiber-polyolefin composite products', *Reinf Plast Comp*, **24**, 121–130.

Sanadi AR, Caulfield DF (2000), 'Transcrystalline interphases in natural fibre–PP composites: effect of coupling agent', *Compos Interface*, **7**(1), 31–43.

Sanadi AR, Caulfield DF, Jacobson RE, Rowell RM (1995), 'Renewable agricultural fibres as reinforcing fillers in plastics: mechanical properties of kenaf fibres–polypropylene composites', *Ind Eng Chim Res*, **34**, 1889–1896.

Sanadi AR, Caulfield DF, Rowell RM (1994a), 'Reinforcing polypropylene with natural fibres', *Plast Eng*, **L**(4), 27–28.

Sanadi AR, Young RA, Clemons C, Rowell RM (1994b), 'Recycled newspaper fibers as reinforcing fillers in thermoplastics. Part I. Analysis of tensile and impact properties in polypropylene', *Reinf Plast Compos*, **13**, 54–67.

Satyanarayana KG, Arizaga GGC, Wypych F (2009), 'Biodegradable composites based on lignocellulosic fibers. An overview', *Prog Polym Sci*, **34**, 982–1021.

Schlößer T, Knothe J (1997), 'Naturfaserverstärkte Fahrzeugteile', *Kunststoffe*, **87**(9), 1148–1152.

Schüssler A (1998), 'Automobilinnenteile aus Naturfaservliesen', *Kunststoffe*, **88**(7), 1006–1008.

Shanks RA, Hodzic A, Wong S (2004), 'Thermoplastic biopolyester natural fiber composites', *J Appl Polym Sci*, **91**, 2114–2121.

Singleton ACN, Baillie CA, Beaumont PWR, Peijs T (2003), 'On the mechanical properties, deformation and fracture of a natural fibre/recycled polymer composite', *Compos Part B Eng*, **34**(6), 519–526.

Sgriccia N, Hawley MC (2007), 'Thermal, morphological, and electrical characterization of microwave processed natural fibre composites', *Compos Sci Technol*, **67**, 1986–1991.

Snijder MHB, Bos HL (2000), 'Reinforcement of polypropylene by annual plant fibers: optimisation of the coupling agent efficiency', *Compos Interface*, **7**(2), 69–75.

Snijder, MHB, Bos, HL, Van Kemenade MJJM (1999), Extruder for continuously manufacturing composites of polymer and cellulosic fibres. European Patent WO9956936, US Patent US 6,565,348 B1.

Snijder MHB, Reinerink EJM, Bos HL (1998), In: *Proceedings, 2nd international symposium on natural polymers and composites*, Brazil.

Son S-J, Lee Y-M, Im S-S (2000), 'Transcrystalline morphology and mechanical properties in polypropylene composites containing cellulose treated with sodium hydroxide and cellulase', *J Mater Sci*, **35**(22), 5767–5778.

Soykeabkaew N, Arimoto N, Nishino T, Peijs T (2008), 'All-cellulose composites by surface selective dissolution of aligned ligno-cellulosic fibres', *Compos Sci Technol*, **68**(10–11), 2201–2207.

Soykeabkaew N, Nishino T, Peijs T (2009a), 'All-cellulose composites of regenerated cellulose fibres by surface selective dissolution', *Compos Pt A: Appl Sci Manuf*, **40**(4), 321–328.

Soykeabkaew N, Sian C, Gea S, Nishino T, Peijs T (2009b), 'All-cellulose nanocomposites by surface selective dissolution of bacterial cellulose', *Cellulose*, **16**(3), 435–444.

Stamboulis A, Baillie CA, Garkhail SK, Van Melick HGH, Peijs T (2000), 'Environmental durability of flax fibres and their composites based on polypropylene matrix', *Appl Compos Mater*, **7**(5–6), 273–294.

Stamboulis A, Baillie CA, Peijs T (2001), 'Effects of environmental conditions on mechanical and physical properties of flax fibers', *Compos Pt A: Appl Sci Manuf*, **32**(8), 1105–1115.

Stamboulis A, Baillie C, Schulz E (1999), 'Interfacial characterization of flax fibre–thermoplastic polymer composites by the pull-out test', *Angew Makromol Chem*, **272**, 117–120.

Thwe MM, Liao K (2003), 'Durability of bamboo–glass fiber reinforced polymer matrix hybrid composites', *Compos Sci Technol*, **63**, 375–387.

Thomason JL, Vlug MA (1996), 'Influence of fibre length and concentration on the properties of glass fibre-reinforced polypropylene: 1. Tensile and flexular modulus'. *Compos Pt A: Appl Sci Manuf*, **27A**, 477–484.

Thomason JL, Vlug MA, Schipper G, Krikor HGLT (1996), 'Influence of fibre length and concentration on the properties of glass fibre-reinforced polypropylene: Part 3. Strength and strain at failure', *Compos Pt A: Appl Sci Manuf*, **27A**, 1075–1084.

Van De Velde K, Kiekens P (2001), 'Influence of fibre and matrix modifications on mechanical and physical properties of flax fibre reinforced poly(propylene)', *Macromol Mater Eng*, **286**(4), 237–242.

Van De Velde K, Kiekens P (2002), 'Biopolymers: overview of several properties and consequences on their applications', *Polym Test*, **21**, 433–442.

Van De Velde K, Kiekens P (2003), 'Effect of material and process parameters on the mechanical properties of unidirectional and multidirectional flax/polypropylene composites', *Compos Struct*, **62**, 443–448.

Van Den Oever M, Peijs T (1998), 'Continuous-glass-fibre-reinforced polypropylene composites II. Influence of maleic-anhydride modified polypropylene on fatigue behaviour', *Compos Pt A: Appl Sci Manuf*, **29**(3), 227–239.

Van Den Oever MJA, Bos HL (1998), 'Critical fibre length and apparent interfacial shear strength of single flax fibre polypropylene composites', *Adv Compos Lett*, **7**(3), 81–85.

Van Den Oever MJA, Bos HL, Van Kemenade MJJM (2000), 'Influence of the physical structure of flax fibres on the mechanical properties of flax fibre reinforced polypropylene composites', *Appl Compos Mater*, **7**(5–6), 387–402.

Van Den Oever MJA, Snijder MHB (2008), 'Jute fiber reinforced polypropylene produced by continuous extrusion compounding, part 1. Processing and ageing properties', *J Appl Polym Sci*, **110**, 1009–1018.

Vila C, Campos AR, Cristovao C, Cunha AM, Santos V, Parajo JC (2008), 'Sustainable biocomposites based on autohydrolysis of lignocellulosic substrates', *Compos Sci Technol*, **68**, 944–952.

Villar MA, Marcovich NE (2003), 'Thermal and mechanical characterization of linear low-density polyethylene/wood flour composites', *J Appl Polym Sci*, **90**, 2775–2784.

Voyutskii S (1963), *Autohesion and adhesion of high polymers*, Wiley, New York.

Wagner HD, Lustiger A, Marzinsky CN, Mueller RR (1993), 'Interlamellar failure at transcrystalline interfaces in glass/polypropylene composites', *Compos Sci Technol*, **48**, 181–184.

Wambua P, Ivens J, Verpoest I (2003), 'Natural fibres: can they replace glass in fibre reinforced plastics?', *Compos Sci Technol*, **63**(9), 1259–1264.

Westerlind BS, Berg JC (1988), 'Surface energy of untreated and surface modified cellulose fibers', *J Appl Polym Sci*, **36**, 523–534.

Wielage B, Lampke Th, Utschick H, Soergel F (2003), 'Processing of natural-fibre reinforced polymers and the resulting dynamic-mechanical properties', *J Mater Process Technol,* **139**, 140–146.

Wood JR, Marom G (1997), 'Determining the interfacial shear strength in the presence of transcrystallinity in composites by the single-fiber microcomposite compressive fragmentation test', *Appl Compos Mater*, **4**(4), 197–207.

Yin S, Rials T, Wolcott M (1999), 'Crystallization behavior of polypropylene and its effect on woodfiber composite properties', In: *Proceedings of the 5th international conference on woodfiber-plastic composites,* Madison, WI, 139–146.

Yu L, Dean K, Li L (2006), 'Polymer blends and composites from renewable resources', *Prog Polym Sci,* **31**, 576–602.

Zafeiropoulos NE, Baillie CA, Matthews FL (2001a), 'A study of transcrystallinity and its effect on the interface in flax fibre reinforced composite materials', *Compos Pt A: Appl Sci Manuf,* **32**(3–4), 527–545.

Zafeiropoulos NE, Baillie CA, Matthews FL (2001b), 'An investigation of the effect of processing conditions on the interface of flax/polypropylene composites', *Adv Compos Lett,* **10**(6), 291–295.

Zafeiropoulos NE, Baillie CA, Matthews FL (2001c), 'The effect of transcrystallinity on the interface of green flax/polypropylene composite materials', *Adv Compos Lett,* **10**(5), 229–236.

Zafeiropoulos NE, Baillie CA, Hodgkinson JM (2002a), 'Engineering and characterisation of the interface in flax fibre/polypropylene composite materials. Part II. The effect of surface treatments on the interface', *Compos Pt A: Appl Sci Manuf,* **33**(9), 1185–1190.

Zafeiropoulos NE, Williams DR, Baillie CA, Matthews FL (2002b), 'Engineering and characterisation of the interface in flax fibre/polypropylene composite materials. Part I. Development and investigation of surface treatments', *Compos Pt A-Appl Sci Manuf,* **33**(8),1083–1093.

Zhang JF, Sun X (2004), 'Mechanical properties of poly(lactic acid)/starch composites compatibilized by maleic anhydride', *Biomacromolecules*, **5**(4), 1446–1451.

Zhang M, Xu J, Zhang Z, Zeng H, Xiong X (1996), 'Effect of transcrystallinity on tensile behaviour of discontinuous carbon fibre reinforced semicrystalline thermoplastic composites', *Polymer*, **37**, 5151–5158.

Preparation of cellulose nanocomposites

A. DUFRESNE, Federal University of Rio de Janeiro, Brazil

Abstract: The preparation of elongated rod-like nanoparticles from lignocellulosic fibers and their use as a reinforcing phase in a polymeric matrix is discussed. The hierarchical structure of natural fibers is described and the expected effects of changing the scale of cellulosic particles from micro to nano are outlined. The preparation methods and morphological features of the ensuing nanofibers are also discussed. The different ways of processing nanocomposite films using a polymer matrix are presented. The main properties of these films, in terms of microstructure, thermal properties, mechanical performances, swelling behavior and barrier properties are reviewed.

Key words: polymer nanocomposites, cellulose whiskers, cellulose nanocrystals, natural nanofibers.

3.1   Introduction

Over the last two decades a good deal of work has been dedicated to the use of lignocellulosic fibers as reinforcing elements in polymeric matrices and for the possibility of replacing conventional fibers such as glass by natural fibers in reinforced composites. However, one of the main drawbacks of lignocellulosic fibers, among others, is the important variation of properties inherent to any natural product. Indeed, their properties are related to climatic conditions, maturity, and type of soil. Disturbances during plant growth also affect the plant structure and are responsible for the wide scatter of mechanical plant fiber properties.

One of the basic approaches to achieving improved fibers and composites is to eliminate the macroscopic flaws by disintegrating the natural grown fibers, and separating the almost defect-free highly crystalline fibrils. This can be achieved by exploiting the hierarchical structure of natural fibers. Aqueous suspensions of cellulose nanoparticles can be prepared by a mechanical treatment or acid hydrolysis of the biomass. The object of this latter treatment is to dissolve away regions of low lateral order so that the water-insoluble, highly crystalline residue may be converted into a stable suspensoid by subsequent vigorous mechanical shearing action. The resulting nanocrystals occur as rod-like particles or whiskers, the dimensions of which depend on the nature of the substrate, but range in the nanometer scale. Because these whiskers contain only a small number of defects, their axial

Young's modulus is close to the one derived from theoretical chemistry and potentially stronger than steel and similar to kevlar. These highly stiff nanoparticles are therefore suitable for the processing of green nanocomposite materials.

Cellulosic nanoparticles, the generally accepted and overused trade name being nanocellulose, have generated interest from the scientific community because of their biodegradability, strength and other characteristics. Sustainability and green issues continue as top priorities for many businesses and individuals, stimulating the search for non-petroleum-based structural materials such as bionanocomposites that are biodegradable, high performance and lightweight.

3.2   Hierarchical structure of natural fibers

Cellulose is one of the most important structural elements in plants and some other living species serving to maintain their structure. It is a ubiquitous structural polymer that confers its mechanical properties to higher plant cells. In nature, cellulose occurs as slender rod-like or threadlike entity, which arises from the linear association of crystallites. This entity is called the microfibril (a collection of cellulose chains) and it forms the basic structural unit of the plant cell wall. Each microfibril can be considered as a string of cellulose crystallites, linked along the chain axis by amorphous domains (Fig. 3.1). Their structure consists of a predominantly crystalline cellulosic core. It is covered with a sheath of paracrystalline polyglucosan material surrounded by hemicelluloses (Whistler and Richards, 1970).

These microfibrils are cemented by other polymers such as lignin and hemicelluloses and aggregate further to form the lignocellulosic fibers. Depending on their origin, the microfibril diameters range from about 2 to 20 nm for lengths that can reach several tens of micrometers. As they are almost defect-free, the modulus of these sub-entities is close to the theoretical limit for cellulose. The potential of cellulose-based composites lies in the Young's modulus of the cellulose crystallite, which was first experimentally studied in 1962 from the crystal deformation of cellulose I using highly oriented fibers of bleached ramie (Sakurada et al., 1962). A value of 137 GPa was reported. This value differs from the theoretical estimate of 167.5 GPa reported by Tashiro and Kobayashi (1991). More recently, Raman spectroscopy technique was used to measure the elastic modulus of native cellulose crystals (Šturcova et al., 2005). A value around 143 GPa was reported. The elastic modulus of single microfibrils from tunicate was measured by atomic force microscopy (AFM) using a three-point bending test (Iwamoto et al., 2009). Values of 145 and 150 GPa were reported for single microfibrils prepared by 2,2,6,6-tetramethylpiperidine–oxyl radical (TEMPO) oxidation and sulfuric acid hydrolysis, respectively.

Microfibril

Fiber

3.1 Schematic diagram of the physical structure of a semi-crystalline cellulose fiber.

These impressive mechanical properties make cellulose nanoparticles ideal candidates for the processing of reinforced polymer composites. Incorporating these nanoparticles in a synthetic or natural polymeric matrix consists therefore in biomimicing nature.

3.3   From micro- to nanoscale

As indicated in section 3.1, a major drawback of natural fibers for composite applications is the big variation of properties inherent to the natural products. The properties are related to climatic conditions, maturity, type of soil. Disturbances during plant growth will also affect the plant structure and are responsible for the enormous scatter of mechanical plant fiber properties. For instance, the number of knobby swellings and growth-induced lateral displacement present in plant fibers influence the tensile strength and elongation at break of the fibers obtained. One of the means to get round this problem involves decreasing the size of the natural particles from the micro to the nanoscale. Conceptually, nanoparticles refer to particles with at least one dimension less 100 nm.

When carrying out this scale shift, exploiting the hierarchical structure of natural fibers, important consequences occur. The first one is obviously an increase of the specific area of the particles, from values around a few m^2 g^{-1} to a few 100 m^2 g^{-1}. This, in turn, results in an increase of the interfacial area with the polymeric matrix and a decrease of the average interparticle distance as the particle size decreases. Some particle–particle interactions can thus be expected. The homogeneous dispersion of nanoparticles in a continuous medium is generally difficult because the surface energy increases when decreasing their dimensions. Another important typical feature of nanoparticles is the possibility of improving properties of the material for low filler content without a detrimental effect on the impact resistance and plastic deformation. A reduction of gas diffusion (the barrier effect) is also likely to occur. Moreover, cellulosic nanoparticles are distinguished by their liquid crystal behavior when suspended in water, presenting birefringence phenomena under polarized light.

Native cellulose fibers are built up by smaller and mechanically stronger entities, the cellulose fibrils. Fibrils contain both crystalline and noncrystalline domains, the latter being located at the surface and along the main axis. Noncrystalline domains form weak spots along the fibrils. The processes for isolating cellulose nanoparticles include simple mechanical methods or a combination of chemical and mechanical methods. The first method involves submitting the natural fibers to multiple mechanical shearing actions. This disintegration process allows the release of the constitutive cellulosic microfibrils. The resulting aqueous suspensions display outstanding properties with a huge increase of the viscosity for very low nanoparticle contents. These micro or nanofibrils occur as very long thin entangled filaments. These nanoelements can also be obtained by a combination of enzymatic hydrolysis that facilitates the fibrillation and mechanical shearing, thus allowing a reduction in the energy consumption involved in the process. The second method involves submitting the natural fibers to a strong acid hydrolysis treatment followed by a sonication treatment to obtain stiff rod-like nanoparticles as explained in section 3.4. These nanocellulosic nanoparticles are easier to disperse in a polymeric matrix than entangled microfibrils and their better defined form is ideal for modeling rheological and reinforcement behaviors. This chapter mainly deals with acid-hydrolyzed nanoparticles but references are also given to mechanically sheared nanocellulose.

3.4   Preparation of cellulose nanocrystals

The extraction or isolation of crystalline cellulosic regions, in the form of nanocrystals, is a simple process based on acid hydrolysis. Various descriptors have been used in the literature to designate these crystalline rod-like nanoparticles. It is mainly referred to as whiskers, nanowhiskers, cellulose nanocrystals, NCC (nanocrystalline cellulose), monocrystals, microcrystals or microcrystallites, despite their nanoscale dimensions. The terms microfibrils, microfibrillated cellulose (MFC), or nanofibrillated cellulose are used to designate cellulosic nanoparticles obtained by a simple mechanical shearing disintegration process (Fig. 3.2).

To prepare cellulose whiskers, the biomass is generally first submitted to a bleaching treatment with NaOH to purify cellulose. After removal of other constituents, such as lignin and hemicelluloses, the bleached material is disintegrated in water, and the resulting suspension is submitted to a hydrolysis treatment with acid. The amorphous regions of cellulose act as structural defects and are responsible for the transverse cleavage of the microfibrils into short nanocrystals under acid hydrolysis. Under controlled conditions, this transformation consists of the disruption of amorphous regions surrounding and embedded within cellulose microfibrils. The crystalline segments remain intact, because of the faster hydrolysis kinetics of amorphous domains compared with crystalline ones. The hydronium ions penetrate the cellulosic material in the amorphous domains promoting the hydrolytic cleavage of the glycosidic bonds to release individual crystallites. The resulting suspension is subsequently diluted with water and washed by successive centrifugations.

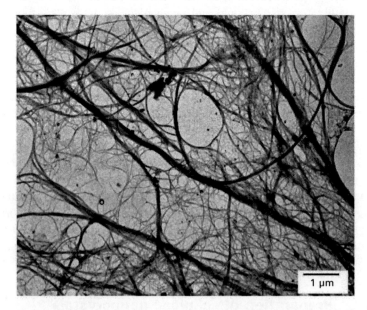

1 µm

3.2 Transmission electron micrograph from a dilute suspension of MFC obtained from *Opuntia ficus-indica* (reprinted with permission from Malainine *et al.,* 2005, copyright Elsevier).

Dialysis against distilled water is then performed to remove free acid in the dispersion. Disintegration of aggregates and complete dispersion of the whiskers is obtained by a sonication step. These suspensions are generally much diluted because of the formation of a gel for low nanoparticle contents. Exact determination of the whisker content can be achieved by weighing aliquots of the solution before and after drying. The dispersions are stored in the refrigerator after filtration to remove residual aggregates. This general procedure has to be adapted depending on the nature of the substrate.

Dong et al. (1998) were among the first researchers studying the effect of hydrolysis conditions on the properties of resulting cellulose nanocrystals. They proved that a longer hydrolysis time leads to shorter monocrystals and also to an increase in their surface charge. The acid concentration was also found to affect the morphology of whiskers prepared from sugar-beet pulp as reported by Azizi Samir et al. (2004b). Beck-Candanedo et al. (2005) reported the properties of cellulose nanocrystals obtained by hydrolysis of softwood and hardwood pulp and investigated the influence of hydrolysis time and acid-to-pulp ratio. It was found that the reaction time is one of the most important parameters to consider in the acid hydrolysis of wood pulp. Moreover, they considered that too long reaction time would digest completely the cellulose to yield its component sugar molecules. On the contrary, a lower reaction time only yields large undispersable fibers and aggregates. It was reported that an increase of the hydrolysis time of pea hull fibers results in a decrease of both length and diameter, whereas the aspect ratio first increases and then decreases (Chen et al., 2009). The effect of the reaction conditions on cellulose nanocrystal surface charge and sulfur content was not significant and it was supposed to be controlled by factors other than hydrolysis conditions. However, chiral nematic pitch decreases when increasing the cellulose concentration and decreasing nanocrystals length. An attempt to find optimized conditions to prepare cellulose nanocrystals from microcrystalline cellulose (MCC) derived from Norway spruce (*Picea abies*) was also reported (Bondenson et al., 2006). The processing parameters have been optimized by using a response surface methodology.

Cellulose fibers can be oxidized by TEMPO-mediated oxidation creating carboxyl groups at the surface. The TEMPO-oxidized cellulose fibers can be converted to transparent and highly viscous dispersions in water, consisting of highly crystalline individual nanofibers (Fukuzumi et al., 2009; Saito et al., 2006; 2007). At pH 10, optimal conditions were reached, giving cellulose nanofibers 3–4 nm wide and a few micrometers long. It was also shown that the hydrolysis of amorphous cellulosic chains can be performed simultaneously with the esterification of accessible hydroxyl groups to produce surface functionalized whiskers in a single step (Braun and Dorgan, 2009). The reaction was carried out in an acid mixture composed of hydrochloric and an organic acid (acetic and butyric). Resulting nanocrystals are of similar dimensions to those obtained by hydrochloric acid hydrolysis alone. Narrower diameter polydispersity indices indicate that surface groups aid the individualization of the nanowhiskers. The resulting surface-modified cellulose whiskers are dispersible in ethyl acetate and toluene indicating increased hydrophobicity and presumably higher compatibility with hydrophobic polymers.

Cellulose nanoparticles are obtained as aqueous suspensions, the stability of which depends on the dimensions of the dispersed species, size polydispersity and surface charge. The use of sulfuric acid to prepare cellulose nanocrystals leads to more stable aqueous suspension than that prepared using hydrochloric acid (Araki *et al.*, 1998). It was shown that the H_2SO_4-prepared nanoparticles present a negatively charged surface whereas the HCl-prepared nanoparticles are not charged. During acid hydrolysis via sulfuric acid, acidic sulfate ester groups are probably formed on the nanoparticle surface. This creates an electric double layer repulsion between the nanoparticles in suspension, which plays an important role in their interaction with a polymer matrix and with each other. The density of charges on the cellulose nanocrystals surface depends on the hydrolysis conditions and can be determined by elementary analysis or conductimetric titration to determine the sulfur content. The sulfate group content increases with acid concentration, acid-to-polysaccharide ratio and hydrolysis time. Based on the density and size of the cellulose whiskers, Araki *et al.* (1998, 1999) estimated for a nanocrystal with dimensions of $7 \times 7 \times 115$ nm^3, that the charge density is 0.155 e nm^{-2}, where e is the elementary charge. With the following conditions: cellulose concentration of 10 wt% in 60% sulfuric acid at 46 °C for 75 min, the charge coverage was estimated at 0.2 negative ester groups per nm (Revol *et al.*, 1992). Other typical values of the sulfur content of cellulose whiskers prepared by sulfuric acid hydrolysis were reported (Marchessault *et al.*, 1961; Revol *et al.*, 1994). It was shown that even at low levels, the sulfate groups caused a significant decrease in degradation temperature and increase in char fraction confirming that the sulfate groups act as flame retardants (Roman and Winter, 2004).

Cellulose nanocrystals can be prepared from any botanical source containing cellulose. In the many published studies, various cellulosic sources have been used as shown in Fig. 3.3. Regardless of the source, cellulose nanocrystals occur as elongated nanoparticles. The persistence of the spot diffractogram when the electron probe is scanned along the rod during transmission electron microscopy (TEM) observation evidences the monocrystalline nature of the cellulosic fragment (Favier *et al.*, 1995a). Therefore, each fragment can be considered as a cellulosic crystal with no apparent defect. Their dimensions depend on several factors, including the source of the cellulose, the exact hydrolysis conditions and ionic strength.

The typical geometrical characteristics for nanocrystals derived from different species are presented in Table 3.1. Even if often composed of a few laterally bound elementary crystallites that are not separated by conventional acid hydrolysis and sonication process (Elazzouzi-Hafraoui *et al.*, 2008), the length and width of hydrolyzed cellulose nanocrystals is generally of the order of a few hundred nanometers and a few nanometers, respectively. It was observed that the length polydispersity has a constant value, whereas the diameter polydispersity depends on the acid used for isolation (Braun *et al.*, 2008). A smaller diameter polydispersity was obtained when using sulfuric

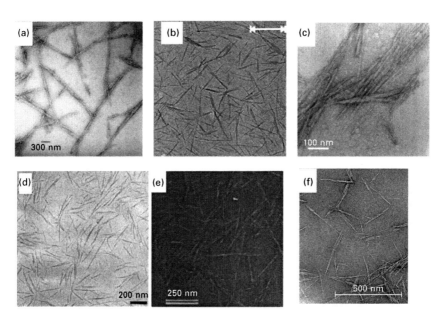

3.3 Transmission electron micrographs from a dilute suspension of (a) bacterial cellulose (reprinted with permission from Grunert and Winter, 2002, copyright Springer), (b) cotton cellulose (reprinted with permission from Fleming *et al.*, 2000, copyright American Chemical Society, the scale bar corresponds to 400 nm), (c) MCC (reprinted with permission from Kvien *et al.*, 2005, copyright American Chemical Society), (d) sugar beet pulp (reprinted with permission from Azizi Samir *et al.*, 2004a, copyright American Chemical Society), (e) wheat straw nanocrystals (reprinted with permission from Helbert *et al.*, 1996, copyright Wiley Interscience), (f) sisal (reprinted with permission from Siqueira *et al.*, 2009, copyright American Chemical Society), (g) tunicin (reprinted with permission from Anglès and Dufresne, 2000, copyright American Chemical Society), (h) acacia pulp (reprinted with permission from Pu *et al.*, 2007, copyright Elsevier), (i) banana rachis (reprinted with permission from Zuluaga *et al.*, 2007, copyright Springer), (j) eucalyptus wood pulp (reprinted with permission from de Mesquita *et al.*, 2010, copyright American Chemical Society), (k) *Luffa cylindrica*, (l) ramie (reprinted with permission from Habibi *et al.*, 2008) copyright The Royal Society of Chemistry), (m) Capim Dourado (reprinted with permission from Siqueira *et al.*, 2010a, copyright Springer).

Continued

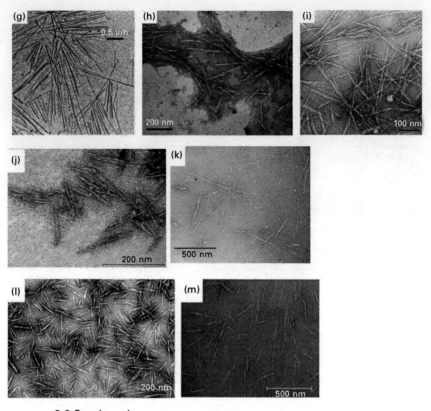

3.3 Continued.

acid instead of hydrochloric acid, because of electrostatic charges resulting from the introduction of sulfate ester groups when using the former.

An important parameter for cellulosic whiskers is the aspect ratio, defined as the ratio of the length to the width (Table 3.1), which determined the anisotropic phase formation and reinforcing properties. The average length ranges between 1 μm for nanocrystals prepared for instance from tunicate and around 200 nm for cotton. The cellulose extracted from tunicate, a sea animal, is referred to as tunicin. The diameter ranges between 15 nm for tunicin and 4–5 nm for sisal or wood. The high value reported for cottonseed linter corresponds to aggregates. The aspect ratio varies between 10 for cotton and 67 for tunicin. Relatively large and highly regular tunicin whiskers are ideal for modeling rheological and reinforcement behaviors and were extensively used in the literature. The shape and dimensions of cellulose whiskers can be measured by microscopic observations or scattering techniques. The cross-sections of microfibrils observed by TEM are square, whereas their AFM topography shows a rounded profile owing to convolution with the shape of the AFM tip (Hanley *et al.*, 1992). AFM images of the surface of highly

Table 3.1 Geometrical characteristics of cellulose nanocrystals from various sources: length (*L*), cross-section (*D*) and aspect ratio (*L/D*) of rod-like particles obtained from acid hydrolysis

Source	*L* (nm)	*D* (nm)	*L/D*	Reference
Acacia pulp	100–250	5–15	–	(Pu *et al.*, 2007)
Alfa	200	10	20	(Ben Elmabrouk, 2009)
Algal (*Valonia*)	>1000	10–20	∞	(Hanley *et al.*, 1992; Revol, 1982)
Bacterial	100 to several 1000	5–10 × 30–50	–	(Grunert and Winter, 2002; Roman and Winter, 2004; Tokoh *et al.*, 1998)
Banana rachis	500–1000	5	–	(Zuluaga *et al.*, 2007)
Capim dourado	300	4.5	67	(Siqueira *et al.*, 2010a)
Cassava bagasse	360–1700	2–11	–	(Morais Teixeira *et al.*, 2009)
Cladophora	–	20 × 20	–	(Kim *et al.*, 2000)
Cotton	100–300	5–15	10	(Dong *et al.*, 1998; Ebeling *et al.*, 1999; Morais Teixeira *et al.*, 2010)
Cottonseed linter	170–490	40–60	–	(Lu *et al.*, 2005)
Date palm tree (rachis/leaflets)	260/180	6.1	43/30	(Bendahou *et al.*, 2009)
Eucalyptus wood pulp	145	6	24	(de Mesquita *et al.*, 2010)
Flax	100–500	10–30	15	(Cao *et al.*, 2007)
Hemp	Several 1000	30–100	–	(Wang *et al.*, 2007)
Luffa cylindrica	242	5.2	47	(Siqueira *et al.*, 2010b)
MCC	150–300	3–7	–	(Kvien *et al.*, 2005)
Mulberry	400–500	20–40	–	(Li *et al.*, 2009)
Pea hull	240–400	7–12	34	(Chen *et al.*, 2009)
Ramie	350–700 150–250	70–120 6–8	12	(de Menezes *et al.*, 2009; Habibi and Dufresne, 2008; Habibi *et al.*, 2008, Lu *et al.*, 2006)
Sisal	100–500 215	3–5 5	60/43	(Garcia de Rodriguez *et al.*, 2006; Siqueira *et al.*, 2009; 2010c)
Sugar beet pulp	210	5	42	(Azizi Samir *et al.*, 2004b)
Tunicin	100 to several 1000	10–20	67	(Favier *et al.*, 1995a; 1995b)
Wheat straw	150–300	5	45	(Helbert *et al.*, 1996)
Wood	100–300	3–5	50	(Araki *et al.*, 1999; Beck-Candanedo *et al.*, 2005 Kim *et al.*, 2000)

crystalline cellulose microfibrils showed periodicities along the microfibril axis of 1.07 and 0.53 nm that were supposed to correspond to the fiber and glucose unit repeat distances, respectively. Scattering techniques include small-angle light (De Souza Lima *et al.*, 2003) and neutron (Orts *et al.*, 1998) scattering.

Layer-by-layer assembled (LBL) tunicin whiskers films show strong antireflection properties having an origin in a novel highly porous architecture reminiscent of a 'flattened matchsticks pile', with film-thickness-dependent porosity and optical properties created by randomly oriented and overlapping whiskers (Podsiadlo *et al.*, 2007). There are potential applications of cellulose nanocrystals in the medical field. Dong and Roman (2007) prepared fluorescently labeled cellulosic nanocrystals to be used as indicators in nanomedicine.

3.5   Processing of cellulose nanocomposites

The main challenge with nanoparticles is related to their homogeneous dispersion within a polymeric matrix. The presence of sulfate groups resulting from the acid hydrolysis treatment when using sulfuric acid to prepare cellulose whiskers or nanocrystals induces the stability of the ensuing aqueous suspension. Water is therefore the initially preferred processing medium. A high level of dispersion of the filler within the host matrix in the resulting composite film is expected when processing nanocomposites in aqueous medium.

3.5.1   Polymer latexes

The first publication reporting the preparation of cellulose nanocrystals reinforced polymer nanocomposites was carried out using a latex obtained by the copolymerization of styrene and butyl acrylate, poly(S-co-BuA), and tunicin whiskers (Favier *et al.*, 1995a). The same copolymer was used in association with wheat straw (Helbert *et al.*, 1996) or sugar beet (Azizi Samir *et al.*, 2004b) cellulose nanocrystals. Other latexes such as poly(β-hydroxyoctanoate) (PHO) (Dubief *et al.*, 1999; Dufresne, 2000; Dufresne *et al.*, 1999), polyvinylchloride (PVC) (Chazeau *et al.*, 1999a; 1999b; 1999c; 2000), waterborne epoxy (Matos Ruiz *et al.*, 2001), natural rubber (NR) (Bendahou *et al.*, 2009; 2010; Siqueira *et al.*, 2010a), and polyvinyl acetate (PVAc) (Garcia de Rodriguez *et al.*, 2006) were also used as matrix. Recently, stable aqueous nanocomposite dispersions containing cellulose whiskers and a poly(styrene-co-hexyl-acrylate) matrix were prepared via miniemulsion polymerization (Ben Elmabrouk *et al.*, 2009). Addition of a reactive silane was used to stabilize the dispersion. Solid nanocomposite films can be obtained by mixing and casting the two aqueous suspensions followed by water evaporation. Alternative methods include freeze-drying and hot-pressing or freeze-drying, extruding and hot-pressing the mixture.

3.5.2   Hydrosoluble or hydrodispersible polymers

The preparation of cellulosic particles reinforced starch (Anglès and Dufresne, 2000; Kvien *et al.*, 2007; Mathew and Dufresne, 2002; Mathew *et al.*, 2008; Orts *et al.*, 2005; Svagan *et al.*, 2009), silk fibroin (Noishiki *et al.*, 2002), poly(oxyethylene) (POE) (Azizi Samir, 2004a; 2004c; 2004f; 2005b; 2006), polyvinyl alcohol (PVA) (Lu *et al.*, 2008; Paralikar *et al.*, 2008; Roohani *et al.*, 2008; Zimmermann *et al.*, 2004; 2005), hydroxypropyl cellulose (HPC) (Zimmermann *et al.*, 2004; 2005), carboxymethyl cellulose (CMC) (Choi and Simonsen, 2006), or soy protein isolate (SPI) (Wang *et al.*, 2006) has been reported in the literature. The hydrosoluble or hydrodispersible polymer is first dissolved in water and this solution is mixed with the aqueous suspension of cellulose nanocrystals. The ensuing mixture is generally evaporated to obtain a solid nanocomposite film. It can also be freeze-dried and hot-pressed.

3.5.3   Nonaqueous systems

In addition to the use of an aqueous polymer dispersion, or latex, an alternative way to process nonpolar polymer nanocomposites reinforced with cellulose nanocrystals involves their dispersion in an adequate (with regard to matrix) organic medium. Coating with a surfactant or surface chemical modification of the nanoparticles can also be considered. The global objective is to reduce their surface energy in order to improve their dispersibility/compatibility with nonpolar media.

Coating of cotton and tunicin whiskers by a surfactant such as a phosphoric ester of polyoxyethylene (9) nonyl phenyl ether was found to lead to stable suspensions in toluene and cyclohexane (Heux *et al.*, 2000) or chloroform (Kvien *et al.*, 2005). Coated tunicin whiskers reinforced atactic polypropylene (aPP) (Ljungberg *et al.*, 2005), isotactic polypropylene (iPP) (Ljungberg *et al.*, 2006), or poly(ethylene-co-vinyl acetate) (EVA) (Chauve *et al.*, 2005) were obtained by solvent casting using toluene. The same procedure was used to disperse cellulosic nanoparticles in chloroform and process composites with polylactic acid (PLA) (Kvien *et al.*, 2005; Petersson and Oksman, 2006). Nanocomposite materials were also prepared by dispersing cellulose acetate butyrate (CAB) in a dispersion of topochemically trimethylsilylated bacterial cellulose nanocrystals in acetone and subsequent solution casting (Grunert and Winter, 2002).

Surface chemical modification of cellulosic nanoparticles is another way to decrease their surface energy and disperse them in organic liquids of low polarity. It generally involves reactive hydroxyl groups from the surface. Experimental conditions should avoid swelling media and the peeling effect of surface-grafted chains inducing their dissolution in the reaction medium.

The chemical grafting has to be mild in order to preserve the integrity of the nanoparticle. Goussé *et al.* (2002) stabilized tunicin microcrystals in tetrahydrofuran (THF) by a partial silylation of their surface. Grunert and Winter (2002) reported the preparation of bacterial cellulose nanocrystals topochemically trimethylsilylated. Resulting nanoparticles were dispersed in acetone to process nanocomposites with a cellulose acetatebutyrate matrix. Araki *et al.* (2001) prepared original sterically stabilized aqueous rod-like cellulose microcrystals suspensions by the combination of HCl hydrolysis, oxidative carboxylation and grafting of poly(ethylene glycol) (PEG) having a terminal amino group on one end using water-soluble carbodi-imide. The PEG-grafted microcrystals displayed drastically enhanced dispersion stability evidenced through resistance to addition of 2 M sodium chloride. They also showed ability to redisperse into either water or chloroform from the freeze-dried state. Alkenyl succinic anhydride (ASA) can be used for acylating the surface of cellulose nanocrystals. Surface chemical modification of tunicin whiskers with ASA was reported by Yuan *et al.* (2006). The acylated whiskers were found to disperse in medium- to low-polarity solvents. It was shown that by controlling the heating time, whiskers with different dispersibility could be obtained. Nogi *et al.* (2006a) and Ifuku *et al.* (2007) were among the first to use acetylated cellulosic nanofibers in the preparation of reinforced clear plastic.

Preparation of stable cellulose whiskers suspensions in dimethylformamide (DMF) (Azizi Samir *et al.*, 2004e; Marcovich *et al.*, 2006), and dimethyl sulfoxide (DMSO) or *N*-methyl pyrrolidine (NMP) (van den Berg *et al.*, 2007) without either addition of a surfactant or any chemical modification was also reported. From DMF, tunicin whiskers reinforced POE plasticized with tetraethylene glycol dimethyl ether (TEGDME) were prepared by casting and evaporation of DMF (Azizi Samir *et al.*, 2004f). Cross-linked nanocomposites were also prepared by dispersing cellulose nanocrystals in a solution of an unsaturated linear polycondensate, addition of a photo-initiator, casting, evaporating the solvent and UV-curing (Azizi Samir *et al.*, 2004d).

3.5.4   Long-chain grafting

Long-chain surface chemical modification of cellulosic nanoparticles involving grafting agents bearing a reactive end group and a long 'compatibilizing' tail was also reported. The general objective was to increase the apolar character of the nanoparticle. In addition, the surface modifications can act as binding sites for active agents in drug delivery systems or for toxins in purifying and treatment systems. These surface modifications may also be able to interdiffuse, upon heating, to form the polymer matrix phase. The covalent linkage between reinforcement and matrix results in near-perfect stress transfer at the interface with exceptional mechanical properties of the composite as a result.

Nanocomposite materials were processed from polycaprolactone (PCL)-grafted cellulose whiskers using the grafting 'onto' (Habibi and Dufresne, 2008) and grafting 'from' (Habibi et al., 2008) approaches. The ensuing nanoparticles were used to process nanocomposites using PCL as matrix and a casting/evaporation technique from dichloromethane. A co-continuous crystalline phase around the nanoparticles was observed. Cellulose whiskers were also surface-grafted with PCL via microwave-assisted ring-opening polymerization yielding filaceous cellulose whisker-graft-PCL nanocrystals which were incorporated into PLA as matrix (Lin et al., 2009). Epoxy functionality was introduced onto the surface of cellulosic nanoparticles by oxidation by cerium (IV) followed by grafting of glycidyl methacrylate (Stenstad et al., 2008). The length of the polymeric chain was varied by regulating the amount of glycidyl methacrylate. The surface of cellulose whiskers was also chemically modified by grafting organic acid chlorides presenting different lengths of the aliphatic chain by an esterification reaction (de Menezes et al., 2009). These functionalized nanoparticles were extruded with low density polyethylene (LDPE) to prepare nanocomposite materials. Cellulose whiskers reinforced waterborne polyurethane nanocomposites were synthesized via in situ polymerization using casting/evaporation technique (Cao et al., 2009). The grafted chains were able to form a crystalline structure on the surface of the nanoparticles and induce the crystallization of the matrix. Cellulose nanoparticles were modified with n-octadecyl isocyanate ($C_{18}H_{37}NCO$) using two different methods with one consisting of an in situ solvent exchange procedure (Siqueira et al., 2010c). Phenol was also enzymatically polymerized in the presence of TEMPO-oxidized cellulosic nanoparticles to prepare nanocomposites under ambient conditions (Li et al., 2010).

3.5.5   Extrusion and impregnation

Very few studies have been reported concerning the processing of cellulose nanocrystals reinforced nanocomposites by extrusion methods. The hydrophilic nature of cellulose causes irreversible agglomeration during drying and aggregation in nonpolar matrices because of the formation of additional hydrogen bonds between amorphous parts of the cellulose nanoparticles. Therefore, the preparation of cellulose whiskers reinforced PLA nanocomposites by melt extrusion was carried out by pumping the suspension of nanocrystals into the polymer melt during the extrusion process (Oksman et al., 2006). An attempt to use PVA as a compatibilizer to promote the dispersion of cellulose whiskers within the PLA matrix was reported (Bondenson and Oksman, 2007). Organic acid chlorides-grafted cellulose whiskers were extruded with LDPE (de Menezes *et al.*, 2009). The homogeneity of the ensuing nanocomposite was found to increase with the length of the grafted chains (Fig. 3.4).

Another possible processing technique of nanocomposites using cellulosic nanoparticles in the dry state involves the filtration of the aqueous suspension to obtain a film or dried mat of particles followed by immersion in a polymer solution. The impregnation of the dried mat is performed under vacuum. Composites were processed by filling the cavities with transparent thermosetting resins such as phenol formaldehyde (Nakagaito and Yano, 2004; 2008; Nakagaito *et al.*, 2005), epoxy (Shimazaki *et al.*, 2007), acrylic (Iwamoto *et al.*, 2008; Nogi *et al.*, 2005; Yano *et al.*, 2005) and melamine formaldehyde (Henriksson and Berglund, 2007). Nonwoven mats of cellulose microfibrils were also used to prepare polyurethane composite materials using a film stacking method (Seydibeyoğlu and Oksman, 2008).

Water-redispersible nanofibrillated cellulose in powder form was recently prepared from refined bleached beech pulp by carboxymethylation and mechanical disintegration (Eyholzer *et al.*, 2010). However, the carboxymethylated sample displayed a loss of crystallinity and strong decrease in thermal stability limiting its use for nanocomposite processing.

3.5.6   Electrospinning

Electrostatic fiber spinning or 'electrospinning' is a versatile method for preparing fibers with diameters ranging from several micrometers down to 100 nm through the action of electrostatic forces. Bacterial cellulose whiskers were incorporated into POE nanofibers with a diameter of less than 1 μm by the electrospining process to enhance the mechanical properties of the electrospun fibers (Park *et al.*, 2007). The whiskers were found to be globally well embedded and aligned inside the fibers, even though they were partially

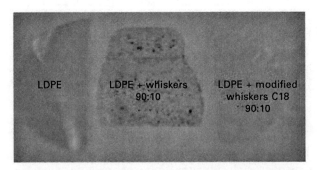

3.4 Photographs of the neat LDPE film and extruded nanocomposite films reinforced with 10 wt% of unmodified and C18 acid chloride-grafted cellulose whiskers (reprinted with permission from de Menezes *et al.*, 2009, copyright Elsevier).

aggregated. Electrospun polystyrene (PS) (Rojas *et al.*, 2009), PCL (Zoppe *et al.*, 2009) and PVA (Peresin *et al.*, 2010) microfibers reinforced with cellulose nanocrystals were obtained by electrospinning. Nonionic surfactant sorbitan monostearate was used to improve the dispersion of the particles in the hydrophobic PS matrix.

3.5.7   Multilayer films

The use of the layer-by-layer (LBL) technique is expected to maximize the interaction between cellulose whiskers and a polar polymeric matrix, such as chitosan (de Mesquita *et al.*, 2010). It also allows the incorporation of high amounts of cellulose whiskers, providing a dense and homogeneous distribution in each layer.

Podsiadlo *et al.* (2005) reported the preparation of cellulose whiskers multilayer composites with a polycation, poly(dimethyldiallylamonium chloride) (PDDA), using the LBL technique. They concluded that the multilayer films presented high uniformity and dense packing of nanocrystals. Oriented self-assembled films were also prepared using a strong magnetic film (Cranston and Gray, 2006a) or spin coating technique (Cranston and Gray, 2006b). The preparation of thin films composed of alternating layers of orientated rigid cellulose whiskers and flexible polycation chains was reported (Jean *et al.*, 2008). Alignment of the rod-like nanocrystals was achieved using anisotropic suspensions of cellulose whiskers. Green composites based on cellulose nanocrystals/xyloglucan multilayers have been prepared using the nonelectrostatic cellulose-hemicellulose interaction (Jean *et al.*, 2009). The thin films were characterized using neutron reflectivity experiments and AFM observations. More recently, biodegradable nanocomposites were obtained by the LBL technique using highly deacetylated chitosan and cellulose whiskers (de Mesquita *et al.*, 2010). Hydrogen bonds and electrostatic interactions between the negatively charged sulfate groups on the nanoparticles surface and the ammonium groups of chitosan were the driving forces for the growth of the multilayered films. A high density and homogeneous distribution of cellulose nanocrystals adsorbed on each chitosan layer, each bilayer being around 7 nm thick, were reported. Self-organized films were also obtained using only charge-stabilized dispersions of cellulose nanoparticles with opposite charges (Aulin *et al.*, 2010) from the LBL technique.

3.6   Properties of cellulose nanocomposites

3.6.1   Microstructure

A visual examination is the simplest way to assess the dispersion of cellulosic nanoparticles within the host polymeric matrix. Because of the nanoscale dimensions of the reinforcing phase the transparency of the nanocomposite film should remain if initially observed for the unfilled matrix. Opacity suggests the presence of aggregates of micrometric sizes (Ljungberg *et al.*, 2005). Optical properties of UV-cured acrylic resin impregnated bacterial cellulose nanofibers were studied by Nogi *et al.* (2006b). Polarized optical microscopy was used by Azizi Samir *et al.* (2004c) to observe and follow the growth of POE spherulites in tunicin whiskers reinforced films. For the unfilled POE matrix, birefringent spherulites were clearly identified through the characteristic Maltese cross pattern indicating a spherical symmetry. For a 10 wt% tunicin whiskers reinforced material, the supermolecular structure was found to be quite different. It was observed that the spherulites exhibited a less birefringent character, most probably owing to a weakly organized structure. It was supposed that the cellulosic filler most probably interfered with the spherulite growth and that during growth the whiskers are ejected and then occluded in interspherulitic regions. The high viscosity of the filled medium most probably restricts this phenomenon and limits the size of the spherulites.

Scanning electron microscopy (SEM) is generally employed for the more extensive morphological inspection of cellulose nanocrystals reinforced polymers by observation of the cryofractured surfaces. By comparing the micrographs showing the surface of fracture of the unfilled matrix and of the composites, the nanoparticles can be easily identified (Fig. 3.5). In fact, they appear as white dots, the concentration of which is a direct function of the particles content in the composite. These shiny dots correspond to the transversal sections of the cellulose whiskers, but their diameter determined by SEM microscopy is much higher than the whiskers diameter. This results from a charge concentration effect owing to the emergence of cellulose whiskers from the observed surface (Anglès and Dufresne, 2000).

The dispersion of nanoparticles in the nanocomposite film strongly depends on the processing technique and conditions. SEM comparison between either cast and evaporated or freeze-dried and subsequently hot-pressed composites based on poly(S-co-BuA) reinforced with wheat straw whiskers, demonstrated that the former were less homogeneous and displayed a gradient of whiskers concentration between the upper and lower faces of the composite film (Dufresne *et al.*, 1997; Helbert *et al.*, 1996). It was suggested that the casting/ evaporation technique results in films that are more homogeneous and in which the whiskers have a tendency to orient randomly into horizontal planes. A two-dimensional in-plane random network of tunicin whiskers reinforced epoxy was also reported from Raman spectroscopy experiments (Šturcová *et al.*, 2005).

TEM observation can also be performed to investigate the microstructure and dispersion of the nanoparticles in the nanocomposite film. Casting/ evaporation was reported to be an efficient processing technique to obtain

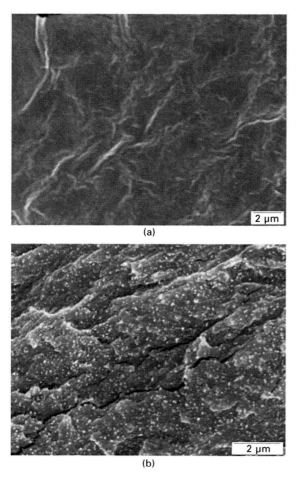

(a)

(b)

3.5 Scanning electron micrographs from the fractured surfaces of (a) unfilled plasticized starch matrix and related composites filled with (b) 25 wt% tunicin whiskers (reprinted with permission from Anglès and Dufresne, 2000, copyright American Chemical Society).

a high dispersion level (Favier *et al.*, 1995a; 1995b; Kvien *et al.*, 2005; Matos Ruiz *et al.*, 2001; Noishiki *et al.*, 2002; Zimmermann *et al.*, 2005). Small angle x-ray scattering (SAXS) and small angle neutron scattering (SANS) are other ways to monitor the dispersion of cellulosic whiskers in the matrix. This latter technique was used to identify an isotropic dispersion of tunicin whiskers in plasticized PVC (Chazeau *et al.*, 1999a). Atomic force microscopy (AFM) imaging has been used to investigate the microstructure of cellulose nanocrystals reinforced polymer nanocomposites (Kvien *et al.*, 2005; Zimmermann *et al.*, 2005). A comparison between field emission (FE) SEM, AFM and bright-field (BF) TEM for structure determination of cellulose whiskers and their nanocomposites with PLA was carried out (Kvien *et al.*, 2005). It was found that AFM overestimated the width of the whiskers owing to the tip-broadening effect. FESEM allowed for a quick examination giving an overview of the sample but with limited resolution for detailed information. Detailed information was obtained from TEM, but this technique requires staining and suffers in general from limited contrast and beam sensitivity of the material.

3.6.2   Thermal properties

The glass–rubber transition temperature, T_g, of cellulose whiskers filled polymer composites is an important parameter, which controls different properties of the resulting composite such as its mechanical behavior, matrix chains dynamics and swelling behavior. Its value depends on the interactions between the polymeric matrix and cellulosic nanoparticles. These interactions are expected to play an important role because of the huge specific area inherent to nanosize particles. For semicrystalline polymers, possible alteration of the crystalline domains by the cellulosic filler may indirectly affect the value of T_g.

No modification of T_g was reported for cellulose whiskers reinforced poly(S-co-BuA) (Dufresne *et al.*, 1997; Favier *et al.*, 1995a; 1995b; Hajji *et al.*, 1996), PHO (Dubief *et al.*, 1999; Dufresne, 2000; Dufresne *et al.*, 1999), PVC (Chazeau *et al.*, 1999a), POE (Azizi Samir *et al.*, 2004a; 2004c), PP (Ljungberg *et al.*, 2005) and NR (Bendahou *et al.*, 2009; Siqueira *et al.*, 2010a). In glycerol plasticized starch based composites, peculiar effects of tunicin whiskers on the T_g of the starch-rich fraction were reported depending on moisture conditions (Anglès and Dufresne, 2000). For low loading level (up to 3.2 wt%), a classical plasticization effect of water was reported (Fig. 3.6). However, an antiplasticization phenomenon was observed for higher whiskers content (6.2 wt% and up). These observations were discussed according to the possible interactions between hydroxyl groups on the cellulosic surface and starch, the selective partitioning of glycerol and water in the bulk starch matrix or at whiskers surface, and the restriction of amorphous starch chains mobility in the vicinity of the starch crystallite coated filler surface. For glycerol plasticized starch reinforced with cellulose crystallites prepared from cottonseed linter (Lu *et al.*, 2005), an increase of T_g with filler content was reported and attributed to cellulose/starch interactions. For tunicin whiskers/sorbitol plasticized starch (Mathew and Dufresne, 2002), T_g increased slightly up to about 15 wt% whiskers and decreased for higher whiskers loading. Crystallization of amylopectin chains upon whiskers addition and migration of sorbitol molecules to the amorphous domains were proposed to explain the observed modifications. When using a PVA matrix, an increase in T_g was reported when the cellulose whiskers

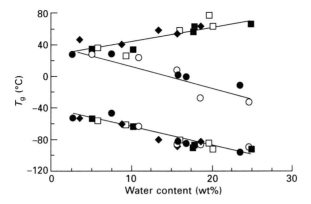

3.6 Glass–rubber transition temperatures associated with the midpoints of the transitions versus water content for glycerol plasticized waxy maize starch filled with 0 (●), 3.2 (○), 6.2 (■), 16.7 (□), and 25 wt% (◆) tunicin whiskers. Solid lines serve to guide the eye (reprinted with permission from Anglès and Dufresne, 2000, copyright American Chemical Society).

content increased (Garcia de Rodriguez *et al.*, 2006; Roohani *et al.*, 2008). A similar observation was reported for CMC reinforced with cotton cellulose whiskers (Choi and Simonsen, 2006).

The melting temperature, T_m, was reported to be nearly independent of the filler content in plasticized starch (Anglès and Dufresne, 2000; Mathew and Dufresne, 2002) and in POE-based materials (Azizi Samir *et al.*, 2004a; 2004c; 2004e) filled with tunicin whiskers. The same observation was reported for CAB reinforced with native bacterial cellulose whiskers (Grunert and Winter, 2002). However, for the latter system, T_m increased when the amount of trimethylsilylated whiskers increased. This difference was attributed to the stronger filler–matrix interaction for chemically modified whiskers. A decrease in both T_m and degree of crystallinity of PVA was reported when adding cellulose nanocrystals (Roohani *et al.*, 2008). However, for electrospun cellulose whiskers reinforced PVA nanofibers, the crystallinity was found to be reduced upon filler addition (Peresin *et al.*, 2010).

A significant increase in crystallinity of sorbitol plasticized starch (Mathew and Dufresne, 2002) was reported when increasing the cellulose whiskers content. This phenomenon was attributed to an anchoring effect of the cellulosic filler, probably acting as a nucleating agent. For POE-based composites the degree of crystallinity of the matrix was found to be roughly constant up to 10 wt% tunicin whiskers (Azizi Samir *et al.*, 2004a; 2004c; 2004e) and to decrease for higher loading level (Azizi Samir *et al.*, 2004c). The nucleating effect of cellulosic nanocrystals appears to be mainly governed by surface chemical considerations (Ljungberg *et al.*, 2006). It was shown from both x-ray diffraction and DSC analysis that the crystallization behavior of films containing unmodified and surfactant-modified whiskers displayed two crystalline forms (α and β), whereas the neat matrix and the nanocomposite reinforced with nanocrystals grafted with maleated polypropylene only crystallized in the α form. It was suspected that the more hydrophilic the whisker surface, the more it appeared to favor the appearance of the β phase. Grunert and Winter (2002) observed from DSC measurements that native bacterial fillers impede the crystallization of the CAB matrix whereas silylated ones help to nucleate the crystallization. The crystallinity of N-octadecyl isocyanate-grafted sisal whiskers reinforced PCL was found to increase upon filler addition, whereas no influence of N-octadecyl isocyanate-grafted sisal MFC was reported (Siqueira et al., 2009). This difference was attributed to the possibility of entanglement of MFC that tends to confine the polymeric matrix and restrict its crystallization.

For tunicin whiskers filled semi-crystalline matrices such as PHO (Dufresne et al., 1999) and glycerol plasticized starch (Anglès and Dufresne, 2000) a transcrystallization phenomenon was reported. For glycerol-plasticized starch-based systems, the formation of the transcrystalline zone around the whiskers was assumed to be caused by the accumulation of plasticizer in the cellulose/amylopectin interfacial zones improving the ability of amylopectin chains to crystallize. This transcrystalline zone could originate from a glycerol–starch V structure. In addition, the inherent restricted mobility of amylopectin chains was proposed to explain the lower water uptake of cellulose/starch composites for increasing filler content. Transcrystallization of PP at cellulose nanocrystal surfaces was identified and found to result from enhanced nucleation owing to some form of epitaxy (Gray, 2008).

The presence of sulfate groups introduced at the surface of the whiskers during hydrolysis with H_2SO_4 promoted their thermal decomposition (Roman and Winter, 2004; Li et al., 2009). Thermogravimetric analysis (TGA) experiments were performed to investigate the thermal stability of tunicin whiskers/POE nanocomposites (Azizi Samir et al., 2004a; 2004c). No significant influence of the cellulosic filler on the degradation temperature of the POE matrix was reported. Cellulose nanocrystals content appeared to have an effect on the thermal behavior of CMC plasticized with glycerin (Choi and Simonsen, 2006) suggesting a close association between the filler and the matrix. The thermal degradation of unfilled CMC was observed from its melting point (270 °C) and it had a very narrow temperature range of degradation. Cellulose nanocrystals degraded at a lower temperature (230 °C) than CMC, but showed a very broad degradation temperature range. The degradation of cellulose whiskers reinforced CMC was observed between these two limits, but of interest was the lack of steps: the composites were reported to degrade as a unit.

3.6.3   Mechanical properties

Nanoscale dimensions resulting in a very high surface area-to-volume ratio and impressive mechanical properties of rod-like cellulose whiskers have attracted significant interest in the last fifteen years. These characteristics along with the remarkable suitability for surface functionalization make them ideal candidates for improving the mechanical properties of the host material.

The first demonstration of the reinforcing effect of cellulose whiskers in a poly(S-co-BuA) matrix was reported by Favier *et al.* (1995a; 1995b), who measured by dynamic mechanical analysis (DMA) a spectacular improvement in the storage modulus after adding tunicin whiskers even at low content into the host polymer. This increase was especially significant above the T_g of the thermoplastic matrix because of its poor mechanical properties in this temperature range. Figure 3.7 shows the isochronal evolution of the logarithm of the relative storage shear modulus (log G'_T/G'_{200}, where G'_{200} corresponds to the experimental value measured at 200 K) at 1 Hz as a function of temperature for such composites prepared by water evaporation. In the rubbery state of the thermoplastic matrix, the modulus of the composite with a loading level as low as 6 wt% is more than two orders of magnitude higher than the one of the unfilled matrix. Moreover, the introduction of 3 wt% or more cellulosic whiskers provides an outstanding thermal stability of the matrix modulus up to the temperature at which cellulose starts to

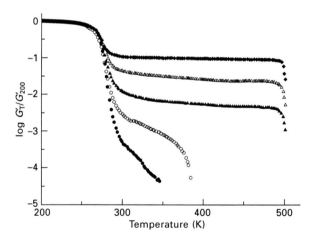

3.7 Logarithm of the normalized storage shear modulus (log G'_T/G'_{200}, where G'_{200} corresponds to the experimental value measured at 200 K) versus temperature at 1 Hz for tunicin whiskers reinforced poly(S-co-BuA) nanocomposite films obtained by water evaporation and filled with 0 (●), 1 (○), 3 (▲), 6 (△) and 14 wt% (◆) of cellulose whiskers (reprinted with permission from Azizi Samir *et al.*, 2005a, copyright American Chemical Society).

degrade (500 K). Since this pioneer work, many studies have reported the reinforcing capability of cellulose nanoparticles.

This outstanding reinforcing effect was ascribed to a mechanical percolation phenomenon (Favier *et al.*, 1995a; 1995b). A good agreement between experimental and predicted data was reported when using the series–parallel model of Takayanagi modified to include a percolation approach. It was suspected that all the stiffness of the material was caused by infinite aggregates of cellulose whiskers. Above the percolation threshold the cellulosic nanoparticles can connect and form a three dimensional continuous pathway through the nanocomposite film. For rod-like particles, such as tunicin whiskers with an aspect ratio of 67, the percolation threshold is close to 1 vol% (Favier *et al.*, 1997a). The formation of this cellulose network was supposed to result from strong interactions between whiskers, such as hydrogen bonds (Favier *et al.*, 1997b). This phenomenon is similar to the high mechanical properties observed for a paper sheet, which result from the hydrogen-bonding forces that hold the percolating network of fibers. This mechanical percolation effect explains both the high reinforcing effect and the thermal stabilization of the composite modulus for evaporated films. Any factor that affects the formation of the percolating whiskers network or interferes with it changes the mechanical performances of the composite (Dufresne, 2006). Three main parameters were reported to affect the mechanical properties of such materials, viz. (i) the morphology and dimensions of the nanoparticles, (ii) the processing method, and (iii) the microstructure of the matrix and matrix/filler interactions.

Morphology and dimensions of the nanoparticles

Cellulose nanocrystals occur as rod-like nanoparticles, the geometrical aspect ratio is an important factor because it determines the percolation threshold value. This factor is linked to the source of cellulose and whiskers preparation conditions. A higher reinforcing effect is obtained for nanocrystals with a high aspect ratio. For instance, the rubbery storage tensile modulus was systematically lower for wheat straw whiskers/poly(S-co-BuA) composites than for tunicin whiskers based materials (Dufresne, 2006). Also, the flexibility and tangling possibility of the nanofibers plays an important role (Azizi Samir *et al.*, 2004b; Bendahou *et al.*, 2009; Siqueira *et al.*, 2009). It was reported that entangled MFC induces a higher reinforcing effect than straight whiskers, whereas the elongation at break was lower.

The processing method

The processing method affects the possibility of formation of a continuous whiskers network and thus the final properties of the nanocomposite material.

Slow processes such as casting/evaporation were reported to give the highest mechanical performance materials compared with freeze-drying/molding and freeze-drying/extruding/molding techniques. During slow water evaporation, because of Brownian motion in the suspension or solution (the viscosity of which remains low up to the end of the process when the latex particle or polymer concentration becomes very high), the rearrangement of the nanoparticles is possible. They have sufficient time to interact and connect to form a continuous network which is the basis of their reinforcing effect. The resulting structure is completely relaxed and direct contacts between nanoparticles are then created. Conversely, during the freeze-drying/hot-pressing process, the nanoparticle arrangement in the suspension is first frozen, and then, during the hot-pressing stage, because of the polymer melt viscosity, the particle rearrangements are strongly limited.

Microstructure of matrix and matrix/filler interactions

The microstructure of the matrix and the resulting competition between matrix/filler and filler/filler interactions also affect the mechanical behavior of cellulose nanocrystals reinforced nanocomposites. Classical composite science tends to favor matrix/filler interactions as a fundamental condition for optimal performance. For cellulose whiskers based composite materials, the opposite trend is generally observed when the material is processed via casting/evaporation method. This unusual behavior is ascribed to the originality of the reinforcing phenomenon of cellulosic nanoparticles resulting from the formation of a H-bonded percolating network. However, when using a processing route other than casting/evaporation in water medium, the dispersion of the hydrophilic filler in the polymeric matrix is also involved (de Menezes *et al.*, 2009) and improved filler/matrix interactions generally lead to higher mechanical properties. In non-percolating systems, for instance for materials processed from freeze-dried cellulose nanocrystals, strong matrix/filler interactions enhance the reinforcing effect of the filler (Ljungberg *et al.*, 2005). The transcrystallization phenomenon reported for PHO (Dufresne *et al.*, 1999) and plasticized starch (Anglès and Dufresne, 2000) on cellulose whiskers resulted in a decrease of the mechanical properties (Anglès and Dufresne, 2001) because of the coating of the nanoparticles with crystalline domains. When using unhydrolyzed cellulose microfibrils extracted from potato pulp rather than cellulose nanocrystals to reinforce glycerol plasticized thermoplastic starch, a completely different mechanical behavior was reported (Dufresne and Vignon, 1998; Dufresne *et al.*, 2000) and a significant reinforcing effect was observed. It was suspected that a tangling effect contributed to this high reinforcing effect (Anglès and Dufresne, 2001).

3.6.4   Swelling properties

Swelling or kinetics of solvent absorption is a method that can highlight specific interactions between the filler and the matrix. It usually involves first drying and weighing the sample, and then immersing it in the liquid solvent or exposing it to the vapor medium. The sample is then removed at specific intervals and weighed until an equilibrium value is reached. The swelling rate of the sample can be calculated by dividing the gain in weight by the initial weight. Generally, the short-term behavior displays a fast absorption phenomenon whereas in the long term, the kinetics of absorption is low and leads to a plateau, corresponding to the solvent uptake at equilibrium. The diffusion coefficient can be determined from the initial slope of the solvent uptake curve as a function of time. For cellulosic particles reinforced composites, it is generally of interest to investigate the water absorption of the material because of the hydrophilic nature of the reinforcing phase. When a nonpolar polymeric matrix is used, the absorption of a nonaqueous liquid can be investigated.

A higher resistance of thermoplastic starch to water was reported when increasing the cellulose nanoparticles content (Anglès and Dufresne, 2000; Lu et al., 2005; Svagan et al., 2009). Both the water uptake and the diffusion coefficient of water decreased upon nanoparticles addition. These phenomena were ascribed to the presence of strong hydrogen bonding interactions between particles and between the starch matrix and cellulose whiskers. The hydrogen bonding interactions in the composites tend to stabilize the starch matrix when it is submitted to a highly moist atmosphere. Moreover, the high crystallinity of cellulose might also be responsible for the decreased water uptake at equilibrium and diffusion coefficient of the material. A lower water uptake and dependence on cellulose whiskers content were reported when using sorbitol rather than glycerol as plasticizer for the starch matrix (Mathew and Dufresne, 2002). An explanation was proposed based on the chemical structure of both plasticizers, more accessible end hydroxyl groups in glycerol being about twice those of sorbitol. Similar results were reported for cellulose whiskers reinforced SPI (Wang et al., 2006) and CMC (Choi and Simonsen, 2006). Sisal whisker addition was found to stabilize PVA-based nanocomposites with no benefit seen when increasing the whisker content beyond the percolation threshold (Garcia de Rodriguez et al., 2006). A lower water uptake was observed when using MFC instead of cellulose whiskers as a reinforcing phase in NR (Bendahou et al., 2010). This observation was explained by the difference in the structure and composition of both nanoparticles and, in particular, by the presence of residual lignin, extractive substances and fatty acids at the surface of MFC that limits, comparatively, the hydrophilic character of the filler. In addition, assuming that the filler/matrix compatibility was consequently lower for whiskers-based nanocomposites, one can imagine that water infiltration could be easier at the filler/matrix interface. For MFC-based nanocomposites, despite higher amorphous cellulose content, the higher hydrophobic character of the filler favors the compatibility with NR and therefore restricts the interfacial diffusion pathway for water.

Swelling experiments of PVC reinforced with tunicin whiskers were conducted in methyl ethyl ketone (MEK) (Chazeau *et al.*, 1999a). A significant decrease in swelling was observed when increasing the cellulose whiskers content, which was assumed to be caused by the existence of an interphase making a link between nanoparticles, thus allowing the formation of a flexible network. The swelling behavior in toluene of poly(S-co-BuA) reinforced with cellulose fibrils from *Opuntia ficus-indica* cladodes was reported by Malainine *et al.* (2005). They observed a strong toluene resistance even at very low filler loading. Although the unfilled matrix completely dissolved in toluene, only 27 wt% of the polymer was able to dissolve when filled with only 1 wt% of cellulose microfibrils. This phenomenon was attributed to the presence of a three-dimensional entangled cellulosic network which strongly restricted the swelling capability and dissolution of the matrix. For higher microfibrils content, no significant evolution was observed because of both the overlapping of the microfibrils restricting the filler–matrix interfacial area and the decrease of the entrapping matrix fraction owing to the densification of the microfibrils network. Similar experiments were conducted with cellulose microfibrils obtained from sugar beet pulp reinforced poly(S-co-BuA) (Dalmas *et al.*, 2006). Toluene resistance of nanocomposites was found to be less significant than for microfibrils from *Opuntia ficus-indica* cladodes and to evolve with filler content. It was observed that the cohesion of composites prepared by evaporation was higher than the one of freeze-dried/hot-pressed materials. This difference was attributed to the presence of a H-bonded network in the former samples. It was concluded that the solvent did not have any effect on the hydrogen bonds of the cellulose network present in evaporated composites. On the contrary, for freeze-dried/hot-pressed materials, the lower interactions created between fibrils and polymer chains meant that they were able to be more easily disentangled and dissolved by the solvent. The swelling behavior of cellulose whiskers reinforced NR in toluene as reported by Bendahou *et al.* (2010) was found to strongly decrease even with only 1 wt% of cellulose nanoparticles and to be almost independent of filler content and nature (MFC or whiskers).

3.6.5   Barrier properties

Petrochemical-based polymers predominate in the packaging of foods because of their ease of processing, excellent barrier properties and low cost. However, there is currently an increasing interest in replacing conventional synthetic polymers by more sustainable materials. One promising application area of cellulosic nanoparticles is barrier membranes, where the nano-sized fillers impart enhanced mechanical and barrier properties. Research in this area is burgeoning and evolving rapidly to enhance the barrier properties and to overcome certain limitations. In addition, it is well known that molecules penetrate with difficulty in the crystalline domains of cellulose microfibrils. Moreover, the ability of cellulosic nanoparticles to form a dense percolating network held together by strong inter-particle bonds suggest their use as barrier films.

The oxygen permeability of paper was considerably reduced when coated with MFC layer (Syverud and Stenius, 2009). The water vapor transmission rate of cotton nanocrystals reinforced CMC (Choi and Simonsen, 2006) and PVA (Paralikar *et al.*, 2008) films were reported to decrease. Cellulose nanoparticles prepared by drop-wise addition of ethanol/HCl aqueous solution into a NaOH/urea/H_2O suspension of MCC were found to decrease the water vapor permeability of glycerol plasticized-starch films (Chang *et al.*, 2010). The water vapor barrier of mango puree based edible (Azeredo *et al.*, 2009) and chitosan films films (Azeredo *et al.*, 2010) were successfully improved by adding cellulose whiskers

3.7   Conclusions and future trends

The potential of nanocomposites in various sectors of research and application is promising and attracting increasing investment. Owing to their abundance, high strength and stiffness, low weight and biodegradability, nano-scale cellulose fiber materials serve as promising candidates for the preparation of bionanocomposites. A broad range of applications of nanocellulose exists even though there are still a high number of unknown possibilities. Tens of scientific publications and experts show its potential even if most of the studies focus on their mechanical properties as reinforcing phase and their liquid crystal self-ordering properties. Packaging is one area in which nanocellulose-reinforced polymeric films can be of interest because of the possibility of producing films with high transparency and improved mechanical and barrier properties. However, although there have been many promising achievements on the laboratory or pilot scale, there are several challenges to overcome in order to be able to produce cellulose-based nanocomposites on an industrial scale.

Two important programs have recently started to produce nanocellulose. One is the creation in 2008 in Finland of the 'Suomen Nanoselluloosakeskus' Centre or 'Finnish Centre for Nanocellulosic Technologies'. One of the challenges is to produce large quantities of microfibrils of uniform quality. The forest industry in Finland is going through a major transition, and the utilisation of new technologies is expected to provide a means for strengthening the competitiveness in the sector. The other program, supported by the

Canadian government and represented by FPInnovations, is the creation of the ArboraNano network. The objective of this program for the valorization of nanocellulose (as cellulose nanocrystals) is also to revive the forestry sectors in Canada, strongly affected by the growing competition from emerging countries in Asia and South America.

However, it is worth noting that there are many safety concerns about nanomaterials, as their size allows them to penetrate into cells and eventually remain in the system. There is no consensus about categorizing nanomaterials as new materials.

3.8   References

Anglès M N, Dufresne A (2000), 'Plasticized starch/tunicin whiskers nanocomposites: 1. Structural analysis', *Macromolecules*, **33**, 8344–8353.

Anglès M N, Dufresne A (2001), 'Plasticized starch/tunicin whiskers nanocomposites: 2. Mechanical behavior', *Macromolecules*, **34**, 2921–2931.

Araki J, Wada M, Kuga S (2001), 'Steric stabilization of a cellulose microcrystal suspension by poly(ethylene glycol) grafting', *Langmuir*, **17**, 21–27.

Araki J, Wada M, Kuga S, Okano T (1998), 'Flow properties of microcrystalline cellulose suspension prepared by acid treatment of native cellulose', *Colloids Surf. A*, **142**, 75–82.

Araki J, Wada M, Kuga S, Okano T (1999), 'Influence of surface charges on viscosity behavior of cellulose microcrystal suspension', *J. Wood Sci.*, **45**, 258–261.

Aulin C, Johansson E, Wågberg L, Lindström T (2010), 'Self-organized films from cellulose I nanofibrils using the layer-by-layer technique', *Biomacromolecules*, **11**, 872–882, doi:10.1021/bm100075e.

Azeredo H M, Mattoso L H C, Avena-Bustillos R J, Filho G C, Munford M L, Wood D, McHugh T H (2010), 'Nanocellulose reinforced chitosan composite films as affected by nanofiller loading and plasticizer content', *J. Food Sci.*, 75, N1–N7, doi:10.1111/j.1750–3841.2009.01386.x.

Azeredo H M, Mattoso L H C, Wood D, Williams T G, Avena–Bustillos R J, McHugh T H (2009), 'Nanocomposite edible films from mango puree reinforced with cellulose nanofibers', *J. Food Sci.*, **74**, N31–N35.

Azizi Samir M A S, Alloin F, Dufresne A (2005a), 'Review of recent research into cellulosic whiskers, their properties and their application in nanocomposite field', *Biomacromolecules*, **6**, 612–626.

Azizi Samir M A S, Alloin F, Dufresne A (2006), 'High performance nanocomposite polymer electrolytes', *Compos. Interfaces*, **13**, 545–559.

Azizi Samir M A S, Alloin F, Gorecki W, Sanchez J Y, Dufresne A (2004a), 'Nanocomposite polymer electrolytes based on poly(oxyethylene) and cellulose nanocrystals', *J. Phys. Chem. B*, **108**, 10845–10852.

Azizi Samir M A S, Alloin F, Paillet M, Dufresne A (2004b), 'Tangling effect in fibrillated cellulose reinforced nanocomposites', *Macromolecules*, **37**, 4313–4316.

Azizi Samir M A S, Alloin F, Sanchez J Y, Dufresne A (2004c), 'Cellulose nanocrystals reinforced poly(oxyethylene)', *Polymer*, **45**, 4033–4041.

Azizi Samir M A S, Alloin F, Sanchez J Y, Dufresne A (2004d), 'Cross-linked nanocomposite polymer electrolytes reinforced with cellulose whiskers', *Macromolecules*, **37**, 4839–4844.

Azizi Samir M A S, Alloin F, Sanchez J Y, El Kissi N, Dufresne A (2004e), 'Preparation of cellulose whiskers reinforced nanocomposites from an organic medium suspension', *Macromolecules*, **37**, 1386–1393.

Azizi Samir M A S, Chazeau L, Alloin F, Cavaillé J Y, Dufresne A, Sanchez J Y (2005b), 'POE–based nanocomposite polymer electrolytes reinforced with cellulose whiskers, *Electrochim. Acta*', **50**, 3897–3903.

Azizi Samir M A S, Montero Mateos A, Alloin F, Sanchez J Y, Dufresne A (2004f), 'Plasticized nanocomposite polymer electrolytes based on poly(oxyethylene) and cellulose whiskers', *Electrochim. Acta*, **49**, 4667–4677.

Beck-Candanedo S, Roman M, Gray D G (2005), 'Effect of reaction conditions on the properties and behavior of wood cellulose nanocrystal suspensions', *Biomacromolecules*, **6**, 1048–1054.

Ben Elmabrouk A, Thielemans W, Dufresne A, Boufi S (2009), 'Preparation of poly(styrene–co–hexylacrylate)/cellulose whiskers nanocomposites via miniemulsion polymerization', *J. Appl. Polym. Sci.*, **114**, 2946–2955.

Bendahou A, Habibi Y, Kaddami H, Dufresne A (2009), 'Physico-chemical characterization of palm from *Phoenix Dactylifera* L, preparation of cellulose whiskers and natural rubber-based nanocomposites', *J. Biobased Mat. Bioenergy*, **3**, 81–90.

Bendahou A, Kaddami H, Dufresne A (2010), 'Investigation on the effect of cellulosic nanoparticles' morphology on the properties of natural rubber based nanocomposites', *Eur. Polym. J.*, **46**, 609–620.

Bondenson D, Mathew A, Oksman K (2006), 'Optimization of the isolation of nanocrystals from microcrystalline cellulose by acid hydrolysis', *Cellulose*, **13**, 171–180.

Bondenson D, Oksman K (2007), 'Polylactic acid/cellulose whisker nanocomposites modified by polyvinyl alcohol', *Composites: Part A*, **38**, 2486–2492.

Braun B, Dorgan J R (2009), 'Single-step method for the isolation and surface functionalization of cellulosic nanowhiskers', *Biomacromolecules*, **10**, 334–341.

Braun B, Dorgan J R, Chandler J P (2008), 'Cellulosic nanowhiskers. Theory and application of light scattering from polydisperse spheroids in the Rayleigh–Gans–Debye regime', *Biomacromolecules*, **9**, 1255–1263.

Cao X, Dong H, Li C M (2007), 'New nanocomposite materials reinforced with flax cellulose nanocrystals in waterborne polyurethane', *Biomacromolecules*, **8**, 899–904.

Cao X, Habibi Y, Lucia L A (2009), 'One-pot polymerization, surface grafting, and processing of waterborne polyurethane-cellulose nanocrystal nanocomposites', *J. Mater. Chem.*, **19**, 7137–7145.

Chang P R, Jian R, Zheng P, Yu J, Ma X (2010), 'Preparation and properties of glycerol plasticized–starch (GPS)/cellulose nanoparticle (CN) composites', *Carbohydr. Polym.*, **79**, 301–305.

Chauve G, Heux L, Arouini R, Mazeau K (2005), 'Cellulose poly(ethylene-co-vinyl acetate) nanocomposites studied by molecular modeling and mechanical spectroscopy', *Biomacromolecules*, **6**, 2025–2031.

Chazeau L, Cavaillé J Y, Canova G R, Dendievel R, Boutherin B (1999a), 'Viscoelastic properties of plasticized PVC reinforced with cellulose whiskers', *J. Appl. Polym. Sci.*, **71**, 1797–1808.

Chazeau L, Cavaillé J Y, Perez J (2000), 'Plasticized PVC reinforced with cellulose whiskers. II. Plastic behavior', *J. Polym. Sci. B: Polym. Phys.*, **38**, 383–392.

Chazeau L, Cavaillé J Y, Terech P (1999b), 'Mechanical behaviour above T_g of a plasticized PVC reinforced with cellulose whiskers, a SANS structural study', *Polymer*, **40**, 5333–5344.

Chazeau L, Paillet M Cavaillé J Y (1999c), 'Plasticized PVC reinforced with cellulose whiskers 1. Linear viscoelastic behavior analyzed through the quasi point defect theory', *J. Polym. Sci. B: Polym. Phys.*, **37**, 2151–2164.

Chen Y, Liu C, Chang P R, Cao X, Anderson D P (2009), 'Bionanocomposites based on pea starch and cellulose nanowhiskers hydrolyzed from pea hull fibre: effect of hydrolysis time', *Carbohydr. Polym.*, **76**, 607–615.

Choi Y J, Simonsen J (2006), 'Cellulose nanocrystal-filled carboxymethyl cellulose nanocomposites', *J. Nanosci. Nanotechnol.*, **6**, 633–639.

Cranston E D, Gray D G (2006a), 'Formation of cellulose-based electrostatic layer-by-layer films in a magnetic field', *Sci. Technol. Adv. Mater.*, **7**, 319–321.

Cranston E D, Gray D G (2006b), 'Morphological and optical characterization of polyelectrolyte multilayers incorporating nanocrystalline cellulose', *Biomacromolecules*, **7**, 2522–2530.

Dalmas F, Chazeau L, Gauthier C, Cavaillé J Y, Dendievel R (2006), 'Large deformation mechanical behavior of flexible nanofiber filled polymer nanocomposites', *Polymer* **47**, 2802–2812.

de Menezes A J, Siqueira G, Curvelo A A S, Dufresne A (2009), 'Extrusion and characterization of functionalized cellulose whisker reinforced polyethylene nanocomposites', *Polymer*, **50**, 4552–4563.

de Mesquita J P, Donnici C L, Pereira F V (2010), 'Biobased nanocomposites from layer-by-layer assembly of cellulose nanowhiskers with chitosan', *Biomacromolecules*, **11**, 473–480.

De Souza Lima M M, Wong J T, Paillet M, Borsali R, Pecora R (2003), 'Translational and rotational dynamics of rodlike cellulose whiskers', *Langmuir*, **19**, 24–29.

Dong S, Roman M (2007), 'Fluorescently labeled cellulose nanocrystals for bioimaging applications', *J. Am. Chem. Soc.*, **129**, 13810–13811.

Dong X M, Revol J F, Gray D G (1998), 'Effect of microcrystallite preparation conditions on the formation of colloid crystals of cellulose', *Cellulose*, **5**, 19–32.

Dubief D, Samain E, Dufresne A (1999), 'Polysaccharide microcrystals reinforced amorphous poly(β-hydroxyoctanoate) nanocomposite materials', *Macromolecules*, **32**, 5765–5771.

Dufresne A (2000), 'Dynamic mechanical analysis of the interphase in bacterial polyester/cellulose whiskers natural composites', *Compos. Interfaces*, **7**, 53–67.

Dufresne A (2006), 'Comparing the mechanical properties of high performances polymer nanocomposites from biological sources', *J. Nanosci. Nanotechnol.*, **6**, 322–330.

Dufresne A, Cavaillé J Y, Helbert W (1997), 'Thermoplastic nanocomposites filled with wheat straw cellulose whiskers. Part II: Effect of processing and modeling', *Polym. Compos.*, **18**, 198–210.

Dufresne A, Dupeyre D, Vignon M R (2000), 'Cellulose microfibrils from potato cells : processing and characterization of starch/cellulose microfibrils composites', *J. Appl. Polym. Sci.*, **76**, 2080–2092.

Dufresne A, Kellerhals M B, Witholt B (1999), 'Transcrystallization in mcl–PHAs/cellulose whiskers composites', *Macromolecules*, **32**, 7396–7401.

Dufresne A, Vignon M R (1998), 'Improvement of starch films performances using cellulose microfibrils', *Macromolecules*, **31**, 2693–2696.

Ebeling T, Paillet M, Borsali R, Diat O, Dufresne A, Cavaillé J Y, Chanzy H (1999), 'Shear-induced orientation phenomena in suspensions of cellulose microcrystals, revealed by small angle x-ray scattering', *Langmuir*, **15**, 6123–6126.

Elazzouzi-Hafraoui S, Nishiyama Y, Putaux J L, Heux L, Dubreuil F, Rochas C (2008), 'The shape and size distribution of crystalline nanoparticles prepared by acid hydrolysis of native cellulose', *Biomacromolecules*, **9**, 57–65.

Eyholzer Ch., Bordeanu N, Lopez-Suevos F, Rentsch D, Zimmermann T, Oksman K (2010), 'Preparation and characterization of water-redispersible nanofibrillated cellulose in powder form', *Cellulose*, **17**, 19–30.

Favier V, Canova G R, Cavaillé J Y, Chanzy H, Dufresne A, Gauthier C (1995a), 'Nanocomposites materials from latex and cellulose whiskers', *Polym. Adv. Technol.*, **6**, 351–355.

Favier V, Canova G R, Shrivastava S C, Cavaillé J Y (1997b), 'Mechanical percolation in cellulose whiskers nanocomposites', *Polym. Eng. Sci.*, **37**, 1732–1739

Favier F, Chanzy H, Cavaillé J Y (1995b), 'Polymer nanocomposites reinforced by cellulose whiskers', *Macromolecules*, **28**, 6365–6367.

Favier V, Dendievel R, Canova G R, Cavaillé J Y, Gilormini P (1997a), 'Simulation and modeling of three-dimensional percolating structures: case of a latex matrix reinforced by a network of cellulose fibers', *Acta Mater.*, **45**, 1557–1565.

Fleming K, Gray DG, Prasannan S, Matthews S (2000), 'Cellulose crystallites: a new and robust liquid crystalline medium for the measurement of residual dipolar couplings', *J. Appl. Polym. Sci.*, **113**, 927–935.

Fukuzumi H, Saito T, Iwata T, Kumamoto Y, Isogai A (2009), 'Transparent and high gas barrier films of cellulose nanofibres prepared by TEMPO-mediated oxidation', *Biomacromolecules*, **10**, 162–165.

Garcia de Rodriguez N L, Thielemans W, Dufresne A (2006), 'Sisal cellulose whiskers reinforced polyvinyl acetate nanocomposites', *Cellulose*, **13**, 261–270.

Goussé C, Chanzy H, Exoffier G, Soubeyrand L, Fleury E (2002), 'Stable suspensions of partially silylated cellulose whiskers dispersed in organic solvents', *Polymer*, **43**, 2645–2651.

Gray D G (2008), 'Transcrystallization of polypropylene at cellulose nanocrystal surfaces', *Cellulose*, **15**, 297–301.

Grunert M, Winter W T (2002), 'Nanocomposites of cellulose acetate butyrate reinforced with cellulose nanocrystals', *J. Polym. Environ.*, **10**, 27–30.

Habibi Y, Dufresne A (2008), 'Highly filled bionanocomposites from functionalized polysaccharides nanocrystals', *Biomacromolecules*, **9**, 1974–1980.

Habibi Y, Goffin A L, Schiltz N, Duquesne E, Dubois P, Dufresne A (2008), 'Bionanocomposites based on poly(ε-caprolactone)-grafted cellulose nanocrystals by ring opening polymerization', *J. Mat. Chem.*, **18**, 5002–5010.

Hajji P, Cavaillé J Y, Favier V, Gauthier C, Vigier G (1996), 'Tensile behavior of nanocomposites from latex and cellulose whiskers', *Polym. Compos.*, **17**, 612–619.

Hanley S J, Giasson J, Revol J F, Gray D G (1992), 'Atomic force microscopy of cellulose microfibrils: comparison with transmission electron microscopy', *Polymer*, **33**, 4639–4642.

Helbert W, Cavaillé J Y, Dufresne A (1996), 'Thermoplastic nanocomposites filled with wheat straw cellulose whiskers. Part I. Processing and mechanical behavior', *Polym. Compos.*, **17**, 604–611.

Henriksson M, Berglund L A (2007), 'Structure and properties of cellulose nanocomposite films containing melamine formaldehyde', *J. Appl. Polym. Sci.*, **106**, 2817–2824.

Heux L, Chauve G, Bonini C (2000), 'Nonflocculating and chiral-nematic self-ordering of cellulose microcrystals suspensions in nonpolar solvents', *Langmuir*, **16**, 8210–8212.

Ifuku S, Nogi M, Abe K, Handa K, Nakatsubo F, Yano H (2007), 'Surface modification of bacterial cellulose nanofibres for property enhancement of optically transparent composites: dependence on acetyl-group DS', *Biomacromolecules*, **8**, 1973–1978.

Iwamoto S, Abe K, Yano H (2008), 'The effect of hemicelluloses on wood pulp nanofibrillation and nanofiber network characteristics', *Biomacromolecules*, **9**, 1022–1026.

Iwamoto S, Kai W, Isogai A, Iwata T (2009), 'Elastic modulus of single cellulose microfibrils from tunicate measured by atomic force microscopy', *Biomacromolecules*, **10**, 2571–2576.

Jean B, Dubreuil F, Heux L, Cousin F (2008), 'Structural details of cellulose nanocrystals/polyelectrolytes multilayers probed by neutron reflectivity and AFM', *Langmuir*, **24**, 3452–3458.

Jean B, Heux L, Dubreuil F, Chambat G, Cousin F (2009), 'Non-electrostatic building of biomimetic cellulose-xyloglucan multilayers', *Langmuir*, **25**, 3920–3923.

Kim U J, Kuga S, Wada M, Okano T, Kondo T (2000), 'Periodate oxidation of crystalline cellulose', *Biomacromolecules*, **1**, 488–492.

Kvien I, Tanem B S, Oksman K (2005), 'Characterization of cellulose whiskers and their nanocomposites by atomic force and electron microscopy', *Biomacromolecules*, **6**, 3160–3165.

Kvien I, Sugiyama J, Votrubec M, Oksman K (2007), 'Characterization of starch based nanocomposites', *J. Mater. Sci.*, **42**, 8163–8171.

Li R, Fei J, Cai Y, Li Y, Feng J, Yao J (2009), 'Cellulose whiskers extracted from mulberry: a novel biomass production', *Carbohydr. Polym.*, **76**, 94–99.

Li Z, Renneckar S, Barone J R (2010), 'Nanocomposites prepared by in situ enzymatic polymerization of phenol with TEMPO-oxidized nanocellulose', *Cellulose*, **17**, 57–68.

Lin N, Chen G, Huang J, Dufresne A, Chang P R (2009), 'Effects of polymer-grafted natural nanocrystals on the structure and mechanical properties of poly(lactic acid): a case of cellulose whisker-graft-polycaprolactone', *J. Appl. Polym. Sci.*, **113**, 3417–3425.

Ljungberg N, Bonini C, Bortolussi F, Boisson C, Heux L, Cavaillé J Y (2005), 'New nanocomposite materials reinforced with cellulose whiskers in atactic polypropylene: effect of surface and dispersion characteristics', *Biomacromolecules*, **6**, 2732–2739.

Ljungberg N, Cavaillé J Y, Heux L (2006), 'Nanocomposites of isotactic polypropylene reinforced with rod-like cellulose whiskers', *Polymer*, **47**, 6285–6292.

Lu Y, Weng L, Cao X (2005), 'Biocomposites of plasticized starch reinforced with cellulose crystallites from cottonseed linter', *Macromol. Biosci.*, **5**, 1101–1107.

Lu Y, Weng L, Cao X (2006), 'Morphological, thermal and mechanical properties of ramie crystallites-reinforced plasticized starch biocomposites', *Carbohydr. Polym.*, **63**, 198–204.

Lu J, Wang T, Drzal L T (2008), 'Preparation and properties of microfibrillated cellulose polyvinyl alcohol composite materials', *Composites: Part A*, **39**, 738–746.

Malainine M E, Mahrouz M, Dufresne A (2005), 'Thermoplastic nanocomposites based on cellulose microfibrils from *Opuntia ficus-indica* parenchyma cell', *Compos. Sci. Technol.*, **65**, 1520–1526.

Marchessault R H, Morehead F F, Joan Koch M (1961), 'Hydrodynamics properties of neutral suspensions of cellulose crystallites as related to size and shape', *J. Colloid Sci.*, **16**, 327–344.

Marcovich N E, Auad M L, Belessi N E, Nutt S R, Aranguren M I (2006), 'Cellulose micro/nanocrystals reinforced polyurethane', *J. Mater. Res.*, **21**, 870–881.

Mathew A P, Dufresne A (2002), 'Morphological investigation of nanocomposites from sorbitol plasticized starch and tunicin whiskers', *Biomacromolecules*, **3**, 609–617.

Mathew A P, Thielemans W, Dufresne A (2008), 'Mechanical properties of nanocomposites from sorbitol plasticized starch and tunicin whiskers', *J. Appl. Polym. Sci.*, **109**, 4065–4074.

Matos Ruiz M, Cavaillé J Y, Dufresne A, Graillat C, Gérard J F (2001), 'New waterborne epoxy coatings based on cellulose nanofillers', *Macromol. Symp.*, **169**, 211–222.

Morais Teixeira E, Corrêa A C, Manzoli A, Leite F L, De Oliveira C R, Mattoso L H C (2010), 'Cellulose nanofibres from white and naturally colored cotton fibers', *Cellulose*, **17**, 595–606, doi: 10.1007/s10570-010-9403-0.

Morais Teixeira E, Pasquini D, Curvelo A A S, Corradini E, Belgacem M N, Dufresne A (2009), 'Cassava bagasse whiskers reinforced plasticized cassava starch', *Carbohydr. Polym.*, **78**, 422–431.

Nakagaito A N, Iwamoto S, Yano H (2005), 'Bacterial cellulose: the ultimate nanoscalar cellulose morphology for the production of high-strength composites', *Appl. Phys. A*, **80**, 93–97.

Nakagaito A N, Yano H (2004), 'The effect of morphological changes from pulp fiber towards nano-scale fibrillated cellulose on the mechanical properties of high-strength plant fiber based composites', *Appl. Phys. A*, **78**, 547–552.

Nakagaito A N, Yano H (2008), 'Toughness enhancement of cellulose nanocomposites by alkali treatment of the reinforcing cellulose nanofibers', *Cellulose*, **15**, 323–331.

Nogi M, Abe K, Handa K, Nakatsubo F, Ifuku S, Yano H (2006a), 'Property enhancement of optically transparent bionanofiber composites by acetylation', *Appl. Phys. Lett.*, **89**, 233123–233125.

Nogi M, Handa K, Nakagaito A N, Yano H (2005), 'Optically transparent bionanofiber composites with low sensitivity to refractive index of the polymer matrix', *Appl. Phys. Lett.*, **87**, 243110–243112.

Nogi M, Ifuku S, Abe K, Handa K, Nakagaito A N, Yano H (2006b), 'Fiber-content dependency of the optical transparency and thermal expansion of bacterial nanofiber reinforced composites', *Appl. Phys. Lett.*, **88**, 133124–133126.

Noishiki Y, Nishiyama Y, Wada M, Kuga S, Magoshi J (2002), 'Mechanical properties of silk fibroin–microcrystalline cellulose composite films', *J. Appl. Polym. Sci.*, **86**, 3425–3429.

Oksman K, Mathew A P, Bondeson D, Kvien I (2006), 'Manufacturing process of cellulose whiskers/polylactic acid nanocomposites', *Compos. Sci. Technol.*, **66**, 2776–2784.

Orts W J, Godbout L, Marchessault R H, Revol J F (1998), 'Enhanced ordering of liquid crystalline suspensions of cellulose microfibrils: a small angle neutron scattering study', *Macromolecules*, **31**, 5717–5725.

Orts W J, Shey J, Imam S H, Glenn G M, Guttman M E, Revol J F (2005), 'Application of cellulose microfibrils in polymer nanocomposites', *J. Polym. Environ.*, **13**, 301–306.

Paralikar S A, Simonsen J, Lombardi J (2008), 'Poly(vinyl alcohol)/cellulose nanocrystals barrier membranes', *J. Membrane Sci.*, **320**, 248–258.

Park I, Kang M, Kim H S, Jin H J (2007), 'Electrospinning of poly(ethylene oxide) with bacterial cellulose whiskers', *Macromol. Symp.*, 249–250, 289–294.

Peresin M S, Habibi Y, Zoppe J O, Pawlak J J, Rojas O J (2010), 'Nanofiber composites of polyvinyl alcohol and cellulose nanocrystals: manufacture and characterization', *Biomacromolecules*, **11**, 674–681.

Petersson L, Oksman K (2006), 'Biopolymer based nanocomposites: comparing layered silicates and microcrystalline cellulose as nanoreinforcement', *Compos. Sci. Technol.*, **66**, 2187–2196.

Podsiadlo P, Choi S Y, Shim B, Lee J, Cuddihy M, Kotov N A (2005), 'Molecularly engineered nanocomposites: layer-by-layer assembly of cellulose nanocrystals', *Biomacromolecules*, **6**, 2914–2918.

Podsiadlo P, Sui L, Elkasabi Y, Burgardt P, Lee J, Miryala A, Kusumaatmaja W, Carman M R, Shtein M, Kieffer J, Lahann J, Kotov N A (2007), 'Layer-by-layer assembled films of cellulose nanowires with antireflective properties', *Langmuir* **23**, 7901–7906.

Pu Y, Zhang J, Elder T, Deng Y, Gatenholm P, Ragauskas A J (2007), 'Investigation into nanocellulosics versus acacia reinforced acrylic films', *Composites Part B: Engineering*, **38**, 360–366.

Revol J F (1982), 'On the cross-sectional shape of cellulose crystallites in Valonia ventricosa', *Carbohydr. Polym.*, **2**, 123–134.

Revol J F, Bradford H, Giasson J, Marchessault R H. Gray D G (1992), 'Helicoidal self-ordering of cellulose microfibrils in aqueous suspension', *Int. J. Biol. Macromol.*, **14**, 170–172.

Revol J F, Godbout L, Dong X M, Gray D G, Chanzy H, Maret G (1994), 'Chiral nematic suspensions of cellulose crystallites: phase separation and magnetic field orientation', *Liq. Cryst.*, **16**, 127–134.

Rojas O J, Montero G A, Habibi Y (2009), 'Electrospun nanocomposites from polystyrene loaded with cellulose nanowhiskers', *J. Appl. Polym. Sci.*, **113**, 927–935.

Roman M, Winter W T (2004), 'Effect of sulfate groups from sulfuric acid hydrolysis on the thermal degradation behavior of bacterial cellulose', *Biomacromolecules*, **5**, 1671–1677.

Roohani M, Habibi Y, Belgacem N M, Ebrahim G, Karimi A N, Dufresne A (2008), 'Cellulose whiskers reinforced polyvinyl alcohol copolymers nanocomposites', *Eur. Polym. J.*, **44**, 2489–2498.

Saito T, Kimura S, Nishiyama Y, Isogai A (2007), 'Cellulose nanofibers prepared by TEMPO–mediated oxidation of native cellulose', *Biomacromolecules*, **8**, 2485–2491.

Saito T, Nishiyama Y, Putaux J L Vignon M R, Isogai A (2006), 'Homogeneous suspensions of individualized microfibrils from TEMPO-catalyzed oxidation of native cellulose', *Biomacromolecules*, **7**, 1687–1691.

Sakurada I, Nukushina Y, Ito T (1962), 'Experimental determination of the elastic modulus of crystalline regions oriented polymers', *J. Polym. Sci.*, **57**, 651–660.

Seydibeyoğlu M Ö, Oksman K (2008), 'Novel nanocomposites based on polyurethane and micro fibrilatted cellulose', *Compos. Sci. Technol.*, **68**, 908–914.

Shimazaki Y, Miyazaki Y, Takezawa Y, Nogi M, Abe K, Ifuku S, Yano H (2007), 'Excellent thermal conductivity of transparent cellulose nanofiber/epoxy resin nanocomposites', *Biomacromolecules*, **8**, 2976–2978.

Siqueira G, Abdillahi H, Bras J, Dufresne A (2010a), 'High reinforcing capability cellulose nanocrystals extracted from *Syngonanthus nitens* (Capim Dourado)', *Cellulose*, **17**, 289–298.

Siqueira G, Bras J, Dufresne A (2009), 'Cellulose whiskers vs. microfibrils: influence of the nature of the nanoparticle and its surface functionalization on the thermal and mechanical properties of nanocomposites', *Biomacromolecules*, **10**, 425–432.

Siqueira G, Bras J, Dufresne A (2010b), '*Luffa cylindrica*: a new lignocellulosic source of fiber, microfibrillated cellulose and cellulose nanocrystals', *BioResources*, **5**, 727–740.

Siqueira G, Bras J, Dufresne A (2010c), 'New process of chemical grafting of cellulose nanoparticles with a long chain isocyanate', *Langmuir*, **26**, 402–411.

Stenstad P, Andresen M, Tanem B S, Stenius P (2008), 'Chemical surface modifications of microfibrillated cellulose', *Cellulose*, **15**, 35–45.

Šturcova A, Davies G R, Eichhorn S J (2005), 'Elastic modulus and stress-transfer properties of tunicate cellulose whiskers', *Biomacromolecules*, **6**, 1055–1061.

Svagan A J, Hedenqvist M S, Berglund L (2009), 'Reduced water vapour sorption in cellulose nanocomposites with starch matrix', *Compos. Sci. Technol.*, **69**, 500–506.

Syverud K, Stenius P (2009), 'Strength and barrier properties of MFC films', *Cellulose*, **16**, 75–85.

Tashiro K, Kobayashi M. (1991), 'Theoretical evaluation of three-dimensional elastic constants of native and regenerated celluloses: role of hydrogen bonds', *Polymer*, **32**, 1516–1526.

Tokoh C, Takabe K, Fujita M, Saiki H (1998), 'Cellulose synthesized by *Acetobacter xylinum* in the presence of acetyl glucomannan', *Cellulose*, **5**, 249–261.

Van den Berg O, Capadona J R, Weder C (2007), 'Preparation of homogeneous dispersions of tunicate cellulose whiskers in organic solvents', *Biomacromolecules*, **8**, 1353–1357.

Wang B, Sain M, Oksman K (2007), 'Study of structural morphology of hemp fiber from the micro to the nanoscale', *Appl. Compos. Mater.*, **14**, 89–103.

Wang Y, Cao X, Zhang L (2006), 'Effects of cellulose whiskers on properties of soy protein thermoplastics', *Macromol. Biosci.*, **6**, 524–531.

Whistler R L, Richards E L (1970), *The carbohydrates*, 2A, Academic Press, New York, p. 447.

Yano H, Sugiyama J, Nakagaito A N, Nogi M, Matsuura T, Hikita M, Handa K (2005), 'Optically transparent composites reinforced with networks of bacterial nanofibers', *Adv. Mater.*, **17**, 153–155.

Yuan H, Nishiyama Y, Wada M, Kuga S (2006), 'Surface acylation of cellulose whiskers by drying aqueous emulsion', *Biomacromolecules*, **7**, 696–700.

Zimmermann T, Pöhler E, Geiger T (2004), 'Cellulose fibrils for polymer reinforcement', *Adv. Eng. Mater.*, **6**, 754–761.

Zimmermann T, Pöhler E, Schwaller P (2005), 'Mechanical and morphological properties of cellulose fibril reinforced nanocomposites', *Adv. Eng. Mater.*, **7**, 1156–1161.

Zoppe J O, Peresin M S, Habibi Y, Venditti R A, Rojas O J (2009), 'Reinforcing poly(ε-caprolactone) nanofibers with cellulose nanocrystals', *ACS Appl. Mater. Interfaces*, **1**, 1996–2004.

Zuluaga R, Putaux J L, Restrepo A, Mondragon I, Gañán P (2007), 'Cellulose microfibrils from banana farming residues: isolation and characterization', *Cellulose*, **14**, 585–592.

4

Characterization of fiber surface treatments in natural fiber composites by infrared and Raman spectroscopy

M. A. MOSIEWICKI, N. E. MARCOVICH and
M. I. ARANGUREN, Institute of Materials Science and
Technology (INTEMA), University of Mar del Plata, Argentina

Abstract: The use of Fourier transform infrared (FTIR) and Raman spectroscopies in the characterization of modified natural fibers and polymer interfaces in composite materials is reviewed. The changes in composition of the fibers owing to leakage of original components and/or incorporation of various species attached or adsorbed on the fiber surfaces, changes of morphology, as well as physical and chemical bonds developed at the polymer composite interfaces have been successfully characterized by these spectroscopic techniques. The potential and complementary use of these techniques is also discussed.

Key words: natural fiber composites, interfacial interactions, Fourier transform infrared spectroscopy, Raman spectroscopy.

4.1   Introduction

The incorporation of natural fibers as filler and/or reinforcing materials in polymeric matrices is an increasingly growing area of importance in industrial and academic fields. One of the major problems found in the research and development of these materials has been to reach good compatibility between the two main phases, fibers and matrices. It has been established that the mechanical performance of the composites depends not only on the properties of the principal components but also on the nature and strength of the interface, that is responsible for the load transfer from the matrix to the fibers. For this reason, numerous strategies have been developed to improve the fillers surface wettability by resins or thermoplastics, to increase the adherence of the matrix to the filler and the corresponding interfacial strength. These strategies include the surface modification of the fillers, the incorporation of coupling agents or, less frequently, the modification of the matrix. Thus, techniques that allow investigation of these changes on filler surfaces or directly (although with increasing difficulties) on the composite interfaces are important tools for establishing correlations between interfaces characteristics and composite properties.

117

Among available surface techniques, probably the most commonly utilized is Fourier transform infrared (FTIR) spectroscopy. All materials absorb infrared radiation and the frequency intervals at which absorption bands appear are associated with the vibrational modes of specific functional groups. Large bodies of infrared data presented in tables and charts of software libraries are accessible to the users and are the results of years of careful and continuous work by a very large number of scientists. For a vibrational motion to be IR active, the dipole moment of the molecule must change, the higher the magnitude of this change, the higher the intensity of the band. Because a functional group is usually associated with more than one absorption band (more than one vibrational mode) and different functional groups may absorb in overlapping regions, it is always wise not to base conclusions on variations of a single band, but to look for confirmation in other non-overlapping FTIR regions or in supporting results from other techniques. Additionally, a group absorbs in a relatively narrow range of frequencies, but the position of a given absorption band can be shifted owing to interferences or perturbations from the surrounding atoms. Although, this may complicate the interpretation of the spectra, it can provide further information regarding possible interactions at the composite interface.

Although at present it is used less frequently, Raman spectroscopy is also based on the vibrational motion of functional groups, detecting changes in polarizability of the molecule. This polarizability decreases with increasing electron density, increasing bond strength and decreasing bond length. This last characteristic has been cleverly utilized to detect extension ratios in fibers/microfibers embedded in different composites, as discussed in detail in chapter 13.

In this way, infrared and Raman spectroscopies are techniques that could be used to provide complementary information about the chemical nature of materials, enabling, in particular, the investigation of the composition of surfaces and/or interactions developed at interfaces of composite materials.

4.2   Methods and techniques

4.2.1   Infrared spectroscopy

Although the infrared portion of the electromagnetic spectrum covers $14000–10$ cm^{-1}, the majority of analytical applications use the mid-infrared range, approximately $4000–400$ cm^{-1}. Many different techniques are currently available and only those related to the scope of this chapter will be summarized below:

Transmission

Even with the development of new FTIR techniques, transmission remains the most popular technique for samples that can be prepared in transparent form (diluted powdery samples pressed in KBr mixtures are included in this definition of transparent samples). In this technique, the sample-beam geometry corresponds to the beam passing through the sample, while the transmitted energy is collected in the detector (Fig. 4.1a). Because of its quantitative capabilities, this technique is generally preferred. Information on a modified surface can be obtained by spectral subtraction of the unmodified sample from the modified sample, to remove absorption from the bulk material.

Attenuated total reflection (ATR)

Attenuated total reflection (ATR) is based on the multiple internal reflection of the beam incident on the sample (Fig. 4.1b). This technique is useful for studying the surface of the materials (up to a depth of 0.5–5 µm) that are put into contact with the crystal (usually ZnSe or monolithic diamond crystals). To obtain good surface contact may be a major problem for some samples. If the materials are rough and hard, the contact with the crystal may only be a point or line; increasing the pressure of the material on the crystal is

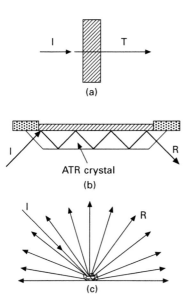

4.1 Schematic representation of the experimental set-up of different FTIR techniques: (a) transmission; (b) attenuated total reflection; (c) diffuse reflectance. I, incident; T, transmitted; R, reflected.

the usual solution although repeatability may become an issue, as well as the useful life of the crystal. Soft flexible materials more readily provide contact and thus repeatability in the measurements.

Diffuse reflectance infrared spectroscopy

In Fourier transform diffuse reflectance infrared spectroscopy (DRIFT), the IR beam is reflected and dispersed on the matt surface of the sample (Fig. 4.1c). The emerging radiation leaves the sample in all directions, but with appropriate optical setups is directed towards the detector. The intensity of the infrared peaks is expressed in Kubelka–Munk units, in which case the spectra look similar to transmission spectra shown as absorbance. The Kubelka–Munk theory was developed for purely diffuse radiation, similarly to Beer's Law for transmission.

Usually, highly absorbing zones are also reflecting regions and some samples may produce regions of flat appearance that correspond to strongly absorbing regions in the transmission spectra. Improving the spectra may require further dilution of the sample in dry KBr powder.

4.2.2   Raman spectroscopy

When a sample is irradiated with a monochromatic radiation of high intensity, such as solid lasers, a fraction of the radiation is scattered and the shifts produced with respect to the wavelength of the incident beam depend on the chemical structure of the molecules. Although Raman spectra and infrared spectra somewhat resemble each other (the peaks corresponding to vibrational modes active in both spectroscopies have the same energy shifts, that is, they appear at the same wavenumbers); the differences are also important, the relative intensity of the peaks differs and in some instances some peaks may not appear. Differences with infrared spectroscopy are not surprising because the mechanisms involved are also different, change in dipolar moment for FTIR-active groups and change in polarizability for Raman-active groups, and these differences make them complementary rather than competitive techniques. In particular, Raman spectroscopy is very useful for samples containing water, because water does not cause interference in Raman, as it strongly does in infrared spectroscopy. Another useful advantage is that glass or quartz cells can be used instead of salt crystals, which are environmentally unstable and fragile. As for FTIR spectroscopy, various methods are available in Raman spectroscopy whose selection will depend on the sample characteristics and the specific aims of the characterization.

4.3   Analysis of natural fibers and surface treatments

4.3.1   Characterization of cellulosic materials

Lignocellulosic fibers consist of bundles of hollow cellulose fibrils. Their cell walls are reinforced with spirally oriented cellulose in a hemicellulose and lignin matrix. Therefore, the cell wall is a composite structure of lignocellulosic material reinforced by helical microfibrillar bands of cellulose. The composition of the external surface of the cell wall is a layer of lignaceous material and waxy substances that bond the cell to its adjacent neighbors (Li *et al.*, 2000; Jacob *et al.*, 2005). These components are the ones to be considered in the study of the interaction with polymer matrices. Thus, to discuss their spectra and modifications or their interactions with different polymers, one should begin by identifying their corresponding absorption bands. Table 4.1 summarizes some FTIR peak assignments for lignocellulosic materials (Marcovich *et al.*, 1996a; Bessadok *et al.*, 2009; Jayaramudu *et al.*, 2009).

On the other hand, Fig. 4.2 is a good example of the information that can be obtained from different techniques and modes applied to a non-treated cellulose paper (Proniewicz *et al.*, 2001). The ATR technique (Fig. 4.2a) is a non-destructive method that requires no sample preparation. The results are collected from a region between the surface and a depth of about 0.5–5 µm depending on the sample characteristics. Intimate contact between the surfaces of the sample and the crystal is required, thus information can vary with sample topology, because of the interfering signal of interstitial trapped air and owing to the beam penetration in the sample.

The older method of transmission FTIR can be used on a neat sample if it is transparent or thin enough to ensure a relatively large percentage of the beam goes through the sample. In Fig. 4.2 b and c the differences found by using pellets prepared from the neat sample and that diluted with KBr pellets are shown. Thus, when the sample is too concentrated or too thick (Fig. 4.2b), some high absorbent regions show essentially zero transmission. That is the case for the 2600–3500 cm^{-1} region, corresponding to the C–H and O–H stretching vibrations, and 1000–1300 cm^{-1} region, which comprises vibrations from the C–C stretching and the COH and CCH deformations. Usually, samples are diluted with KBr and pressed into a pellet to eliminate the problem of highly absorbent samples. However, this method is destructive and time consuming because the cellulose has to be in the form of a fine powder and well dispersed into the salt in order to avoid spectral artifacts. When sampling plant fibers, the pressing step can also have the added disadvantage of destroying the cell walls, exposing internal groups, in particular hydroxyl groups that were originally not available. Thus, a difference can be expected in the -OH bands if other not so invasive methods are utilized.

Table 4.1 Peak assignments of the infrared spectra of lignocellulosic materials

Band position (cm^{-1})	Assignment
3550–3650	O–H stretching in free or weakly H-bonded hydroxyls
3200–3400	O–H stretching in H-bonded hydroxyls
2840–2940	C–H stretching region
2725	Overtone of interacting C–O stretch and O–H deformation
2568	Same
1720–1740	C=O stretching in carbonyl
1625–1660	Adsorbed water molecules in noncrystalline cellulose
~1600	Aromatic skeleton ring vibration and vibrations owing to adsorbed water
~1505	Aromatic skeleton ring vibration
1450–1475	C–H deformation and CH$_2$ (sym.) + OH deformation
1400–1430	C–H deformation (methoxyl group in lignin)
~1370	C–H deformation (symmetric)
~1327	Syringyl ring breathing with C–O stretching (lignin) and CH$_2$ wagging in cellulose
1250–1260	Guaiacyl ring breathing with C–O stretching (lignin)
1240–1245	C–O bond of the acetyl group in xylan and hemicellulose
~1230	Phenolic O–H deformation (lignin) – syringyl structure
1160–1230	C–O stretching of ester groups
1150–1160	C–O–C stretching (antisymmetrical) in cellulose and aromatic C–H CH$_2$ wagging in cellulose
1098–1120	Skeletal vibration involving C–O stretching of the β glycosidic linkages
~1060	C–OH stretching vibration
1036	Aromatic C–H in plane deformation, guaiacyl and C–O deformation primary alcohol in lignin and C–O stretching in cellulose
1003	Skeletal vibration and C–O stretching in cellulose
890–900	Antisymmetrical stretching owing to β linkage in cellulose
830	Aromatic C–H out of plane vibration owing to lignin

Similarly, Marcovich *et al.* (1996a) discussed the effect of sample concentration when using transmission and DRIFT modes for studying sawdust samples. They found that the opacity of the pellets limited the use of the transmission technique to weight concentrations equal or lower than 2% by weight in KBr. However, the DRIFT technique applied to neat samples gave good spectra with detailed band structure. Moreover, Fig. 4.3 illustrates that the frequency at which the bands appear is independent of the methods used, but the intensity depends strongly on the technique selected for collecting the spectra or preparing the sample. For example, the strong absorption band at 3000–3500 cm^{-1} (corresponding to –OH absorption) is highly dependent on the sample preparation, as well as the spectroscopic method selected: it appears with higher intensity in the transmission spectra, for which a KBr pellet was prepared by pressing. Although DRIFT requires

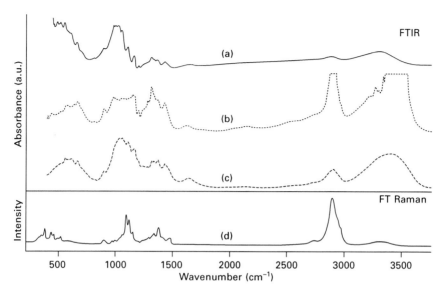

4.2 Vibrational spectra of non-treated cellulose in the range 200–3800 cm^{-1}: (a) ATR technique; (b) transmission through sample; (c) transmission through KBr pellet; (d) FT Raman. Reprinted from Proniewicz *et al.*, 2001, with permission from Elsevier.

also grinding of the sample (as in transmission), it is not necessary to apply pressure hence the band appears with much lower intensity. Similarly, the peak at about 1650 cm^{-1} corresponding to absorbed water appears clearly in transmission but not so much in DRIFT. Finally, the region between 900 and 1250 cm^{-1} looks similar for the diluted samples under the two FTIR methods, although the DRIFT trace of the neat sample shows distortion of the main absorption band (around 1100 cm^{-1}).

The FT Raman technique displays a more direct correlation between band and bonds. However, it is not very sensitive to hydrogen bonds, as can be seen from the very low intensity of the band in the region of 3000 to 3500 cm^{-1} (Fig. 4.2d). Similarly the band at around 1625 cm^{-1}, which appears in the infrared spectra and is assigned to the bending motions of water molecules, is almost missing in the Raman spectra.

4.3.2   Characterization of modified fibers (and nanofibers)

As already mentioned, the high-modulus, low-cost, lightweight characteristics associated with natural fibers make them attractive as polymer reinforcements. However, the performance of a composite depends not only on the properties of the reinforcement and matrix, but also on the bonding between them. A poor fiber–matrix interaction derived from polar fibers and, typically,

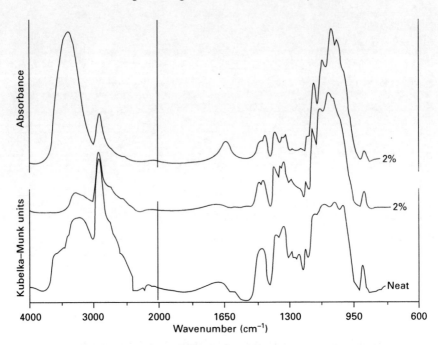

4.3 Transmission and DRIFT spectra of cellulose. Transmission spectrum was obtained at 2 (wt%) of cellulose in KBr. DRIFT spectra were obtained from 2% and neat samples. Reprinted from Marcovich *et al.,* 1996a, with permission from Brill.

non-polar polymers leads to composites with poor mechanical properties and low durability when exposed to aggressive environments. Thus, to improve interfacial adhesion between reinforcement and polymer, the fiber surface is frequently modified or coupling agents are included in the composite formulation. The hydrophilization of non-polar polymeric matrices is also possible, but most often treatments are performed on the reinforcements (Gauthier *et al.,* 1998a; Nuñez *et al.,* 2003). Lignocellulosic materials contain surface hydroxyl groups that can be bonded to molecules that have functional groups selected to improve polymer compatibility. The use of FTIR and Raman spectroscopies in the analysis of the most frequently used chemical modifications is summarized below.

4.4   Chemical treatments

4.4.1   Alkaline treatment

The alkaline treatment, also called mercerization, involves the immersion and soaking of the fibers in an alkaline medium, typically a rather concentrated NaOH aqueous solution. This treatment removes practically all non-cellulose components except waxes (Idicula *et al.*, 2006; Luz *et al.*, 2008). Owing to the dissolution of lignin by alkali, some pores are formed on the fiber surface. The loss of the cementing material of the vegetable fibers produces some fibrillation because of the breakage and separation of the fiber bundles (Aranguren and Reboredo, 2007). These changes increase the contact area between the fiber and the matrix (Idicula *et al.*, 2006).

The effects of the alkaline treatment on fabrics of *Polyalthia cerasoides* tree, palm fibers, sisal fibers, hemp and kenaf fibers, henequen fibers and wood flour were addressed by Alawar *et al.* (2009), Jayaramudu (2009), Marcovich *et al.* (2005) and Rahman (2009), Rong *et al.* (2001), Sgriccia *et al.* (2008), among others. Using FTIR and chemical analysis, these studies confirmed the reduction of the hemicellulose and lignin contents after fiber mercerization. The major change in the FTIR spectra of the fibers owing to the alkaline treatment was the reduction (or sometimes the complete disappearance) of the peak centered at 1730–1740 cm^{-1}, attributed to the carbonyl group (C=O) stretching. This disappearance is a consequence of the extraction of hemicellulose and lignins and/or the formation of ionic carboxylates in the incompletely extracted samples, in which instance the corresponding peak appears at lower frequencies (1590 cm^{-1}). It was also established that the peak located at 1240 cm^{-1}, assigned to the C–O bond of the acetyl group in xylan, separated into two smaller peaks at 1260 and 1230 cm^{-1}. The first one corresponds to vibrations in the guaiacyl structure of the lignin and the second to the syringyl structure (Marcovich *et al.*, 2005; Reddy *et al.*, 1990; Roy *et al.*, 1991). Since NaOH modification happens to increase the amount of accessible polar hydroxyl groups from the cellulosic fibers (Marcovich *et al.*, 1998; 1999), the water sensitive regions of the FTIR spectrum become useless to detect other changes. In this sense, FT Raman spectroscopy results are very attractive because water does not interfere with the sample signal. This advantage was interestingly used by Jähn and coworkers (2002) to study the extent of the polymorphic transformation of cellulose I into cellulose II as a function of the alkali concentration. They detected the polymorphic transformation of the cellulosic fine structure of the flax fibers *in vivo* by analyzing the FT Raman spectra at frequencies below 1500 cm^{-1}, as denoted in Fig. 4.4. The Raman lines characterizing the cellulose modification I are denoted by an asterisk whereas the typical Raman lines of modification II are denoted by a cross. Particularly, in the range of the methylene bending vibrations (1400–1475 cm^{-1}) the spectra show the differences between the conformational arrangements of the side chains of the anhydroglucopyranose residues of cellulose. They indicate the simultaneous presence of two stereochemically non-equivalent –CH$_2$OH groups in cellulose I (1476 and 1455 cm^{-1}). In cellulose II, only one type of –CH$_2$OH group is present (1460 cm^{-1}) and the two scissoring vibrations of the methylene groups merge into one signal. On the other hand, the Raman

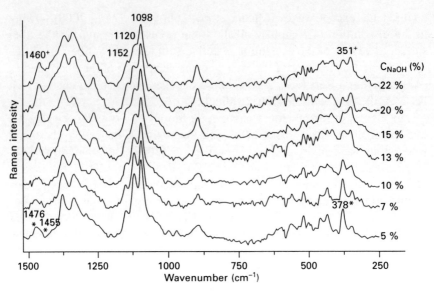

4.4 FT Raman spectra of a series of NaOH treated flax fiber bundles in the frequency range below 1500 cm^{-1}. Raman lines characterizing the most common polymorphic modifications are shown: cellulose I (*) and cellulose II (+). Reprinted from Jähn *et al.*, 2002, with permission from Elsevier.

lines at 1120 and 1098 cm^{-1} assigned to the skeletal vibrational modes v_s (C–O–C) and v_{as} (C–O–C) of the β(1→4) glycosidic linkages of the β-D-glucopyranosyl units of cellulose can serve as characteristic marker bands for multicomponent systems such as cellulosic plant tissues. In particular, the vibrational mode v_s (C–O–C) at 1120 cm^{-1} is superimposed on skeletal modes of non-cellulosic carbohydrates (Himmelsbach and Akin, 1998) and lignin components (Agarwal and Atalla, 1986; Agarwal *et al.*, 1995). For that reason, the intensity and the band shape of this cellulosic Raman line, which is also coupled with the C–C stretching mode of the breathing vibration of the glucopyranose rings at 1153 cm^{-1}, is strongly affected. In contrast, the Raman line at 1098 cm^{-1} assigned to the asymmetric stretching mode v_{as} (C–O–C), is not influenced by non-cellulosic carbohydrates. Taking into account the previous considerations, the changes in the C–H stretching region could be studied and related to the effect of the alkali treatment on the molecular structures of cellulosic plant fibers.

4.4.2   Esterification

The esterification of natural fibers using various anhydrides has been studied extensively. The type of lignocellulosic material used as a substrate in the esterification reaction is of vital importance, because its three main components, i.e. lignin, hemicelluloses and cellulose exhibit different reactivity towards the anhydrides. The acetylation reaction is one of the most frequently informed in literature (Adebajo and Frost, 2004; Bessadok *et al.*, 2009; Liu *et al.*, 1995; Luz *et al.*, 2008; Rong *et al.*, 2001; van Hazendonk *et al.*, 1996; Zafeiropoulos *et al.*, 2002). For example, Luz *et al.* (2008) reported the changes observed in the FTIR spectra for cellulose from sugarcane bagasse and sugarcane bagasse without hemicellulose after attaching acetyl groups to the OH groups of the fibers. The authors reported the appearance of a band at 1758 cm^{-1}, corresponding to the C=O carbonyl peak from bonded acetyl groups. Furthermore, after fiber acetylation, there was a decrease in OH band intensity, which was shifted from 3400 to 3700 cm^{-1} owing to the reduction of hydroxyl groups available to form hydrogen bonds and the higher energy of the remaining free-OH groups. Similar effects were found by Zhang and coworkers (2008), who performed a mechanochemical acetylation of cellulose powder (solid-state at ambient temperature) using acetic anhydride. They demonstrated that the reaction occurred under those conditions and essentially on the cellulose surface.

Tserkia and coworkers (2005) modified flax, hemp and wood fibers by acetylation and propionylation, and confirmed the esterification reaction by ATR-FTIR. The treatment with both anhydrides led to the appearance/increase of the absorbance in the regions 1735–1737 cm^{-1} (stretching vibration of the C=O group in ester bonds) and 1162–1229 cm^{-1} (C–O stretching) in all cases. A reduction of the strong absorption between 3400 and 3600 cm^{-1} (OH groups) was also noticed after fiber esterification.

Marcovich and coworkers (1996a, 1996b, 1997, 2005) showed that the DRIFT technique was effective in assessing surface modifications on wood flour resulting from esterification with maleic anhydride (MAN), even under mild conditions (room temperature without catalyst). The formation of new ester bonds owing to the reaction of the OH groups in the wood with the anhydride was confirmed by the increased intensity of the 1710–1740 cm^{-1} band (attributed to both the C=O groups in acid and ester groups attached to the filler) and the appearance of a new peak at 850 cm^{-1} related to the absorption of *cis* C=C conjugated to the carbonyl group in MAN (Bellamy, 1975). The similar intensity of the peaks corresponding to ester and acid carbonyls indicated that only one of the formed acid groups reacts with the filler surface and the other remains unreacted, in agreement with other studies (Ly *et al.*, 2008; Kishi *et al.*, 1988; Felix and Gatenholm, 1991; Maldas and Kokta, 1992). Samples esterified under stronger conditions (xylene reflux temperatures) showed an increase in the absorbance of the band at 1740 cm^{-1} that was also a function of the reaction time. This observation corresponds to the increase in the concentration of ester bonds.

Wood flour was also esterified with a commercial alkenyl succinic anhydride (ASA) (Acha *et al.*, 2003; Marcovich *et al.*, 2005, Fig. 4.6). The FTIR

4.5 Expected reaction between wood flour and ASA.

spectrum of the treated sample revealed the usual increase in the intensity of the carbonyl absorption at 1735 cm^{-1}, owing to ester formation, and the absence of the anhydride absorption bands at 1600–1800 cm^{-1}. An increase in the CH stretching bands (2900–3000 cm^{-1}) was also observed owing to the CH groups of the attached ASA molecule. Additionally, a new peak at 1650 cm^{-1} appeared, which corresponds to the unsaturation of the attached ASA. As is the case with many unsaturated fatty acids, in which the double bond is located approximately in the middle of the molecule, the 1650 cm^{-1} absorption peak, characteristic of C=C absorption, did not appear in the FTIR spectrum of the neat ASA and had to be identified using Raman spectroscopy (strong peak). However, it clearly appeared in the spectra of the modified wood flour, because the reaction broke the symmetry of the molecule.

The attachment of longer chains on cellulose whiskers and flax strands is also well exemplified by Junior de Menezes *et al.* (2009) and Cañigueral *et al.* (2009), respectively. The cellulose was modified by grafting organic acid chlorides with aliphatic chains of different lengths, and thus, their spectra showed the expected peak at 1737 cm^{-1} and an increase in the signals at 2953, 2919 and 2850 cm^{-1} ascribed to the presence of grafted alkane chains (Fig. 4.6). The concomitant decrease of the magnitude of the broad band around 3300 cm^{-1} for modified whiskers compared with unmodified ones is attributed to the partial disappearance of OH groups, confirming the success of the grafting reaction with organic acid chlorides. On the other hand, the flax strands were modified with alkyl ketene dimer and the resulting spectrum showed the same signals discussed previously added to the peak at 1700 cm^{-1}, which corresponds to the ketone group of the hydrolyzed reactive.

Regarding commercial coupling agents, the use of maleated polypropylene (MAPP) or maleated polyethylene (MAPE) copolymers to improve the interface between cellulosic fibers and PP or PE matrices is a common practice and has been frequently reported in the scientific literature (Acha *et al.*, 2006; Bledzki *et al.*, 2005; Felix and Gatenholm, 1991; Gassan and Bledzki, 1997; Hristov *et al.*, 2004; Ichazo *et al.*, 2001; Joly *et al.*, 1996; Lu *et al.*, 2005; Miguez Suarez *et al.*, 2003). Just to include an example, Lu *et al.* (2005) demonstrated by FTIR spectroscopy and ESCA (electron spectroscopy for chemical analysis) the chemical linkage formed between

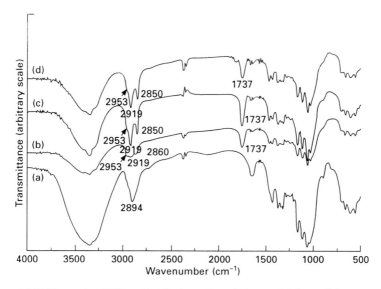

4.6 FTIR spectra (KBr pellets) of ramie cellulose whiskers: (a) unmodified; surface modified with (b) hexanoyl chloride; (c) lauroyl chloride; and (d) stearoyl chloride, after Soxhlet extraction with acetone. Reprinted from Junior de Menezes *et al.*, 2009, with permission from Elsevier.

the wood fibers and several MAPE coupling agents through esterification and proposed that succinic and half succinic esters were the two primary covalent bonding products of the reaction. Their results are discussed further in 4.4.5.

4.4.3   Silanization

Researchers at Grenoble have been involved in a series of studies (Abdelmouleh *et al.*, 2002, 2004, 2005 and 2007) aiming at understanding the silane–cellulose system. The main advantages of using silane coupling chemicals is that at one end, they bear alkoxysilane groups capable of reacting with the OH-rich surface of the fibers, whereas the other end can be tailored to suit a specific matrix. One of the studies (Abdelmouleh *et al.*, 2007) illustrates the grafting of the coupling agent through initial physical adsorption of the hydrolyzed silane followed by a curing process at 120 °C under an inert atmosphere. Figure 4.7 presents the DRIFT spectra corresponding to the cellulose samples treated with γ-methacryloxypropyltrimethoxy silane (MPS), before and after heat treatment. In both cases, the spectrum corresponding to untreated cellulose was subtracted to investigate the band issuing from the silane moiety. The two spectra show different bands at 1713 and 1636 cm^{-1}, which are associated with the stretching vibrations of the C=O and

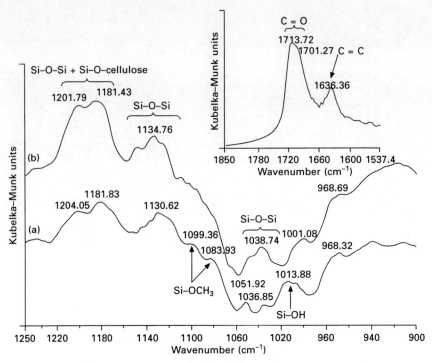

4.7 DRIFT subtraction spectra (MPS treated fibers – untreated fibers): (a) before and (b) after treatment (900–1250 cm⁻¹ region). Inset: 1850–1540 cm⁻¹ region after treatment. Reprinted from Abdelmouleh *et al.*, 2007, with permission from Elsevier.

C=C groups in the acrylic moieties, respectively. The broad shape of the C=O peak suggests that this group is interacting by hydrogen bonding with the surface hydroxyl groups. The broad intense bands around 1200 and 1135 cm⁻¹ were assigned to the stretching of the –Si–O–cellulose and –Si–O–Si– bonds, respectively (Felix and Gatenholm, 1991; Navoroj *et al.*, 1984). The strong increase in the intensity of these bands after the heat treatment suggested that the grafting of silane onto cellulose as well as the intermolecular condensation between adjacent adsorbed –Si–OH groups were substantially enhanced. The peaks near 1100 and 1080 cm⁻¹ are related to residual unhydrolyzed Si–OCH₃ groups and their small intensity indicates that most of the silane adsorbed under these conditions was hydrolyzed. The large band around 1015 cm⁻¹, present in the spectrum of the uncured sample, was attributed to –Si–OH groups. This band disappeared after the heat treatment and was replaced by a wide band around 1040 cm⁻¹, characteristic of –Si–O–Si– moieties. These peak assignments are in agreement with those reported in other studies dealing with glass surfaces treated with the same coupling agents (Navoroj *et al.*, 1984). However, FTIR spectroscopy applied to the study of silanized natural fibers has not always been reported as a successful analysis technique. Sgriccia *et al.* (2008) treated hemp and kenaf fibers with silane using gentler conditions than the Grenoble group and were unable to identify the characteristics of silane moieties on the surface of the fibers by FTIR spectroscopy.

Other studies (Herrera-Franco and Aguilar-Vega, 1997; Valdez-Gonzalez *et al.*, 1999a, 1999b) have also used FTIR spectroscopy and ESCA to confirm the reaction of a silane coupling agent with henequen fibers and found that the fiber–matrix interaction became stronger when the fibers were previously treated with alkali. The pretreatment eliminated cementitious components and lead to a rougher topology and larger accessible surface for reaction.

4.4.4   Urethane bonds

The reactivity of lignocellulosic fibers towards isocyanates has been frequently used as a compatibilization procedure in polymeric based composites (Kokta *et al.*, 1990; Ly *et al.*, 2008; Maldas *et al.*, 1989a,b). As an example, Joly and coworkers (1996) demonstrated the success of the chemical modification by showing the presence of two new bands at 3343 and 1705 cm^{-1}, attributed to the formation of urethane groups and associated with their NH and CO vibrations, respectively, after modifying Avicel cellulose, flax and ramie fibers with aliphatic isocyanates. More recently, Ly and coworkers (2008) modified cellulose fibers and kraft softwood pulps with both 1,4-phenylene diisocyanate (PDI) and methylene-bis-diphenyl diisocyanate (MDI). The FTIR spectra of the Soxhlet extracted modified fibers revealed a new band at 2275 cm^{-1}, corresponding to the isocyanate function and a peak at 831 cm^{-1}, associated with the presence of disubstituted aromatic ring. The small size and rigidity of the isocyanate molecules selected exclude the possibility of bridging two different fibers, which explained the presence of isocyanate functions. However, the expected urethane linkage signal was masked by a large broad signal from absorbed water ($\sim$1650 cm^{-1}). Gandini *et al.* (2001) used TDI (toluene di-isocyanate) to treat cellulose and also confirmed the reaction by FTIR spectroscopy.

In a comprehensive experiment, Marcovich *et al.* (2006) treated cellulose nanocrystals (prepared from acid hydrolysis of microcrystalline cellulose Avicel) with a polymeric MDI, followed by Soxhlet extraction with toluene to remove unreacted isocyanate. Because of the large area available for reaction in the cellulose nanocrystals, the modifications were distinctively observed in the FTIR spectra. The spectrum of the treated (unwashed) crystals showed new peaks corresponding to the absorption of isocyanate (2270 cm^{-1}) and urethane (1720 cm^{-1}). The spectrum of the extracted crystals showed that there was no change in the 1720 cm^{-1} peak, because the urethane bonds were formed by reaction with the OH in the surface of the cellulose crystals.

On the other hand, a large, but incomplete reduction of the isocyanate band was reported, which results from the removal of non-attached MDI, and the presence of unreacted isocyanate in the attached moieties.

4.4.5   Acrylate derivatives

Kalia and coworkers (2008) reported the preparation of graft copolymers of flax fibers with methyl acrylate (MA) using a redox system. FTIR spectrum of original flax showed a broad peak at 3422.8 cm^{-1} owing to bonded –OH groups and at 2918.8, 1653.5 and 1058.7 cm^{-1} arising from –CH$_2$, C–C and C–O stretching, respectively. However, for the graft copolymer, an additional peak at 1731.2 cm^{-1} owing to the >C=O group of MA was observed, suggesting MA graft copolymerization onto flax fiber through covalent linkages.

Bessadok *et al.* (2009) treated agave (*Americana* L.) fibers with acrylic acid (AA), among other reactives. Owing to the introduction of new CH and CH$_2$ groups of the acrylic acid, the spectra of the treated agave fibers presents stronger bands at 2960, 2900 and 2855 cm^{-1} than that of the untreated fibers. At the same time the new ester groups resulting from the acidic AA treatment are highlighted by the band at 1730 cm^{-1} characteristic of the carbonyl (C=O) group. Analagously, Sangthong *et al.* (2009) treated sisal by admicellar polymerization with a poly(methyl methacrylate) film coating (Fig. 4.8). Figure 4.9 shows the FTIR spectra of pure PMMA, untreated sisal fiber, and admicellar-treated sisal fiber. The admicellar-treated sisal fiber spectrum, shows the peaks at 1734 cm^{-1} (C=O) and 1457 cm^{-1} (–O–CH$_3$), which are the characteristic peaks of pure PMMA whereas the spectrum of bare sisal fiber, shows the characteristic peaks of cellulose at 3421 cm^{-1} (–OH) and

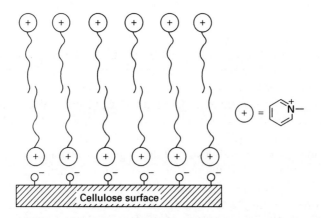

4.8 The ion pairing mechanism between the cellulose anions and the pyridinium cations in admicelle formation. Reprinted from Sangthong *et al.*, 2009, with permission from Elsevier.

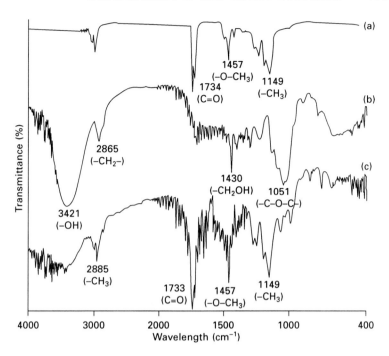

4.9 FTIR spectra of (a) PMMA, (b) untreated sisal fiber, and (c) treated sisal fiber. Reprinted from Sangthong, 2009, with permission from Elsevier.

1430 cm^{-1} (–CH$_2$OH). These results confirm that the sisal fiber surface was successfully coated with PMMA by admicellar polymerization.

4.4.6   Enzymatic treatment

In some instances, the outermost surface of a lignocellulosic reinforcement is covered with hydrophobic noncellulosic components, i.e. lipophilic extractives and silica. This is true for wheat straws (WSs), stalks, husks, grasses, cereals and others. These materials impede uniform and efficient fiber–matrix adhesion and are responsible for the nonwetting behavior of these kinds of fillers when they are used for the production of particleboards (Han *et al.*, 2001). Jiang and co-workers (2009) demonstrated that lipases from *Candida cylindracea* could effectively remove the hydrophobic lipophilic extractives and silica from the outer surface of untreated wheat straw, and increased the hydroxyl group content in the outer surface. They found, in the FTIR spectrum of the wheat straw outer surfaces, two strong and sharp peaks at 2916 and 2850 cm^{-1} assigned to the asymmetric and symmetric stretching, respectively, of the CH$_2$ group comprising the majority of the aliphatic fractions of waxes. Meanwhile, the small sharp band at 718 cm^{-1}

being characteristic of the $-(CH_2)_n - (n \geq 6)$ in plane deformations rocking (Bio-Rad, 2005) was observed. After enzymatic treatment, the sharp peaks at 2916 and 2850 cm^{-1} significantly decreased and they turned into a weak rounded band, now associated with the contribution of the CH_2- group in lignin and polysaccharides (Fang *et al.*, 2002; Moran *et al.*, 2008). The peak at 718 cm^{-1}, meantime, was reduced to a very weak shoulder. The two peaks at 790 and 970 cm^{-1}, attributed to the Si–C stretching vibration and Si–O vibration, respectively (Das *et al.*, 2007; Frost and Mendelovici, 2006), were reduced to two corresponding weak shoulders after the lipase treatment indicating that the silicon-containing components were also removed concurrently during the modification. Additionally, the intensity of several typical peaks assigned to lignin and polysaccharides (cellulose and hemicellulose) increased significantly after treatment: the band at 896 cm^{-1}, characteristic of β-glucosidic linkages between the sugar units (Gupta *et al.*, 1987), the infrared bands at 1422, 1600, 1502 cm^{-1}, assigned to methoxyl group in lignin, the aromatic skeletal vibrations and the aromatic skeletal vibrations coupled with C–H in plane deformations, and the peaks at 1245 cm^{-1} (C–O stretching of acetyl group present in the lignin moiety as well as in hemicellulose, Subramanian *et al.*, 2005) and 1160 cm^{-1} (C–O–C antisymmetric bridge in hemicelluloses and cellulose and to aromatic C–H deformation of the syringyl and guaiacyl units in lignin) were revealed. Moreover, the broad band associated with hydrogen bonded hydroxyl groups at 3320 cm^{-1} and the band at 1648 cm^{-1}, attributable to the bending of adsorbed water (Brandrup and Immergut, 1989; Lojewska *et al.*, 2005), increased significantly, illustrating that the lipase treatment increased the amount of accessible hydroxyl groups and thus, of water absorption owing to the higher hydrophilicity of the WS outer surface.

4.4.7   Irradiation

Kato *et al.* (1999) treated different types of cellulose fibers with vacuum ultraviolet (VUV) irradiation in order to ascertain the efficiency of this technique in terms of oxidation capacity and compared their results with other common oxidation techniques. They monitored the evolution of the surface chemical composition by FTIR spectroscopy through the intensity variations of the band at 1720–1740 cm^{-1} (stretching vibration of carbonyl) and showed that VUV irradiation induced surface oxidation as efficiently as chromic and nitric acids, with the additional advantage of neither causing losses in bulk mechanical properties nor coloring the cellulose fibers. Strlic *et al.* (2003) studied the effect of neodymium-doped yttrium aluminium garnet (Nd:YAG) laser cleaning at 1064 nm on the surface of Whatman cellulose, rag paper, cotton, linen, silk and wool. They used FTIR diffuse reflectance spectroscopy in order to establish the chemical changes occurring at the surface of these materials. The peaks affected by laser treatment were the vibration at 1063 cm^{-1}, associated with C–O stretching, the band at 1111 cm^{-1} related with skeletal ring asymmetric stretching and that at 1413 cm^{-1} associated with scissoring, whereas the bands at 1492 and 875 cm^{-1} are new and cannot be associated with any of the usual bands appearing in FTIR spectra of pure cellulose. They concluded that chemical modifications of the surface were typical of the thermal degradation of cellulosic materials.

The application of γ-irradiation to cellulose and its derivatives has been extensively studied, but mostly in terms of the degradation mechanisms taking place (Takács et al., 1999). The effects of this treatment are shown in FTIR spectra by the appearance of a carbonyl band with increasing intensity as the duration of the treatment increased, attributed to terminal oxidized functions (carbonyl and carboxyl moieties) borne by the degraded cellulose fragments (Földváry et al., 2003; Takács et al., 1999).

4.5   Interfaces in polymer composites

Following the development of the field of composite materials, the need to assess and understand interface interactions has become increasingly important as the information can be further used in tailoring and improving material properties. The previously discussed studies utilized FTIR and Raman spectroscopies to investigate the efficiency of fiber modifications, which were aimed to affect interface interactions (De la Orden et al., 2007). However, although covalent bonding between filler and matrix has been frequently invoked in the literature (Bledzki and Gassan, 1999; Gauthier et al., 1998b; Kamdel et al., 1991; Li et al., 2000) to explain changes in the macroscopic properties, most frequently the interfacial reaction was not directly confirmed, because of the obvious complexity of accessing the composite interface. To the best of our knowledge, the first study that proved an interfacial reaction between fibers and polymeric matrix was that by Zadorecki and Flodin (1985). They synthesized two coupling agents based on trichloro-s-triazine with various terminal unsaturated groups in order to improve the bonding between cellulose fibers and an unsaturated polyester matrix. They proved both the covalent reaction of the coupling agent onto cellulose and the copolymerization of the treated fibers with styrene by FTIR and XPS (x-ray photoelectron spectroscopy) techniques. In the same context, De la Orden et al. (2007) presented an interesting FTIR spectroscopic analysis of composites made from bleached eucalyptus kraft pulp (92% cellulose) sprayed with polyethylenimine (PEI) as reactive coupling agent and PP matrix. Figure 4.10 shows the DRIFT spectra corresponding to composites with and without PEI as coupling agent (graphs a and b, respectively). The difference spectrum (curve c) reveals a negative band centered at 1715 cm^{-1} (C=O stretching vibration). The study suggested that during the high temperature processing

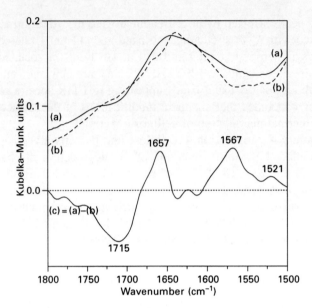

4.10 DRIFT spectra of PP composites with PEI-treated fibers (a) and untreated fibers (b). Graph (c) is the difference spectrum, (a – b). Reprinted from De la Orden *et al.*, 2007, with permission from Elsevier.

of the polyolefin composites, carbonyl and carboxyl groups are generated owing to oxidation of cellulose and PP. These groups further react with the amine groups of the coupling agent (PEI) present in the composites. Thus, the absorption of the C=O band disappears, and the formation of secondary amides explains the appearance of new bands, amide I at 1657 cm^{-1} and amide II at 1567 cm^{-1}. The formation of Schiff bases (imines) absorbing at 1660 cm^{-1} was also considered as a possible contribution to the bands observed in this region for the PEI-containing composites.

In order to investigate the participation of cellulose in the formation of the observed covalent bonds, composite samples were subjected to Soxhlet extraction with refluxing xylene to extract the fibers and remove all the components not covalently bonded to them. Again, spectral subtraction was performed between the spectra of the fibers extracted from the composites with and without coupling agent. The results agree with the observations made on the bulk composites, that is, the absence of the carbonyl/carboxyl band and the presence of amide and/or imine groups in the extracted fibers from PEI-containing composites. At the same time, these results also confirm that the PEI became covalently attached to the fibers.

Analogously, Demjén *et al.* (1999) also reported a covalent reaction occurring between amine functionalities of aminosilanes and carboxyl groups produced in the thermal oxidation of PP during processing.

Methacrylic anhydride was used by Hill and Cetin (2000) to attach wood flour to polystyrene. Using FTIR spectroscopy they demonstrated that methacrylic anhydride modified wood flour was able to form covalent bonds with styrene monomers via free radical polymerization. The authors reported the appearance of a weak bond at 1657 cm^{-1} in the FTIR spectrum of the grafted wood flour, owing to the attached unsaturation of the anhydride. This band substantially decreased after reaction with styrene, and weak absorption peaks appeared at 700–760 cm^{-1} owing to the phenyl group of styrene. The reactions were also confirmed by weight gain, saponification of the grafted flour and ^{13}C NMR.

Further to this, Marcovich *et al.* (2005) investigated the co-reaction between maleic anhydride previously attached to wood flour and an unsaturated polyester resin, through their co-reaction with styrene. They used a DRIFT technique in order to enhance the signal from surface chemical groups. They investigated the co-reaction of the grafted wood flour with styrene by mixing treated and untreated wood flour with styrene and hot-pressing the mixture (as was done with the styrene–unsaturated polyester composites). The untreated flour–styrene mixture led to a material that could be easily converted into powder by hand pressure. This material was extracted with refluxed toluene leaving a solid residue undistinguishable by DRIFT analysis from the original flour. However, the material obtained from MAN-treated wood flour and styrene could only be broken using a hammer. DRIFT spectra of the toluene-extracted material (Fig. 4.11) clearly showed the disappearance of the 1650

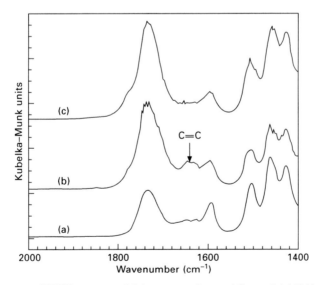

4.11 DRIFT spectra of (a) untreated wood flour, (b) MAN-esterified wood flour and (c) the same esterified wood flour after being treated with styrene and extracted with toluene.

cm^{-1} band corresponding to the anhydride unsaturations, which reacted with styrene. This study as well as that of Hill and Cetin (2000) showed that the unsaturated anhydrides grafted to wood flour by esterification are active for radical co-reaction with styrene and styrene containing resins.

More recently, Krouit and coworkers (2008) reported the grafting of polycaprolactone chains onto the surface of cellulose fibers via click-chemistry under heterogeneous conditions. Cellulose esters were prepared by reacting Avicel cellulose with undecynoic acid, in order to prepare cellulose substrate bearing multiple C≡C-terminated hairs and in parallel, polycaprolactonediol (PCL) was converted to an azidoderivative. Finally, modified cellulose was reacted with azido-PCL under heterogeneous conditions through 'click chemistry', which was followed by a Soxhlet extraction. After the initial modification, the FTIR spectrum of the treated cellulose fibers (Fig. 4.12) showed a new band at 1740 cm^{-1}, characteristic of the ester O–C=O moiety. The fingerprint of C≡C bands at 2100–2260 cm^{-1} was not observed owing to the low concentration of these surface groups with respect to the absorption of the bulk cellulose. Moreover, the C≡C–H signal expected to appear at 3300 cm^{-1}, was found to be overlapped by the peak corresponding to hydroxyl group. However, the reaction could be confirmed by elemental analysis and XPS, which indicated a low degree of substitution, because the reaction was mainly confined to the surface of the cellulose particles.

The click-chemistry step reaction with the azido-PCL was monitored *in situ* by FTIR spectroscopy through the reduction of the azido band (~2100 cm^{-1}). The FTIR analysis also revealed an increasing intensity of the ester band around 1740 cm^{-1}, which indicated that the grafting of the matrix on the cellulosic fibers was successfully achieved (Fig. 4.12c). In this case, XPS and weight gain (20%) confirmed the FTIR analysis.

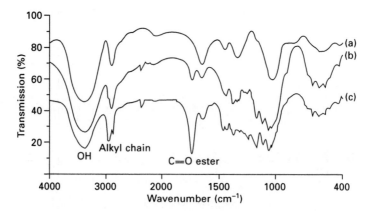

4.12 Comparative FTIR spectra of (a) cellulose, (b) cellulose undecynoate, and (c) cellulose grafted with PCL. Reprinted from Krouit *et al.*, 2008, with permission from Elsevier.

In polyolefin composites, the use of coupling agents is the usual method for improving compatibility. Lu *et al.* (2005) investigated the interface in composites made from wood fiber and high density polyethylene (HDPE), in the presence of different coupling agents, a maleated polyethylene (MAPE) and a maleated polypropylene (MAPP). The composites were prepared in a one step reactive process in which the wood fiber, polyolefin, coupling agent and peroxide initiator were fed into an intensive mixer. The melt was cooled, ground and molded in a heated press. To investigate the composition of the interface, the prepared composites were extracted by a Soxhlet procedure with hot xylene and dried before being analyzed by transmission FTIR, using KBr pellets at approximately 1 wt% dilution. The most interesting features in the FTIR spectra obtained from the unextractable composite samples (solid residue consisting of the fibers and attached polymer/copolymer) are the shoulders appearing in the ester carbonyl region and the large increase of the absorption bands in the C-H region ($\sim$2900 cm^{-1}). These characteristics are an indication that the coupling agent remained attached to the surface of the fiber after the xylene extraction. The authors additionally proposed that succinic and half succinic ester were the two types of covalent products that bonded the wood fiber to the polymer matrix at the interface. The formation of a half ester was also reported as the main product of reaction in the modification of wood flour with maleic anhydride according to Marcovich *et al.* (2005), who characterized the treated wood flour using DRIFT.

Analogously, Hristov and Vasileva (2003) used also FTIR to investigate the effect of a maleated PP, added as coupling agent, on the interface of a wood fiber–PP composite. They also reported that the coupling agent was located at the interface and covalently attached to the wood fibers by esterification.

Reduced hydrophilicity was the main effect reported by Colom *et al.* (2003). They studied two different coupling agents added to aspen fibers–HDPE composites (a maleated polyethylene and a γ-methacryloxypropyl trimethoxy silane). The effect of the chemicals on the interface was inferred from the FTIR spectra of composites prepared with 40 wt% of aspen fibers. Comparison of the spectra, showed that the peak at ca. 1635 cm^{-1} (corresponding to the vibrational absorption of water) appeared with a relatively lower intensity in the composites that incorporated coupling agents. This peak appeared in the spectra of the composites because of the hydrophilic nature of the incorporated fibers, but its intensity was much reduced by the protective effect of the coupling agents that surround the lignocellulosic reinforcement.

4.6   Summary

The various examples presented in this chapter on the use of FTIR and Raman spectroscopies applied to vegetable-based composites are not meant to provide an exhaustive review of the subject, but rather to give a glimpse into the different approaches to identify interfacial interactions.

FTIR spectroscopy is by far the most used of the two techniques, because of the easiness of use, the amount of information returned and also the long tradition in analytical chemistry. Different modes allow different geometries of the samples to be used and thus access to slightly different aspects of the chemistry being analyzed, (for example, different depths of beam incidence into the samples) or simply allow selection of the most suitable sample preparation through the selection of the FTIR mode to be used (transmission, attenuated or diffuse reflectance and others).

Because a functional group is associated with multiple vibrational absorptions, the spectra of composites, whose individual components can absorb also in overlapping regions, are difficult to analyze. Given the complexity of the spectra of unmodified lignocellulosic materials, it is not wise to base conclusions on the variations of a single band, but to look for confirmation in other non-overlapping regions or, as has been shown to be the case in many of the presented examples, in supporting results from other techniques.

Although less used up to the present, Raman was shown to be a valuable technique in confirming chemical changes occurring during modification of vegetable materials. In this respect, the fact that water does not interfere with Raman absorption is presented as a characteristic to take advantage of when working with natural fibers, which are well known for being very hygroscopic in nature.

It is clear that many interesting contributions to the identification of interfacial interactions in natural fiber composites are the result of ingenious experiments by many researchers, who utilized model/simplified matrices to identify a particular covalent interfacial bond or took advantage of extracting fibers from soluble matrix-composites to investigate the nature of any attached material. However, the direct observation of the interfacial bonds by FTIR and Raman spectroscopy is not possible, at least at the moment. The ever increasing improvement of dedicated equipment and the possibility of using microscopy associated to these techniques (for example, confocal Raman microscopy) may offer in a near future new tools to enable direct access closer to the interface region of lignocellulosic composites.

4.7   References

Abdelmouleh M, Boufi S, Belgacem M N, Duarte A P, Ben Salah A, Gandini A (2004), 'Modification of cellulosic fibres with functionalised silanes: development of surface properties', *Int J Adh Adhes*, **24**, 43–54.

Abdelmouleh M, Boufi S, Belgacem M N, Dufresne A (2007), 'Short natural-fibre reinforced polyethylene and natural rubber composites: effect of silane coupling agents and fibres loading', *Comp Sci Technol*, **67**, 1627–1639.

Abdelmouleh M, Boufi S, Belgacem M N, Dufresne A, Gandini A (2005), 'Modification of cellulose fibres with functionalized silanes: effect of the fiber treatment on the performance of cellulose-thermoset composites', *J Appl Polym Sci*, **98**, 974–984.

Abdelmouleh M, Boufi S, Ben Salah A, Belgacem M N, Gandini A (2002), 'Interaction of silane coupling agents with cellulose', *Langmuir*, **18**, 3203–3208.

Acha B A, Aranguren M I, Marcovich N E and Reboredo M M (2003), 'Composites from PMMA modified thermosets and woodflour', *Polym Eng Sci*, **43**(5), 999–1010.

Acha B A, Reboredo M M, Marcovich N E (2006), 'Effect of coupling agents on the thermal and mechanical properties of PP–jute fabric composites', *Polym Int*, **55**(9), 1104–1113.

Adebajo M O, Frost R L (2004), 'Infrared and 13C MAS nuclear magnetic resonance spectroscopic study of acetylation of cotton', *Spectrosc Acta Part A: Mol Biomol Spectrosc*, **60**(1–2), 449–453.

Agarwal U P, Atalla R H (1986), 'In-situ Raman microprobe studies of plant cell walls: macromolecular organization and compositional variability in the secondary wall of *Picea mariana* (Mill.) B.S.P.', *Planta*, **169**, 325–332.

Agarwal U P, Atalla R H, Forsskahl I (1995), 'Sequential treatment of mechanical and chemimechanical pulps with light and heat: a Raman spectroscopic study', *Holzforschung*, **49**, 300–312.

Alawar A, Hamed A M, Al-Kaabi K (2009), 'Characterization of treated date palm tree fiber as composite reinforcement', *Composites: Part B*, **40**, 601–606.

Aranguren M I, Reboredo M M (2007), Plant based reinforcements for thermosets matrices, processing and properties, in *Engineering biopolymers: homopolymers, Blends and Composites*, Editors: D Bhattacharyya, S Fakirov, Cincinnati, OH, Hanser Gardner Publications.

Bellamy L J (1975), *The infrared spectra of complex molecules*, 3rd Ed, Vol 1, Chapman and Hall, UK.

Bessadok A, Langevin D, Gouanvé F, Chappey C, Roudesli S, Marais S (2009), 'Study of water sorption on modified agave fibres', *Carb Polym*, **76**, 74–85.

Bio-Rad Laboratories Informatics Division (2005), *The Sadtler handbook of infrared spectra*, Bio-Rad Laboratories, Inc.

Bledzki A K, Gassan J (1999), 'Composites reinforced with cellulose based fibres', *Prog. Polym. Sci.* **24**, 221–274.

Bledzki A K, Letman M, Viksne A, Rence L (2005), 'A comparison of compounding processes and wood type for wood fibre-PP composites', *Composites Part A: Appl Sci Manuf*, **36**(6), 789–797.

Brandrup J, Immergut E H (1989), *Polymer handbook*, 3rd Ed. Wiley & Sons, New York, USA.

Cañigueral N, Vilaseca F, Méndez J A, López J P, Barberà L, Puig J, Pèlach M A, Mutjé P (2009), 'Behavior of biocomposite materials from flax strands and starch-based biopolymer', *Chem Eng Sci*, **64**, 2651–2658.

Colom X, Carrasco F, Pagès P, Canavate J (2003), 'Effects of different treatments on the interface of HDPE/lignocellulosic fiber composites', *Comp Sci Technol*, **63**(2), 161–169.

Das G, Bettotti P, Ferraioli L, Raj R, Mariotto G, Pavesi L, Soraru G D (2007), 'Study of the pyrolysis process of an hybrid CH3SiO1.5 gel into a SiCO glass', *Vib Spectrosc*, **45**, 61–68.

De la Orden M U, González Sánchez C, González Quesada M, Martínez Urreaga J (2007), 'Novel polypropylene–cellulose composites using polyethylenimine as coupling agent', *Composites: Part A*, **38**, 2005–2012.

Demjén Z, Pukánszky B, Nagy Jr J (1999), 'Possible coupling reactions of functional silanes and polypropylene', *Polymer*, **40**, 1763–1773.

Fang J M, Fowler P, Tomkinson J (2002), 'Preparation and characterization of methylated hemicelluloses from wheat straw', *Carbohydr Polym*, **47**, 285–293.

Felix J M, Gatenholm P (1991), 'The nature of adhesion in composites of modified cellulose fibers and polypropylene', *J Appl Polym Sci*, **42**, 609–620.

Földváry C, Takács E, Wojnárovits L (2003), 'Effect of high-energy radiation and alkali treatment on the properties of cellulose', *Rad. Phys. Chem.*, **67**, 505–508.

Frost R L, Mendelovici E (2006), 'Modification of fibrous silicates surfaces with organic derivatives: an infrared spectroscopic study', *J Colloid Interface Sci*, **294**, 47–52.

Gandini A, Botaro V, Zeno E, Bach S (2001), 'Activation of solid polymer surfaces with bifunctional reagents', *Polym Int*, **50**(1), 7–9.

Gassan J, Bledzki A K (1997), 'The influence of fiber-surface treatment on the mechanical properties of jute-polypropylene composites', *Composites Part A* **28A**, 1001–1005.

Gauthier H, Coupas A-C, Villemagne P and Gauthier R (1998a), 'Physicochemical modifications of partially esterified cellulose evidenced by inverse gas chromatography', *J. Appl. Polym. Sci.* **69**, 2195–2203.

Gauthier R, Joly C, Coupas A C, Gauthier H, Escoubes M (1998b), 'Interfaces in polyolefin/cellulosic fiber composites: chemical coupling, morphology, correlation with adhesion and aging in moisture', *Polym Comp*, **19**(3), 287–300.

Gupta S, Madan R N, Bansal M C (1987), 'Chemical composition of *Pinus caribaea* hemicellulose', *Tappi J*, **70**, 113–116.

Han G, Umenura K, Zhang M, Honda T, Kawai S (2001), 'Development of high-performance UF-bonded reed and wheat straw medium-density fiberboard', *J Wood Sci*, **47**, 350–355.

Herrera-Franco P J, Aguilar-Vega M (1997), 'Effect of fiber treatment on the mechanical properties of LDPE–henequen cellulosic fiber composites', *J. Appl. Polym. Sci.* **65**, 197–207.

Hill C A S, Cetin N S (2000), 'Surface activation of wood for graft polymerization', *Int J Adhes Adhes*, **20**, 71–76.

Himmelsbach D S, Akin D E (1998), 'Near-infrared Fourier-transform Raman spectroscopy of flax (*Linum usitatissimum* L.) stems', *J Agric Food Chem*, **46**, 991–998.

Hristov V N, Lach R, Grellmann W (2004), 'Impact fracture behavior of modified polypropylene/wood fiber composites', *Polym Test*, **23**(5), 581–589.

Hristov V, Vasileva S (2003), 'Dynamic mechanical and thermal properties of modified poly(propylene) wood fiber composites', *Macrom Mater Eng*, **288**(10), 798–806.

Ichazo M N, Albano C, González J, Perera R, Candal M V (2001), 'Polypropylene/wood flour composites: treatments and properties', *Comp Struct*, **54**(2–3), 207–214.

Idicula M, Boudenne A, Umadevi L, Ibos L, Candau Y, Thomas S (2006), 'Thermophysical properties of natural fibre reinforced polyester composites', *Comp Sci Technol*, **66**, 2719–2725.

Jacob M, Joseph S, Pothan L A, Thomas S (2005), 'A study of advances in characterization of interfaces and fiber surfaces in lignocellulosic fiber-reinforced composites', *Compos Interfaces*, **12**(1–2), 95–124.

Jähn A, Schröder M W, Füting M, Schenzel K, Diepenbrock W (2002), 'Characterization of alkali treated flax fibres by means of FT Raman spectroscopy and environmental scanning electron microscopy', *Spectrosc Acta Part A*, **58**, 2271–2279.

Jayaramudu J, Jeevan Prasad Reddy D, Guduri B R, and Varada Rajulu A (2009), 'Properties of natural fabric polyalthia cerasoides', *Fibers Polymers*, **10**(3), 338–342.

Jiang H, Zhang Y, Wang X (2009), 'Effect of lipases on the surface properties of wheat straw', *Ind Crops Prod*, **30**, 304–310.

Joly C, Kofman M, Gauthier R (1996), 'Polypropylene/cellulosic fiber composites chemical treatment of the cellulose assuming compatibilization between the two materials', *JMS: Pure Appl. Chem. A*, **33**, 1981–1996.

Junior de Menezes A, Siqueira G, Curvelo A P S, Dufresne A (2009), 'Extrusion and characterization of functionalized cellulose whiskers reinforced polyethylene nanocomposites', *Polymer*, **50**, 4552–4563.

Kalia S, Kaith B S, Sharma S, Bhardwaj B (2008), 'Mechanical properties of flax-g-poly(methyl acrylate) reinforced phenolic composites', *Fibers and Polymers*, **9**(4), 416–422.

Kamdel D P, Riedel B, Adnot A, Kaliaguine S (1991), 'ESCA spectroscopy of poly(methyl methacrylate) grafted onto wood fibers', *J. Appl. Polym. Sci.* **43**, 1901–1912.

Kato K, Vasilets V N, Fursa M N, Meguro M, Ikada Y, Nakamae K (1999), 'Surface oxidation of cellulose fibers by vacuum ultraviolet irradiation', *J Polym Sci A: Polym Chem*, **37**, 357–361.

Kishi H, Yoshioka M, Yamanoi A, Shiraishi N (1988), 'Composites of wood and polypropylenes I', *Mokuzai Gakkaishi* **34**(2), 133–139.

Kokta B V, Maldas D, Daneault C, Beland P (1990), 'Composites of polyvinyl chloride-wood fibers. I. Effect of isocyanate as a bonding agent ', *Polymer Plast Technol Eng*, **29**(1–2), 87–118.

Krouit M, Bras J, Belgacem M N (2008), 'Cellulose surface grafting with polycaprolactone by heterogeneous click-chemistry', *Eur Polym J*, **44**, 4074–4081.

Li Y, Mai Y-W, Ye L (2000), 'Sisal fibre and its composites: a review of recent developments', *Comp Sci Technol*, **60**, 2037–2055.

Liu, F P, Wolcott M P, Gardner D J, Rials T G (1995), 'Characterization of the interface between cellulosic Ffibers and a thermoplastic matrix', *Compos Interfaces*, **2**(6), 419–432.

Lojewska J, Miskowiec P, Lojewski T, Pronienwicz L M (2005), 'Cellulose oxidative and hydrolytic degradation: *in situ* FTIR approach', *Polym Degrad Stab*, **88**, 512–520.

Lu J Z, Negulescu I I, Wu Q (2005), 'Maleated wood-fiber/high-density-polyethylene composites: coupling mechanisms and interfacial characterization', *Compos Interfaces*, **12**(1–2), 125–140.

Luz S M, Del Tio J, Rocha G J M, Gonçalves A R, Del'Arco Jr A P (2008), 'Cellulose and cellulignin from sugarcane bagasse reinforced polypropylene composites: effect of acetylation on mechanical and thermal properties', *Composites: Part A*, **39**, 1362–1369.

Ly B, Thielemans W, Dufresne A, Chaussy D, Belgacem M N (2008), 'Surface functionalization of cellulose fibres and their incorporation in renewable polymeric matrices', *Comp Sci Technol*, **68**, 3193–3201.

Maldas D, Kokta B V (1992), 'Performance of hybrid reinforcements in PVC composites. IV. Use of surface-modified glass fiber and different cellulosic materials as reinforcements', *Int J Polym Mater*, **17**, 205–214.

Maldas D, Kokta B V, Daneault C (1989a), 'Influence of coupling agents and treatments on the mechanical properties of cellulose fiber–polystyrene composites', *J Appl Polym Sci*, **37**(3), 751–775.

Maldas D, Kokta B V, Daneault C (1989b), 'Thermoplastic composites of polystyrene: effect of different wood species on mechanical properties', *J Appl Polym Sci*, **38**(3), 413–439.

Marcovich N E, Aranguren M I, Reboredo M M (1997), Chemical modification of lignocellulosic materials. The utilization of natural fibers as polymer reinforcement, in *Lignocellulosic–plastics composites*, Editors: Leão A L, Carvalho F X, Frollini E, San Pablo, Universidade de Sao Paulo – Universidade Estadual Paulista.

Marcovich N E, Bellesi N E, Auad M L, Nutt S R, Aranguren M I (2006), 'Cellulose micro/nanocrystals reinforced polyurethane', *J Mater Res*, **21**(4), 870–881.

Marcovich N E, Reboredo M M, Aranguren M I (1996a), 'FTIR spectroscopy applied to woodflour', *Composite Interfaces*, **4**(3), 119–132.

Marcovich N E, Reboredo M M, Aranguren M I (1996b), 'Sawdust modification: maleic anhydride treatment', *Holz als Roh- und Werkstoff European Journal for Wood and Wood Industries*, **54**(3), 189–193.

Marcovich N E, Reboredo M M, Aranguren M I (1998), 'Mechanical properties of woodflour unsaturated polyester composites', *J Appl Polym Sci*, **70**, 2121–2131.

Marcovich N E, Reboredo M M, Aranguren M I (1999), 'Moisture diffusion in polyester–woodflour composites', *Polymer*, **40**(26), 7313–7320.

Marcovich N E, Reboredo M M, Aranguren M I (2005), 'Lignocellulosic materials and unsaturated polyester matrix composites: interfacial modifications', *Compos Interfaces*, **12**(1–2), 3–24.

Miguez Suarez J C, Coutinho F M B, Sydenstricker T H (2003), 'SEM studies of tensile fracture surfaces of polypropylene–sawdust composites', *Polym Test*, **22**(7), 819–824.

Moran J I, Alvarez V A, Cyras V P, Vazquez A (2008), 'Extraction of cellulose and preparation of nanocellulose from sisal fibers', *Cellulose*, **15**, 149–159.

Navoroj S, Culler R, Koenig J L, Ishida H (1984), 'Structure and adsorption characteristics of silane coupling agents on silica and E-glass fiber; dependence on pH', *J Colloid Interf Sci*, **97**(2), 309–317.

Nuñez A J, Sturm P C, Kenny J M, Aranguren M I, Marcovich N E, Reboredo M M (2003), 'Mechanical characterization of PP – woodflour composites', *J App Polym Sci*, **88**(6), 1420–1428.

Proniewicz L M, Paluszkiewicz C, Weselucha-Birczynska A, Majcherczyk H, Baranski A, Konieczna A (2001), 'FT-IR and FT-Raman study of hydrothermally degradated cellulose', *J Mol Struct*, **596**, 163–169.

Rahman M M (2009), 'UV-cured henequen fibers as polymeric matrix reinforcement: studies of physico-mechanical and degradable properties', *Mater Design*, **30**, 2191–2197.

Reddy S S, Bhaduri S K, Sen S K (1990), 'Infrared spectra of alkali treated jute stick', *J Appl Polym Sci*, **41**, 329–336.

Rong M Z, Zhang M Q, Liu Y, Yang G C, Zeng H M (2001), 'The effect of fiber treatment on the mechanical properties of unidirectional sisal-reinforced epoxy composites', *Comp Sci Technol*, **61**, 1437–1447.

Roy A K, Sen S K, Bag S C, Pandey S N (1991), 'Infrared spectra of jute stick and alkali-treated jute stick', *J Appl Polym Sci*, **42**, 2943–2950.

Sangthong S, Pongprayoon T, Yanumet N (2009), 'Mechanical property improvement of unsaturated polyester composite reinforced with admicellar-treated sisal fibers', *Composites: Part A*, **40**, 687–694.

Sgriccia N, Hawley M C, Misra M (2008), 'Characterization of natural fiber surfaces and natural fiber composites', *Composites: Part A*, **39**, 1632–1637.

Strlic M, Kolar J, Vid-Simon S, Marincek M (2003), 'Surface modification during Nd:YAG (1064 nm) pulsed laser cleaning of organic fibrous materials', *Appl Surface Sci*, **207**, 236–245.

Subramanian K, Kumar P S, Jeyapal P (2005), 'Characterization of ligno-cellulosic seed fiber from *Wrightia tinctoria* plant for textile applications: an exploratory investigation', *Eur Polym J*, **41**, 853–861.

Takács E, Wojnárovits L, Borsa J, Földváry C, Hargittai P, Zöld O (1999), 'Effect of γ-irradiation on cotton-cellulose', *Rad Phys Chem*, **55**, 663–666.

Tserkia V, Matzinosa Kokkoub P, Panayiotoua C (2005), 'Novel biodegradable composites based on treated lignocellulosic waste flour as filler. Part I. Surface chemical modification and characterization of waste flour', *Composites: Part A*, **36**, 965–974.

Valdez-Gonzalez A, Cervantes-Uc J M, Olayo R, Herrera-Franco P J (1999a), 'Chemical modification of henequén fibers with an organosilane coupling agent', *Compos, Part B: Eng* **30**, 321–331.

Valdez-Gonzalez A, Cervantes-Uc J M, Olayo R, Herrera-Franco P J (1999b), 'Effect of fiber surface treatment on the fiber–matrix bond strength of natural fiber reinforced composite', *Compos, Part B: Eng* **30**, 309–320.

van Hazendonk J M, Reinerink E J M, de Waard P, van Dam J E G (1996), 'Structural analysis of acetylated hemicellulose polysaccharides from fibre flax (*Linum usitatissimum* L.)', *Carbohydr Res*, **291**, 141–154.

Zadorecki P, Foldin P (1985), 'Surface modification of cellulose fibers. I. Spectroscopic characterization of surface-modified cellulose fibers and their copolymerization with styrene', *J Appl Polym Sci*, **30**(6), 2419–2429.

Zafeiropoulos N E, Williams D R, Baillie C A, Matthews F L (2002), 'Engineering and characterization of the interface in flax fibre/polypropylene composite materials. Part I. Development and investigation of surface treatments', *Compos, Part A: Appl Sci Manuf*, **33**(8), 1083–1093.

Zhang W, Zhang X, Liang M, Lu C (2008), 'Mechanochemical preparation of surface-acetylated cellulose powder to enhance mechanical properties of cellulose-filler-reinforced NR vulcanizates', *Compos Sci Technol*, **68**, 2479–2484.

5

Testing the effect of processing and surface treatment on the interfacial adhesion of single fibres in natural fibre composites

A. ARBELAIZ and I. MONDRAGON, University of the
Basque Country, Spain

Abstract: The main methods for mechanical characterization of the single fibre–polymer matrix interfacial adhesion are described: fibre pull-out, single fibre fragmentation and microdebonding tests. The basis of each method, assumptions and main advantages and drawbacks are discussed. A review of lignocellulosic polymer fibre–matrix interface adhesion data is presented. The effect of different surface treatments and also the effect of different processing parameters are also shown.

Key words: lignocellulosic fibres, single fibre polymer–matrix composites, interfacial shear strength, surface treatments, processing parameters.

5.1   Introduction

Fibre-reinforced polymer matrix composites are multi-phase systems. They are important mainly because of their high strength and modulus, and light weight. The superior performance of such composites depends not only on the strength of the reinforcement material and the matrix, but also on the fibre–matrix interface/interphase.[1,2] One of the most typical types of failure in composite materials is interfacial debonding. It is usually a result of poor bonding at the interface between the fibre and the matrix.[3] To overcome this poor bonding when lignocellulosic fibres are used, various fibre-surface treatments have been used.[4–8] The determination of τ, interfacial shear strength (IFSS), between fibre and matrix in composite materials is possible with a variety of methods ranging from tests with a single fibre embedded in the matrix to composite specimens. The objective of these methods is to predict interfacial shear strength and to relate interfacial shear strength to fibre, matrix and interface/interphase properties.[9]

According to its definition, interface is an infinitesimally thin section between two phases, whereas interphase is an interfacial region with finite volume and distinct physical properties or gradients in properties.[10] An interphase is formed if either of two homogeneous phases influences a region of the other phase in a manner that alters its chemical or physical structure.

146

From the schematic drawing of the structure of a fibre-reinforced material, shown in Fig. 5.1, it can be seen that two interfaces do exist, one between the interphase and the fibre and another one between the interphase and the bulk matrix.[10]

A prerequisite to form an adhesive bond between two materials, but not a sufficient condition for good adhesion, is that they have to come into close molecular contact. For fibre-reinforced polymer composites, the ability of a given polymer melt to wet a fibre surface depends on the surface tension of both materials. A general condition that must be fulfilled for wetting is that the surface tension of the polymer melt must be less than that of the fibre. In addition, optimum conditions for wetting exist when the polarities of the two phases match each other.[10] The mechanisms that cause two materials to adhere one to another are not well understood and various mechanisms for stress transfer across the interface are possible. The most accepted theories are adsorption and bonding, diffusion, electrostatic attraction and mechanical interlocking (Fig. 5.2).

Adsorption (Fig. 5.2a) is the process whereby a molecule is attracted to a specific site on a solid surface. The attraction may be the result of relatively weak interactions or primary bonds such as ionic, covalent or metallic can be formed across the interface. Primary bonds represent one of the strongest types of interaction and thus an extremely important adhesion mechanism.[10] Electrostatic attraction bonding (Fig. 5.2b) is the result of concentration on each surface of particles charged with opposite polarities that attract one to another.[11] Diffusion or interpenetrating network (Fig. 5.2c) involves the simple concept that when two polymers are in close contact at temperatures above their glass transition temperatures and are partly soluble, then the long chain molecules, or at least segments of them, will interdiffuse and form entanglements.[10] Mechanical interlocking adhesion (Fig. 5.2d) is the

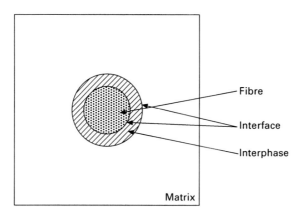

5.1 Schematic drawing of a fibre-reinforced material.

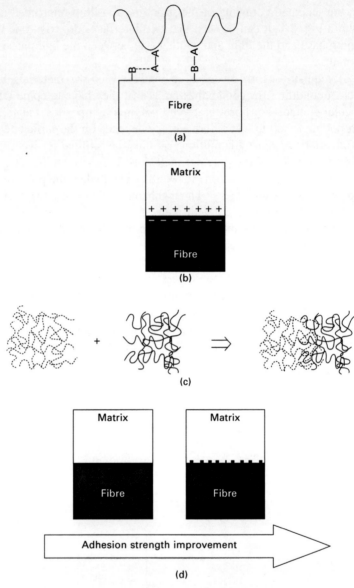

5.2 Schematic drawings of the principles of the most accepted adhesion mechanism theories: (a) adsorption and bonding, (b) electrostatic attraction, (c) diffusion or interpenetrating network, (d) mechanical interlocking.

result of mechanical interlocking and consequently is strongly influenced by surface roughness. A prerequisite of this kind of bonding is that matrix can introduce itself into the microvoids and microcavities of fibre. The strength is transmitted between two bodies because of the shear stresses.[11] Adhesion can be the result of combined effects of more than one of these mechanisms, consequently interfacial bond strength interpretation is complicated.[2,11] In the following, the main methods for characterization of the single fibre–polymer matrix interface adhesion are described and a review of lignocellulosic-polymer fibre–matrix interface adhesion data is presented.

5.2   Methods for characterization of single-fibre–polymer matrix interfacial adhesion

There is no generally accepted physical model to describe adhesion at the interface and the process of debonding. Several studies used different approaches for interfacial shear strength calculations.[12] Interfacial fibre–matrix adhesion tests can be classified into two general categories: multiple-fibre or composite tests and single-fibre tests. Multiple-fibre or composite tests give a very good indication of the expected composite performance. These tests have the disadvantages of requiring expensive and time-consuming sample preparation, requiring large quantities of material.[13] On the other hand, single-fibre tests are easy to perform and can be done on a small quantity of fibre.

Sample preparation is generally inexpensive. However, the single-fibre tests are possibly not indicative of the performance of an actual composite since the microstructure of a composite deviates significantly from single-fibre tests as there are defects of various types, misaligned fibres, fibre-poor and fibre-rich zones, fibre spacings, fibre diameter, and material properties such as the fibre strength and the interfacial strength follow random statistical distributions. Methods for characterization of the single-fibre–polymer matrix interfacial adhesion are mainly suitable for studying various surface treatments in a given fibre–matrix system. In the following, the most often used methods for characterization of the single-fibre–polymer matrix interfacial adhesion are presented.

5.2.1   Single-fibre fragmentation test

The single-fibre fragmentation test (SFFT) method was originally used with metals by Kelly and Tyson[14] who observed a multiple-fibre fracturing phenomenon in a system consisting of a low concentration of brittle tungsten fibres embedded in a copper matrix.[2] The main assumptions made by Kelly and Tyson were that the matrix yields plastically, the fibre–matrix adhesion is perfect and the shear stress along the fibre is constant and equal to the shear yield strength of the matrix.[15] In addition, a single fragmentation test to measure the interfacial shear strength is available for those composite systems where the ultimate elongation of the matrix is higher than the fibre elongation. The test states that the strain at matrix fracture must be at least three times higher than that of the fibre.[16] The method involves completely embedding a single fibre along the centreline of a relatively large dog-bone shaped specimen of matrix material (Fig. 5.3a–b). The specimen is then pulled in tension uniaxially along the fibre axis. Stress is transferred to the fibre by the matrix. As the strain is increased, the fibre will break. When the fibre first breaks, the matrix restrains the fibre fragments for returning to their unstressed dimension. However, the axial stress in the fibre at a distance from the break increases until it reaches the original stress in the fibre before the first break. Further increase in the specimen strain results in another break in the fibre (Fig. 5.4). This fibre breakage continues with increased strain until the fibre fragments become so small that the matrix can no longer transfer

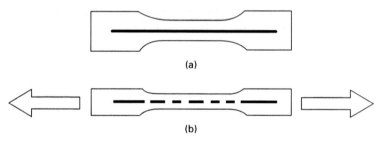

(a)

(b)

5.3 Specimen consisting of a fibre embedded in the matrix (a) before stretching; and (b) after stretching.

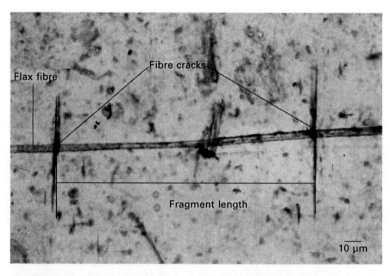

5.4 A typical optical micrograph showing fibre breaks after the fragmentation test for MAPP–flax specimen (from reference 67, copyright Koninklijke Brill NV, Leiden, 2008, with permission).

stress over a long enough distance to break the fibre.[13] At this point, as no more breaks can occur with increased strain, the fragmentation process is saturated and the test is completed.[17]

When a matrix is transparent, fibre fracture phenomena can be easily observed through an optical microscope and fragment lengths can be determined. When the matrix used is a semicrystalline thermoplastic, after saturation observation of the fibre fragments by optical microscopy is difficult. In order to properly locate the fibre break points and measure the fragment lengths, the specimens can be melted.[18]

Some studies incorrectly assumed that the length of the broken fibre fragments, referred to as the fibre critical length l_c, is an indication of the ability of the polymeric matrix to transfer stress to the fibre.[13] Kelly and Tyson defined the critical fragment length as the maximum fragment length over which no further fibre fracture occurs.[19] However, the average of the measured fragmentation length, $\bar{l}_f$ is not equal to l_c [3,20] because, when the fibre length is just above l_c, the fibre can break into two pieces both having lengths less than l_c. When the fibre length is just below l_c, it can not break any further. Therefore, the measured fragmentation length, $\bar{l}_f$ should be distributed between $\frac{1}{2}l_c$ and l_c.[3]

The axial stress acting in a fibre fragment of a single fibre composite specimen varies along the fragment length. The two fibre fragment ends are unloaded: axial stress in the fibre at those points is zero. Moving along the fibre from the end towards the fragment centre, stress increases. This fibre stress build-up occurs owing to stress transfer by the surrounding matrix. For a perfectly plastic matrix ideally adhering to the fibre, the stress build-up in the fibre is rectilinear from the fibre ends to the centre,[3] as represented schematically in Fig. 5.5 and 5.6, where $\sigma_{f_3} > \sigma_{f_2} > \sigma_{f_1}$ are the fibre tensile strengths at different fibre lengths. For the case represented in Fig. 5.5a, the tensile stress that supports the fibre increases linearly from the fibre ends but the highest stress value is not enough to break the fibre. On the other hand, for the case represented in Fig. 5.5b, the fibre length is sufficient to reach the tensile strength value of embedded fibre and the fibre breaks.

The simplest way to relate average fibre length and critical length is the approximation[3] shown in equation 5.1,

$$\bar{l}_f = \frac{3}{4} l_c \qquad\qquad [5.1]$$

Several studies[3,9,18,21] assume a distribution function to determine $\bar{l}_f$. The most commonly used distribution functions are the Gaussian and Weibull distributions.

Based on a force balance in a micromechanical model for a system of elastic fibre embedded in a plastic matrix[17] and assuming that the stress transferred across the interface acts over a length l_c, a force balance can be

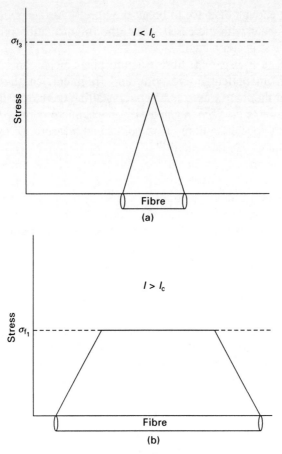

5.5 Scheme of stress distribution along the fibre axis for fibre length (a) lower than l_c and (b) higher than l_c.

obtained by setting the total force transferred across the interface equal to the breaking force of the fibre (Fig. 5.6a–b), where σ_{f_2} is the fibre strength at fibre length l_c and r is the fibre radius. Behind the apparent simplicity of this expression, there is a great deal of complexity because σ_{f_2} refers to the ultimate strength of very short lengths of the fibre and τ includes a combination of different modes of stress transfer.[22] To correctly calculate τ, valid values of σ_{f_2}, r and l_c must be determined. The fibre tensile strength depends on the distribution of flaws along the length, thus it is strongly influenced by the gauge length. Hence it is incorrect to use the tensile strength values obtained for macroscopic gauge lengths.[12] On the other hand, the value of σ_{f_2} is too complicated to be experimentally measured because testing the fibre at very short gauge lengths of the order of l_c is impossible in most cases[23] or at least it can entail experimental difficulties.[12] Sometimes this

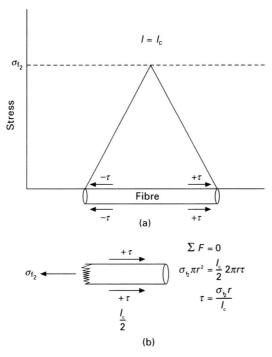

5.6 (a) Distribution of interfacial shear stress and fibre stress in a short fibre ($l = l_c$); (b) force balance in the micromechanical model.

strength value was obtained through a linear extrapolation of experimental data measured in single fibre tensile tests at longer gauge lengths.[24] Many studies obtained this strength by extrapolation using the two-parameter Weibull distribution function.[12,25] However, for several reasons, this value of fibre strength obtained may differ from the *in situ* fibre strength,[23] equation 5.2,

$$\sigma_{f_2} = \sigma_{f_0} \left(\frac{l_c}{l_0} \right)^{-\frac{1}{\rho}}$$

[5.2]

where σ_{f_0} is the fibre strength at a gauge length of l_0 and ρ is the shape parameter of the Weibull distribution function for fibre strength.

The main difficulty in measuring the single-fibre diameter of natural fibres is that it can vary widely between two ends and there can also be differences in diameter among filaments.[26,27] The cross-section of natural fibres is polygonal or oval, and somewhat irregularly ellipsoidal.[28–30] As a result, any determination assuming a round cross-section is expected to give a higher fibre diameter, particularly by an optical method, and thus a lower value of τ.[26]

As mentioned above, different methods are used to relate average fibre length and critical length. Therefore the calculation of τ depends on the values of σ_{f_2}, r and l_c used. Calculation of τ gives an indication of the ability of the matrix to transfer stress to the fibre. Because there is normally no interfacial failure between fibre and matrix, it is incorrect to call τ the interfacial shear strength. Rather, τ is a measure of the stress transfer across the interface, and it should be referred to as the stress transfer or stress transfer coefficient.[13]

Even with the Kelly–Tyson model different approaches can be encountered for calculating the variables in the expression determining the stress transfer coefficient, i.e. the critical length l_c and tensile strength σ_{f_2} of the fibre at the critical length. Besides, the method assumes a constant shear stress across the interface but ignores the matrix properties, conditions that can hardly be met in real fibre–matrix systems. Therefore, the single-fibre fragmentation test method seems inappropriate for comparison between different systems even on a qualitative basis. However, the main advantage of this method is that it is applicable over a wide range of interfacial bond strengths and it can provide qualitative indications of interfacial bond strength in a given system.[2]

5.2.2   Fibre pull-out test

Although the fibre pull-out test is popular because of its conceptual simplicity, this test, however, presents some difficulties.[16] For testing, a fibre is embedded in a very thin film or bead of polymer (Fig. 5.7–5.8) and the specimen is gripped in a tensile testing machine. The fibre is pulled out of the polymer and the force required to remove the fibre is measured.[13] The adhesive strength is calculated by dividing the measured maximum load by the area of contact of the fibre with the polymer. The area of contact can be measured using optical microscopy. Because of difficulties in measuring the exact embedded surface area for each fibre, it is usually assumed that the fibres are perfectly cylindrical with a smooth surface, which may not be completely valid and it therefore affects the accuracy of calculation.[31] The small diameter of the fibre requires that the area of interfacial contact be

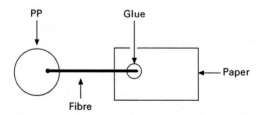

5.7 Schematic drawing of pull-out sample.

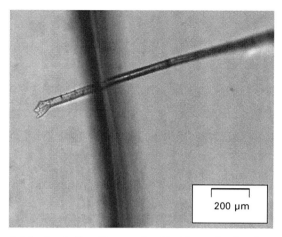

5.8 A single flax fibre embedded in melted polypropylene matrix.

small, otherwise the fibre may break rather than pulling-out, thus altering the data analysis.[13] Interfacial shear strength values can be obtained according to the Kelly–Tyson equation[7,28,32,33] by using the assumptions that the shear stress at the interface is uniformly distributed along the embedded length,[26,32] the interfacial loading is in shear mode and the friction between debonded fibre and matrix is negligible,[34] as shown by equation 5.3:

$$\tau = \frac{F_{max}}{\pi dl} \qquad\qquad [5.3]$$

where F_{max} is the maximum force recorded by the load cell and d and l are fibre diameter and embedded length, respectively. For the fibre to be pulled out rather than broken, test conditions require that tensile stress in the pulled-out fibre is less than its ultimate tensile strength. Taking into account the Kelly–Tyson equation, the maximum embedded fibre length l_{max} is given by $l_{max} < \frac{\sigma r}{2\tau}$. As l_{max} is usually low, it causes experimental difficulties and a large data scatter.[2]

Figure 5.9 shows a typical pull-out test curve of a flax fibre–polypropylene system. This graph can be separated into three parts corresponding to the different stages involved in pull-out. During the first part of testing, the graph is considered to represent the rectilinear elastic behaviour of the fibre–matrix system and the fibre–matrix interface remains intact. For the second stage, after initiation, debonding occurs by means of crack propagation along the embedded fibre length. The third part occurs after complete debonding has taken place, where the remaining force is the result of frictional interactions between the fibre and the matrix.[26,35]

Herrera-Franco *et al.*[36] studying high-density polyethylene (HDPE)–henequen fibre composite noted that pull-out graphs for various surface

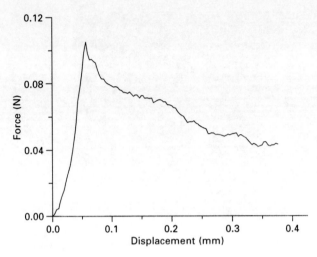

5.9 A typical pull-out test curve of flax fibre–PP system (from reference 26 with permission, copyright Wiley, 2010).

treatments exhibited the nonlinear behaviour characteristic of a ductile matrix. However, after the load reached its maximum value the graphs showed significant differences. For the native henequen, they observed that the load increased gradually and when it reached a maximum value there was a smooth transition and it began to decrease in a linear fashion until the total embedded length of the fibre was pulled-out. They mentioned that this behaviour agreed well with the behaviour of a poor interphase that resulted because of the incompatibility between the hydrophilic fibre and the hydrophobic matrix. However, they suggested that the load–displacement graph for a fibre treated with an aqueous NaOH solution and then pre-impregnated with dissolved HDPE seemed to be a very strongly bonded interphase because the pull-out force was higher and the pull-out process occurred catastrophically.

The fibre pull-out test is mainly suitable for studying various surface treatments in a given fibre–matrix system of strong fibres and low to medium interfacial bond strength. Comparison between different systems may be misleading. The fibre pull-out test is inadequate for weak fibres because the test gives a significant data scatter owing to the permissible embedded fibre lengths being extremely short.[2] Pull-out tests entail experimental difficulties like sample clamping and embedded fibre alignment problems, besides the meniscus at the point of a fibre entry into matrix sometimes makes it difficult to determine the exact embedded length of the fibre.[16] As shown in equation [5.3], τ is defined as the ratio of the maximum force measured in a pull-out test to the contact area. This approach is very simplistic and, not surprisingly, gives a large scatter in the experimental data.[37] In addition, τ underestimates the interfacial shear strength because the interface fails where the maximum stress is located and the interfacial shear stress is not distributed evenly along the fibre. Moreover, τ decreases as the embedded length increases because of the uneven distribution of shear stress at the interface.[38] Interfacial shear stresses calculated from pull-out tests can be used only to estimate the bonding strength and compare the effect of the treatment on the fibre–matrix interfacial bonding strength.[31]

As a result, the apparent IFSS values obtained in various studies can not be compared with each other.[37,39] Pisanova *et al.*[37] proposed to characterize the interfacial bond strength in terms of the local interfacial shear strength. The local IFSS is a semi-empirical parameter that can be determined from experimental data by extrapolating τ to zero embedded length, equation [5.4],

$$\tau_{loc} = \lim_{l_e \to 0} \tau \qquad\qquad [5.4]$$

5.2.3   Microdebonding test

The microdebonding test is a modified version of the pull-out test, proposed by Miller *et al.*[40] The test was developed for characterising the single-fibre–matrix interface in order to eliminate any meniscus effect.[41] In this test a small drop of matrix is deposited onto the fibre at some point. The fibre with the microbead is mounted in a micro-device and then the fibre is pulled out (Fig. 5.10a–b).[42] The test has been widely used to evaluate τ for

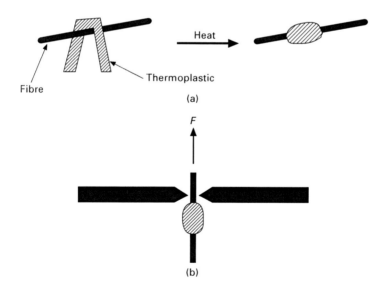

5.10 (a) Procedure for forming microdebonding samples with thermoplastic polymer; (b) schematic diagram of microdebonding test.

both thermoset and thermoplastic matrix composites[40,41] because sample preparation is simplified. Although this test is easy to perform, there are several concerns that have to be taken into account such as stress concentrations during specimen loading, non-uniform shear stress distribution along the fibre–matrix interface, geometry of the polymer drop, and the effect of the strain rate. All these factors significantly affect the test results and their scatter. The reliability of the data is dependent on the shape of the drop. Symmetric, round drops are easier to test and analyze than drops with flat surfaces. If the length of drop in contact with the fibre surface exceeds a critical value, the fibre will break before debonding and pull-out.[43] Miller *et al.*[40] noted a reduced variability in shear debonding results when the drop-sizes were larger, they suggested that the variability occurred because localized surface differences were more likely to be averaged out with a greater contact area.

Results obtained by various methods for characterization of the single fibre–polymer matrix interface adhesion values are not similar. However, several studies[5,36] observed that when different methods were used for the same system, the results were similar for the different tests.

5.3   Review of lignocellulosic polymer fibre–matrix interfacial adhesion data

5.3.1   Introduction

As mentioned above, for measuring interfacial shear strength there are many approaches. However, the experimental τ should be regarded as a mere adhesion parameter and not as the actual stress acting at the fibre–matrix interface.[18] Apart from the intrinsic problems of each method for measuring interfacial shear strength, the use of natural fibres adds another problem associated with them, the highly anisotropic nature of this kind of fibres. The lack of homogeneity in these fibres makes the interpretation of interfacial adhesion test results difficult[34] and a wide scattering of results is seen. The fibres are highly anisotropic especially when the diameter is considered because their diameters vary from fibre to fibre.[44] In addition, several studies[5,45,46] have shown that the cross-section and apparent diameter of natural fibres vary considerably along their length. Some studies[28,29,30] observed that the cross-section of natural fibres is not round but somewhat irregularly ellipsoidal and fibres show a hollow multicellular nature. As a result, any determination assuming a round cross-section is expected to give an inaccurate value.[28]

The situation of single-fibre composites of natural fibre is complex because fibres themselves could be considered composite materials. In addition, if bundles of natural fibres are used instead of a single fibre, two types of interfaces should be considered, one between the fibre bundle and the matrix and the other between the cells.[11] Much information is available concerning the lignocellulosic fibre–polymer matrix interfacial adhesion (Table 5.1). An increase in interfacial shear strength for composites based on treated fibres is considered to be evidence of the treatment efficiency. Published results should be treated carefully owing to the assumptions made for data analysis and the lack of homogeneity of natural fibres properties. The obtained results are not immediately representative of the actual composite where performance increase is generally much lower.[6]

5.3.2   Effect of fibre surface modification

The adhesion between two solids is directly related to their surface energies and wetting is a requisite to get a good interfacial adhesion. Moreover, the main disadvantage of natural fibres is their hydrophilic nature, which lowers the compatibility with hydrophobic polymer matrices during composite fabrication.[4] The effect of different chemical and physical treatments on natural fibre–polymer matrix interfacial shear strength has been investigated.[7,8,16,18,26] Fibre–matrix adhesion can be the result of combined effects of more than one adhesion mechanism, as mentioned above. Therefore, the study of adhesion of single-fibre composites of natural fibres is complex. Usually lignocellulose fibres after treatments become more hydrophobic indicating that treated fibres can be better wetted out by a hydrophobic matrix. A good adhesion is also desirable to prevent, or inhibit, environmental agents from invading and destroying the interface. For lignocellulosic fibres, the degradation caused by water at the interface is of primary concern because the fibres are highly hygroscopic.[27]

Fibre surface treatments based on the use of sodium hydroxide, silane and maleic acid anhydride (MA) based copolymer are the most common.[5,7,26,47] Figures 5.11–5.14 show possible mechanisms of adhesion improvement for maleic anhydride polypropylene copolymer (MAPP), silane and sodium hydroxide fibre surface treatments,[48–50] respectively. The reaction between cellulose and copolymer can be divided into two main steps, as shown in Fig. 5.12. In the first step, the copolymer is converted into the more reactive anhydride form and sterification of the cellulose fibres takes place in the second step.[48] As shown in Fig. 5.13, the silane can hydrolyse to some extent to form silanols, thus providing a link to cellulose through their -OH groups by the formation of hydrogen bonds. Mwaikambo et al.[50] suggested that alkalization of lignocellulosic fibres removed lignin, pectin, waxy substances, and natural oils covering the external surface of the fibre cell wall (Fig. 5.14). This treatment can reveal the fibrils, and gives a rough surface topography to the fibre, as shown by Arbelaiz et al.[26] (Figure 5.15a–b). Mwaikambo et al.[50] also mentioned that alkalization depolymerizes the native cellulose I molecular structure producing short crystallites (Fig. 5.14b).

Table 5.1 Interfacial shear strength of lignocellulosic fibre–polymer matrix composites

Matrix	Lignocellulosic fibre	Best fibre surface/ matrix modification	τ (MPa) unmodified	τ (MPa) modified	Test method	Reference
LDPE	Jute	Mercerized	3.8	3.6	SFFT	18
PP	Jute	Mercerized	11	12	SFFT	18
HDPE	Henequen	NaOH + silane agent + preimpregnation with matrix dilute solution	~2.5*	9.0	Pull-out	5
HDPE	Henequen	NaOH + silane agent + preimpregnation with matrix dilute solution	~4*	20.0	SFFT	5
PHBV	Henequen	–	5.24	–	Microdebonding	28
PHBV	Henequen	–	6.97	–	SFFT	28
PP	Ramie	MAPP	9.6	~15*	Microdebonding	42
Epoxy	Jute	Bleached	83	140	SFFT	27
LDPE	Regenerated cellulose	–	~4.8*	~8.5*	SFFT	51
PP	Ramie	–	~5.0*	–	Pull-out	52
PP	Regenerated cellulose	–	~4.2*	–	Pull-out	52
PP/EPDM	Flax	MAPP	1.2	8.9	Pull-out	53
Acrylic	Hemp	SMA	~2.3*	~4*	Pull-out	31
Epoxy	Pineapple leaf	Combination NaOH + DGEBA	26.53	72.40	SFFT	54
PP-MAPP	Jute	5% MAPP	~4.6*	~6.5*	Microdebonding	55
PP-MAPP	Hemp	5% MAPP	~4.8*	~6.3*	Microdebonding	55
PS	Rayon	Energy glow discharge plasma	7.0	11.9	Microdebonding	56
HDPE	Henequen	NaOH + silane agent	5.4	16	SFFT	36
HDPE	Henequen	NaOH + silane agent	2.5	5.0	Pull-out	36
PS	EFB	Acetylation	~1.1*	~2.0*	Pull-out	41

Matrix	Fiber	Treatment			Method	Ref.
PS	Coir	Acetylation	~1.3*	~2.3*	Pull-out	41
Epoxy	EFB	–	~1.9*	~2.5*	Pull-out	41
Epoxy	Coir	–	~2.2*	~2.7*	Pull-out	41
UP	EFB	–	~1.5*	~2.0*	Pull-out	41
UP	Coir	–	~1.9*	~2.4*	Pull-out	41
PP	Hemp	White rot fungi treatment	3.26	5.84	Pull-out	35
HDPE	Sisal	Permanganate oxide	1.6	3.2	Pull-out	57
Phenol–formaldehyde	Banana	–	44	–	Pull-out	58
PP	Flax	MAPP	14.85	19.64	Pull-out	7
MAPP	Flax	–	–	25.14	Pull-out	7
Polyester	Flax	Depolymerization of lignin and hemicelluloses into lower molecular aldehyde and phenolic functionalities followed by a subsequent curing	17.1	11.7	Pull-out	62
PP	Flax	Acetylation	6.33	11.61	SFFT	62
Polyester	Sisal	NaOH	2.6	6.9	Pull-out	63
PP	Henequen	Electron beam	3.9	5.3	Microdebonding	64
PLA	Jute	Water + ultrasonication	~5.5*	~13*	Microdebonding	65
PLA	Kenaf	Water	~10.5*	~11.5*	Microdebonding	65
PP	Henequen	Water + ultrasonication	~4*	~6*	Microdebonding	65
Polyester	Henequen	Water + ultrasonication	~5.5*	~8.7*	Microdebonding	65
HDPE	Agave americana	Boiled in water + NaOH	~0.55*	~0.7*	Pull-out	66
MAPP	Flax	–	–	12.03	SFFT	67
MAPP	Hemp	–	–	14.33	SFFT	67
MAPP	Cotton	–	–	0.66	SFFT	67
PP	Henequen	NaOH + ultrasonication	4.05	9.62	Microdebonding	68

Continued

Table 5.1 Continued

Matrix	Lignocellulosic fibre	Best fibre surface/ matrix modification	τ (MPa) unmodified	τ (MPa) modified	Test method	Reference
PS	Rayon	Heat	3.06	3.72	Microdebonding	33
PS	Cotton	Acetylation	3.36	6.46	Microdebonding	33
PS	Wood	Acetylation	3.06	10.0	Microdebonding	33
Soy protein isolate	Ramie	–	29.8	–	Microdebonding	30
PP	Flax	–	3.39	–	Microdebonding	69
PP	Hemp	–	5.07	–	Microdebonding	69
PP	Sisal	–	4.6	–	Microdebonding	69
MaterBi	Flax	–	4.18	–	Microdebonding	69
MaterBi	Hemp	–	2.97	–	Microdebonding	69
MaterBi	Sisal	–	3.22	–	Microdebonding	69
Lactide–glycolide copolymer	Flax	–	8.98	–	Microdebonding	69
Lactide–glycolide copolymer	Hemp	–	11.33	–	Microdebonding	69
Lactide–glycolide copolymer	Sisal	–	14.26	–	Microdebonding	69
PE	Sisal	Stearic acid	2.16	2.66	SFFT	71
Polyester	Date palm	NaOH + silane agent	~11*	~16*	Pull-out	72
Epoxy	Ramie	–	~16*	–	Microdebonding	73
Epoxy	Kenaf	–	~10*	–	Microdebonding	73
Vinyl ester	Bamboo	–	10.82	–	Pull-out	75
HDPE	Flax	Duralin	–	10.1	Pull-out	34
LDPE	Flax	Duralin	–	6.2	Pull-out	34

PP	Flax	Dew–retted	–	10.6	Pull-out	34
MAPP	Flax	Dew–retted	–	11.4	Pull-out	34
Vinyl ester	Flax	Acrylic acid	28	31	SFFT	8
Polyester	Flax	Acrylic acid	13	15	SFFT	8
Epoxy	Flax	–	33	–	SFFT	8

*Values taken from figures in the references.

5.11 Schematic representation of interactions between polymer matrix and surface-modified lignocellulosic fibre.

5.12 Model of reaction between cellulosic fibres and PP–MAPP copolymer.

In the following, a review of obtained interfacial adhesion data for different surface treatments of lignocellulosic fibre-polymer matrix interface adhesion data is presented. Arbelaiz et al.[26] studied flax fibres as reinforcement for polypropylene (PP) matrix composites. For improving compatibility between flax fibre bundles and PP matrix, fibre surface treatments with MA, MAPP

5.13 Model of reaction between cellulosic fibres and silane agent.

copolymer, vinyltrimethoxysilane and alkali were carried out. The effect of surface modification on fibre–matrix interfacial shear strength was analyzed. The polar component of the surface energy was lowered by treatment and the treated fibres were better wetted out by the polypropylene matrix than

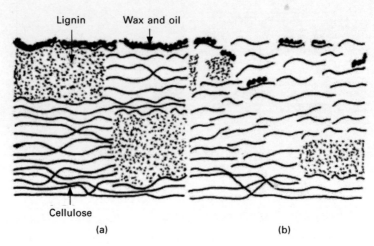

Lignin   Wax and oil

Cellulose

(a)   (b)

5.14 Typical structure of (a) untreated and (b) alkalized cellulosic fibre (from reference 50 with permission, copyright Wiley, 2010).

the untreated ones. The apparent interfacial strength, as determined by pull-out tests, for a surface-modified single flax fibre showed that only MAPP and alkali treatment led to improvement of fibre–matrix adhesion increasing the IFSS by around 15 and 8%, respectively. Therefore, besides wetting characteristics, other surface properties should be taken into account for improving the interfacial strength. Arbelaiz et al.[26] suggested that for silane and MA treated fibres, the vinyl group, which should interact with the thermoplastic matrix, could not create entanglements, as the compound used for treatment is not long enough. Therefore, stresses could not be transferred from PP chains, through fibre bonded coupling agent, to the fibre. Therefore, to stress transfer, a minimum chain length is necessary.

Karlson et al.[51] studied low-density polyethylene (LDPE) regenerated cellulose system by single fragmentation test. They determined the fibre surface perimeter by the Wilhelmy plate method using a nonpolar liquid that had a lower total surface free energy than regenerated cellulose in order to wet the fibre completely. They found that when regenerated cellulose fibres were mechanically treated, long and numerous twisted fibrils were observed on the fibre surface. The surface area of treated fibres was increased after treatment and, consequently, the interfacial shear strength increased significantly as a result of surface fibrillation. They mentioned that fibrils present on the surface could provide numerous mechanical anchoring sites for the polyethylene matrix.

Adusumalli et al.[52] determined the interfacial shear strength of different fibre–polypropylene composites. They used regenerated cellulose, ramie and glass fibres. They showed that both cellulose fibres had lower interfacial shear strength than sized glass fibre, presumably because of the different chemical

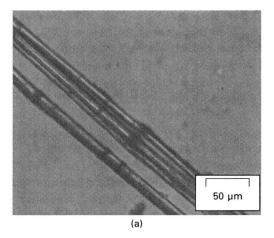

(a)

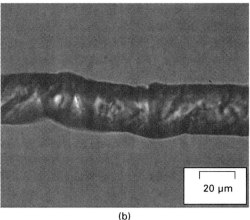

(b)

5.15 Optical microscopy photographs for (a) untreated flax fibre with smooth surface (from reference 26 with permission, copyright Wiley, 2010). (b) Sodium hydroxide treated fibre with rough surface (from reference 26 with permission, copyright Wiley, 2010).

character of cellulose (polar, hydrophilic) and PP (nonpolar, hydrophobic). Suitable surface chemical properties led to good wetting and development of adhesion forces. In spite of regenerated cellulose and ramie being chemically very similar as they consist almost exclusively of cellulose, ramie fibres showed higher interfacial shear strength than regenerated cellulose, presumably because of the higher surface roughness observed for these fibres.

García-Hernández *et al.*[44] modified sugar cane bagasse fibres by surface treatments in order to improve fibre adhesion to polystyrene (PS) matrices. The fibres were modified by alkaline treatment, silane coupling agents, fibre coating with PS and grafting of PS on the fibre (with and without crosslinker). According to pull-out tests, the IFSS increased for all treated fibres, as compared with untreated ones. The highest IFSS improvement was obtained for grafting of PS on the fibres in the presence of a crosslinker. The improvement obtained with this treatment was partly attributed to a good interdiffusion between chains of PS in the matrix and the grafted PS with crosslinker. Scanning electronic microscopy (SEM) analysis revealed that after treatment PS aggregations distributed as dots appeared along the fibre surface, resulting in more interlocking sites that contributed to the adhesion.

López Manchado et al.[53] investigated the enhancement of interfacial adhesion of PP/ethylene propylene diene monomer (EPDM)–flax fibre composites using MA as compatibilizer. The interface between the fibre and polymer matrix was characterized by pull-out tests. The addition of small amounts of MA-modified matrices significantly increased the interfacial shear strength. This effect was more evident when the matrix was PP modified with MAPP addition. This improvement was attributed to the presence of MA functional groups causing esterification of flax fibres, and thereby increasing the surface energy of the fibres to a value much closer to the surface energy of the matrix. Thus, a better wettability and a higher interfacial adhesion was obtained.

Behzad et al.[31] modified hemp fibres applying a paper sizing technique using a copolymer of styrene and dimethylaminopropylamine maleimide (SMA) as a surface-modifying agent in order to improve the performance of acrylic matrix composites. The estimated average stress to pull out the treated fibres was 71% higher than that measured for untreated fibres. This improvement was attributed to the presence of the nonpolar phenolic groups in the SMA component increasing the dispersive component of the fibre. Therefore, better spreading and wetting was expected, and, thus, the number of voids and bubbles at the interface, commonly present in all types of composites, would be reduced.

Lopattananon et al.[54] studied the influence of fibre modification on interfacial adhesion of pineapple leaf fibre–epoxy composites by single-fibre fragmentation tests. They verified that the interfacial shear strength of modified fibres was substantially higher than that of untreated ones. Sodium hydroxide and diglycidyl ether of bisphenol A (DGEBA), and the combination of both, were employed as reagents to modify the fibres. The adhesion between alkali-treated fibre and epoxy matrix increased by mechanical anchoring effects because alkalization made the fibre surface rough. The modification of fibres with epoxy solution resulted in grafting of the epoxy resin molecule at OH sites of the fibre. Therefore, the high interfacial shear strength of epoxy-treated composites might result from higher fibre surface affinity with the epoxy matrix. However, the strongest interfacial adhesion was obtained for the fibres that were treated by alkalization combined with DGEBA deposition. In this instance, the number of grafting sites at OH functional groups of the cellulose fibres increased as a result of both treatments, leading to more reactivity of the fibre with the matrix.

Interfacial evaluation of the various untreated and treated jute and hemp fibres reinforced PP–MAPP matrix composites was investigated by a micromechanical technique combined with acoustic emission (AE) and dynamic contact angle measurements by Park *et al.*[55] They modified the fibre surface with alkali and silane treatments and the PP matrix was modified by the addition of different amounts of MAPP. They observed that the IFSS of the natural fibre–MAPP/PP composites significantly increased with increasing content of MAPP in the mixture as well as after treating with alkaline solution and silane coupling agent. They suggested that chemical bridges could be created in the interfacial area between the fibre surface and the matrix for MAPP and silane coupling agent. Moreover, the advancing contact angle of silane-treated fibres was slightly lower than that of untreated fibres because the silane coupling agent blocked the high energy sites. Although the advancing contact angle of alkali treated fibres increased owing to the more hydrophilic fibre surface, the IFSS improvement was the result of weak boundary layers (wax, lignin and pectin) of natural fibres being removed leading to the increase of the surface area.

Morales *et al.*[56] used low-energy glow-discharge plasmas to functionalize cellulose fibres. PS films were synthesized by plasma (PPS) on the surface of cellulose fibres to obtain a good compatibility between the fibres and the PS matrix. The average interfacial shear strength was obtained from microdebonding tests for each plasma treatment applied to the fibres. Results showed that the adhesion in the fibre–matrix interface increased upon time for the first 4 min of treatment. However, at longer plasma exposures, fibres might be degraded, thus reducing the adhesion with the matrix. The greatest increase in the interfacial strength after plasma treatment was about 70%. They observed that the surface of the untreated cellulose fibres was smooth and uniform which would make the adhesion to the matrix difficult and produce a very weak interface. On the other hand, micrographs of fibres, exposed to 2 min of periodic glow discharges and 4 min of continuous glow discharges, respectively, showed that small fragments of PS adhered to the fibre at several points, which was an indication that the low-energy glow-discharge plasma modification could create different activation sites that promoted the formation of chemical bonds between the polymeric matrix and the fibres. Consequently fibre–matrix adhesion was improved.

Herrera-Franco *et al.*[36] studied the degree of fibre–matrix adhesion and its effect on the mechanical reinforcement of short henequen fibres and a polyethylene matrix. Henequen fibres were modified by various treatments, including an alkali treatment, a silane coupling agent and a pre-impregnation process with HDPE/xylene solution. The fibre–matrix interface shear strength calculated by both single-fragmentation and pull-out tests, was used as an indicator of the fibre–matrix adhesion improvement. They observed that all surface treatments improved the IFSS. All values obtained by fragmentation tests were higher than pull-out values ones. Although the IFSS values obtained by both methods were different, the lowest and highest values were obtained for the same systems, composites with untreated fibres and fibres treated with NaOH aqueous solution and then a silane coupling agent, respectively. The initial treatment with the aqueous solution of NaOH was thought to remove some lignin and hemicelluloses from the fibre surface, thus increasing the fibre surface area. Such a fibre surface increase resulted in a larger area of contact between the fibres and the matrix. Therefore, the hydroxyl groups on the cellulose fibres could better react with the silane-coupling agent because a larger number of possible reaction sites were available.

Khalil et al.[41] studied the effect of acetylation on the interfacial shear strength of oil palm empty fruit bunch (EFB) and coconut fibre (coir) with various commercial matrices such as epoxy, unsaturated polyester and PS. Pull-out tests showed that composites based on acetylated fibres had a higher interfacial shear strength. They mentioned that IFSS values increased after acetylation because the hydrophobicity of natural fibres increased. In general, coir fibres exhibited higher IFSS values than EFB ones. This was attributed to the higher lignin content of coir.

Li et al.[35] determined the interfacial shear strength of untreated and white rot fungi-treated hemp fibre reinforced PP/maleated PP matrix. They observed that the IFSS value of white rot fungi-treated fibre composite was 40% higher than for the untreated fibre one. They suggested that the treatment could remove noncellulosic compounds, thus increasing the surface area and roughness. Therefore, the increase in surface area increased the potential for interaction between hydroxyl sites and the MAPP coupling agent. On the other hand, the higher roughness increased mechanical interlocking between fibres and matrix.

Li et al.[57] also investigated interfacial properties of untreated and treated sisal fibres reinforced high-density polyethylene composites. Four fibre surface treatments were applied to fibres in order to improve the interfacial bonding properties between sisal fibre and HDPE matrix. Two types of silanes, 3-aminopropyltriethoxysilane (silane 1) and γ-methacryloxypropyltrimethoxysilane (silane 2), were used as coupling agents to modify the surface of the sisal fibres. On the other hand, permanganate ($KMnO_4$) and dicumyl peroxide (DCP) were selected as oxidants to treat the fibre surface. The highest IFSS value was obtained after permanganate treatment, the IFSS improvement was 100% from 1.6 MPa (untreated fibre) to 3.2 MPa. The silane 2 also improved the IFSS to 2.9 MPa, whereas the IFSS for sisal fibre treated with silane 1 was similar to that for the untreated fibre composite. They suggested that permanganate could etch the sisal fibre surface and make it rougher, so that mechanical interlocking could be introduced between the sisal fibre and the HDPE matrix which was in accordance with the pull-out curve presented. Comparing the results obtained by silane-treated fibres, it was suggested that because silane 2 had a carbon main chain, it could form van der Waals bonds with the HDPE matrix, which has a similar chemical structure, thus increasing the IFSS. However, as silane 1 did not have a long carbon chain structure, no van der Waals bonding with the HDPE matrix could occur between the matrix and treated sisal fibres. Therefore, the IFSS was almost the same as that for the untreated composites.

Joseph et al.[58] compared the interfacial shear strength values obtained from single-fibre pull-out tests of composites fabricated using banana fibres and glass fibres with a phenol/formaldehyde (PF) matrix. The results obtained revealed that the adhesion between banana fibres and PF matrix was much higher than that between glass and PF matrix. They suggested that this was the result of the hydrophilic nature of banana fibres, caused by the hydroxyl groups of lignin and cellulose, which could easily form hydrogen bonds with methylol and phenolic hydroxyl groups of the resol resulting in a strong interlocking between banana fibres and resol matrix. However, in glass–PF composites the interfacial shear strength value was very small compared with that of banana fibre composites, suggesting that there was no interaction between fibre and matrix.

Van de Velde et al.[7] studied the influence of surface treatments and the influence of matrix modification on the apparent interfacial shear strength of flax fibre reinforced polymer matrix. Flax fibre was treated with propyltrimethoxysilane, phenylisocyanate and MA-grafted polypropylene. The matrices used for this study were PP (two types having different molecular weights) and MAPP (two types differing in molecular weight and the content of grafted MA). The first, MAPP1, had a higher molecular weight than the second, MAPP2. However, MAPP1 had a lower grafted MA content than MAPP2 did. The single-fibre pull-out test was used for IFSS measurement, the results of which showed that treatment of flax fibre with propyltrimethoxysilane reduced fibre–matrix adhesion. On the other hand, small improvements were found for the phenylisocyanate-treated flax fibre PP interfacial strength. The presence of the alkyl chain MAPP on the coupling agent could possibly further increase the fibre–matrix adhesion. Comparing the values of systems obtained with different MAPP matrices, the results showed that although the MA content of MAPP1 was lower than for MAPP2, the apparent interfacial shear strength was almost double. It was suggested that an optimum MA content existed, for which the interfacial properties were optimum. Moreover, the higher molecular weight of MAPP1 compared with MAPP2 could also create better entanglements between fibre-bonded MAPP polymer chains and the matrix polymer chains resulting in the improvement in the interfacial shear strength. On the other hand, the difference in interfacial adhesion could also be attributed to the difference in crystallinity between the two MAPP used. Results obtained by differential scanning calorimetry showed that MAPP1 presented higher crystallinity than MAPP2. It was also suggested that crystallites could be formed around the fibre to act like cross-links by tying many molecules together, thus creating a possible positive effect on the interfacial adhesion.

Snijder et al.[59] studied the coupling efficiency of various MAPP grades with varying MA content grafted per polymeric PP chain and different molecular weights. It was concluded that the molecular weight was a more important factor than the MA content. Felix et al.[60] studied the mechanical properties of cellulose–PP composites. They used three different MA-based compatibilizing agents: alkylsuccinic anhydride, and two MAPP with the same MA content but different molecular weights. They found that the molecular weight of the compatibilizing agent was crucial for improving the mechanical properties of cellulose–PP composites. The mechanical properties increased with an increased molecular weight of MA-based compatibilizing agent. Similar results were obtained by Arbelaiz et al.[61] for flax–PP composites modified with various MAPP. It was suggested that MAPP with higher molecular weight led to more entanglements between the MAPP chains and the PP matrix and, consequently, improved the mechanical properties of flax–PP composites.

The work done by Eichhorn et al.[62] involved the modification of the interface between flax fibres and a variety of matrices. The interfacial shear strength was measured by pull-out and single fragmentation tests. As reinforcements, green flax (flax as received from the fields), dew retted (green flax after bacterial treatment on the fields) and a commercial fibre Duralin (flax treated by a method developed by CERES BV, Netherlands, which involves the depolymerization of lignin and hemicelluloses into lower molecular aldehyde and phenolic functionalities followed by a subsequent curing that hydrophobizes the fibre surface) were chosen. The following polymers were used as matrices: unsaturated polyester, epoxy, LDPE, HDPE, isotactic PP and MA-modified PP. Results showed that the greatest average interfacial shear strength was obtained with the Duralin–epoxy system, and the lowest was for green flax and LDPE composite.

On the other hand, more methods such as acetylation and stearic acid sizing have also been used to promote a better interface between dew retted flax and isotactic PP. The aim of both treatments was to treat the hydroxyl groups of the fibres with acid groups and, subsequently, to hydrophobize the fibre surface in order to obtain a better compatibility with PP. SFFT test results showed that acetylation slightly improved the interface for dew retted flax and caused a significant improvement for green flax. Stearic acid sizing caused an improvement in the interface only for low reaction times, whereas for longer reaction times there was a deterioration of the interface.

Joly et al.[42] used microdebonding tests with ramie–PP composites. The IFSS results confirmed an improvement of 60% when MAPP treated fibres were used. They suggested that the improvement was the result of entanglements or interdiffusion of the PP chains grafted on the cellulosic fibres with the chains of the matrix. When fibres were treated with small alkyl chains (C_8–C_{18}), the IFSS value showed that the length of these alkyl chains had almost no effect on the IFSS. They suggested that the chain length was probably too short to create entanglements with matrix chains and also some degradation of the fibres owing to the swelling medium might mask a positive effect of the compatibilizing agent.

Sydenstricker et al.[63] evaluated treated and untreated sisal-reinforced polyester biocomposites. Sisal fibres were treated with NaOH and N-isopropyl acrylamide. The sisal–polyester interface was investigated through pull-out tests in order to find the optimal sisal chemical treatment conditions. Pull-out results for sisal fibre–polyester composites showed that all treatments improved the interfacial adhesion between fibres and matrix. The highest IFSS values were obtained for treatment with 2 wt% NaOH and 2 wt% acrylamide. Moreover, these concentrations also showed the highest fibre tensile strength.

Valadez-Gonzalez et al.[5] explored the possibility of improving the effective mechanical properties of a HDPE matrix reinforced with henequen fibres by enhancing the fibre–matrix interface physicochemical interactions. Such interactions were characterized using two micromechanical methods, single-fibre fragmentation and pull-out tests. Three different treatments were applied to the henequen fibres: a treatment with an aqueous solution, a treatment first with alkaline solution and then with a silane coupling agent and the last treatment with alkali and then impregnated with a dilute matrix solution. The interfacial shear strength between natural fibres and thermoplastic matrices improved by fibre surface modification. They mentioned that the alkaline treatment increased the surface roughness resulting in a better mechanical interlocking and, moreover, for fibres treated after alkalization with a silane coupling agent and impregnated with a dilute matrix solution, NaOH treatment increased the amount of cellulose exposed on the fibre surface, thus increasing the number of possible reaction sites. The treatment with NaOH removed some lignin and hemicelluloses from the surface of the fibres and, consequently, the hydroxyl groups on the cellulose fibres were able to better interact with the silane-coupling agent because of the availability of a larger number of possible reaction sites. After treating henequen fibres with NaOH and combining silane surface treatment modification with preimpregnation process, the IFSS value was the highest owing to synergistic effect of both treatments.

Luo et al.[28] measured the IFSS of henequen fibre–poly(hydroxybutyrate-co-hydroxyvalerate) resin (PHBV) using single-fibre fragmentation and microdebonding tests. The IFSS of henequen fibre–PHBV measured using SFFT and microdebonding tests showed close agreement in spite of the different stresses experienced during each test. They attributed these findings mainly to the mechanical interlocking resulting from the fibre surface roughness, because no hydrogen bonding was possibile for PHBV. However, they suggested that fibre–matrix mechanical interlocking could be limited because of the high viscosity of the resin, which limited its ability to penetrate into and around the cells.

Tripathy et al.[18] studied the interfacial adhesion between four different forms of jute fibres (sliver, bleached, mercerized and untreated) and polyolefinic matrices (LDPE and PP). The fibre–matrix adhesion was estimated by means of single-fibre fragmentation tests. Results indicated a low adhesion between fibres and polyolefins, that confirmed that nonpolar matrices can not develop good adhesion with the polar fibre surfaces. With both thermoplastic matrices, sliver jute fibres and untreated jute fibres showed the highest values of stress transfer ability, which was attributed to the fact that sliver jute and untreated jute fibres have a high lignin content whose aromatic nature might be responsible for somewhat better wettability and interfacial interaction with polyolefins as a result of the higher dispersive forces. On the other hand, the bleached jute fibres and mercerized jute fibres showed the lowest interfacial interaction with polymer matrix. The bleached jute fibres and mercerized jute fibres were chemically treated with acid and acid + base solutions, respectively. As a considerable amount of lignin was removed, the concentration of polar groups at the surface probably increased, and therefore the adhesion to nonpolar polymer was consequently reduced.

Tripathy et al.[27] also investigated interfacial strength of untreated jute fibres and fibres treated with three different treatments with an epoxy resin as matrix. They carried out single-fibre composite tests in order to determine the critical fragment length and interfacial strength. The bleached jute fibres showed the highest IFSS value of about 140 MPa. They suggested that the high adhesion for bleached jute fibres might be caused by a greater mechanical anchorage of the epoxy resin on the more regular, clean surface with high microporosity. The possibility of chemisorption of the epoxy resin with hydroxyl groups of the cellulose in the absence of lignin might be another reason for improved adhesion.

Cho et al.[64] demonstrated that the intensity of electron beam (e-beam) irradiated on henequen fibres strongly influenced the interfacial adhesion between the fibres and a PP matrix. The interfacial shear strength was measured by single-fibre microdebonding tests. Henequen fibre surfaces modified with an appropriate dosage of e-beam, a low e-beam intensity of 10 kGy, resulted in the improvement of the interfacial properties. However, the e-beam irradiation at intensities higher than 10 kGy led to the reduction of the properties investigated. It was suggested that at 200 kGy the interfacial and mechanical properties were increased to the level of the composite with the henequen fibres treated at 10 kGy but the fibre surfaces were likely to be damaged.

Cho et al.[65] studied natural fibres (jute, kenaf and henequen) reinforced thermoplastic (PLA and PP) and thermosetting (unsaturated polyester) matrix composites. They observed that the interfacial shear strength of untreated kenaf–jute, henequen–PP and henequen–unsaturated polyester green composites was significantly improved by water treatment, particularly with a dynamic ultrasonication method. They suggested that the water treatment removed not only the surface impurities but also some waxy components present on the fibre surfaces. As a result, the treated fibres had rougher surfaces with more crevices than the untreated ones. They concluded that water treatment of natural fibres may be practically favourable with many advantages for improving the properties of composites.

Thamae et al.[66] studied the influence of the fibre extraction method, alkali and silane treatment on the interface of Agave americana waste–HDPE composites using pull-out tests. They observed that mercerization of agave americana fibres could improve the interfacial shear strength value by about 104% compared with untreated fibre ones. They stated that mercerization removed impurities, exposed hydrophobic lignin, and increased fibre surface area by exposing some ultimate fibres. However, the IFSS measurements reported in this work were very low, which they attributed to the lack of compatibility between natural fibres and HDPE. Silane treatment did not have any influence on the interfacial shear strength between agave americana fibres and HDPE.

Huber et al.[67] studied natural fibres (cotton, flax and hemp) polymeric matrix adhesion using single-fibre fragmentation tests. As polymeric matrices PP with 2% MA and PLA were used. Although SFFT tests have to be performed with a single fibre, in this study flax and hemp fibre bundles were used. The IFSS measurement for flax–MAPP was about 12 MPa and for hemp–MAPP was higher but the value was not given. Finally, the IFSS for the cotton–MAPP system was very low, about 0.7 MPa, which was attributed to be a consequence of the difference in chemical composition of the cotton fibres. Taking into account that cotton is a fibre made up of only cellulose and hemicellulose compared with the bast fibres hemp and flax, which contain lignin and pectin, these results indicated that fibre composition influences the fibre–matrix adhesion. The SFFT results reported for flax–PLA composites showed a poor adhesion between fibre and matrix because no fragments of flax fibre in the PLA matrix could be generated during the tests. The adhesion between fibre and PLA was not strong enough to transmit forces from the matrix to the fibre. This result was in accordance with SEM micrographs presented.

Joffe *et al.*[8] studied flax fibres with vinylester (VE), unsaturated polyester (UP) and epoxy matrices adhesion by SFFT. Fibres were treated with acrylic acid and vinyltrimethoxysilane. They observed that the IFSS of epoxy and VE matrices was somewhat higher than that for UP. The surface treatments used did not lead to significant variation in the IFSS. Hence, they concluded that the adhesion of flax fibres and thermoset matrices did not benefit from the common surface treatments.

Lee *et al.*[68] studied the effect of fibre surface treatments on the interfacial properties of henequen–PP composites using single-fibre microdebonding tests. The surfaces of henequen fibres were treated with two different media, tap water and sodium hydroxide, that underwent both soaking and ultrasonic methods. The soaking and ultrasonic treatments with tap water and sodium hydroxide at different concentrations and treatment times significantly influenced the interfacial properties. The greatest improvement in the interfacial shear strength was achieved by ultrasonic alkalization performed with 6 wt% NaOH for 60 min. The IFSS improved by about 138% compared with that of the untreated fibre. The water treatment removed not only the surface impurities but also some waxy components present on the fibre surfaces. As a result, the treated fibres had rougher surfaces with more crevices than the untreated ones.

Liu *et al.*[33] characterized the interface between cellulosic fibres and PS matrix by microdebonding tests. They found that no consistent relationship existed between IFSS results and the embedded length of the microdrop in the range examined. Moreover, the results showed that the acetylated wood fibres had the highest shear strength. Acetylation also improved the IFSS for cotton, although to a lesser degree. No differences were found in the IFSS with fibre treatments for rayon. The high IFSS between the acetylated wood fibre and the PS matrix was attributed to the improved wetting and spreading of the molten PS on the acetylated wood fibre surfaces.

Lodha *et al.*[30] fabricated composites using ramie fibres and soy protein isolate (SPI). They characterized the interfacial shear strength using microdebonding tests. The average IFSS value was 29.8 MPa. Although the SPI–polymer resin contained several amino acids with polar groups, it also contained amino acids, such as alanine, glycine, leucine and isoleucine, which had nonpolar groups and, therefore, these amino acids could not be hydrogen bonded with cellulose molecules in the fibres.

Czigany *et al.*[69] examined the interfacial adhesion of three different natural reinforcing fibres and three different commercial matrix materials using microdebonding tests. The natural fibres used were flax, hemp and sisal. As polymeric matrices, PP, biodegradable MaterBi (a thermoplastic biodegradable polymer made of starch) and PuraSorb (a biodegradable lactide–glycolide copolymer) were used. Based on microdebonding tests, it was revealed that among the examined specimens the strongest adhesion existed in the

PuraSorb–sisal combination. Purasorb matrix is highly polar owing to the oxygen atoms present in the polymer chain. They suggested that very strong hydrogen bridges were formed with the –OH groups on the surface of the fibres, resulting in superior fibre–matrix interfacial adhesion.

Zafeiropoulos[70] studied flax fibre–PP composites and a sensitivity analysis was presented for single-fibre fragmentation tests. The results indicated a large standard deviation for the IFSS measurements, which was partly attributed to the significant standard deviation of the fibre strength as well to the significant standard deviation of the fibre diameter. Traditional stress analysis failed to correctly assess the interface, whilst a statistically based data analysis could overcome the fibre heterogeneity problem. He suggested that it was not possible to assess the effect of the fibre surface treatments at the interface in flax fibre–PP composites.

Torres *et al.*[71] studied the interfacial properties of sisal-reinforced polyethylene composites by means of single-fibre fragmentation tests. Fibre treatment with stearic acid increased the interfacial shear strength by 23% with respect to untreated fibres. The improvements in IFSS found for the treated specimens were consistent with observations from SEM micrographs. They suggested that the variability of the results was relatively high as indicated by a high standard deviation (21%), which might be attributed to experimental errors (opacity of the matrix), as well as the inherent characteristics of the fibres and the fibre–matrix interface.

Wazzan[72] modified date palm (*Phoenix dactylifera* L.) fibre surface using different treatments in order to improve their adhesion to polyester matrices. Three fibre surface treatments were used, an aqueous alkaline solution, silane coupling agent, and a combination of both treatments. He investigated the improvement in adhesion through single-fibre pull-out tests. IFFS values increased for all treated fibres compared with nontreated fibres. In particular, the combination of alkaline and silane coupling agents resulted in the best adhesion improvement to the polyester matrix. The alkali treatment gave up to 18% increase in the shear strength. Wazzan mentioned that this improvement was because of the removal of pectins. A higher shear strength was obtained when using the fibres treated with the silane coupling agent. However, the best effect was reached with a combination of both treatments, which increased the average value of fibre–matrix shear strength by more than 40% with respect to the composite made with untreated fibres. The IFSS varied with the fibre embedded for all date palm fibre–polyester systems. The apparent IFSS increased as the embedded length decreased.

Park *et al.*[73] evaluated the IFSS values of various ramie and kenaf fibre–epoxy composites using the combination of microdebonding test and nondestructive acoustic emission. They observed that the IFSS was higher for ramie fibre than for kenaf fibres. Kenaf fibres contain a higher proportion of hemicelluloses, lignin, and waxes than ramie fibres and these compounds are responsible for poor adhesion between the fibre surface and epoxy matrix because of their different inherent properties.

Park et al.[55] studied jute and hemp fibres and noticed that the IFSS of a given natural fibre–matrix system increased when the critical embedded area decreased. The critical embedded area was used instead of the conventional critical embedded length because of the different diameter for both fibres. The critical embedded area was obtained by the intersection of two linear regression lines: one is the fibre pull-out linear regression line, whereas the other is the fibre fracture linear regression line. From the experimental plots, they observed that the critical embedded area decreased with an increase in MAPP content in the PP–MAPP matrix material or after treating the natural fibres with alkaline solution and silane coupling agent.

5.3.3   Effect of processing parameters

The presence of moisture has a negative effect on the performance of lignocellulosic fibre based composites.[74] Water molecules absorbed in natural fibres can reduce the intermolecular hydrogen bonding between cellulose molecules in the fibre and establish intermolecular hydrogen bonding between cellulose molecules in the fibre and water molecules, thereby reducing the interfacial adhesion between the fibre and the matrix and, as a result, decreasing the mechanical properties of the composites.[61]

Chen et al.[75] fabricated bamboo strips–vinyl ester resin composites to investigate the influence of moisture on interfacial shear strength by pull-out tests. The objective was to clarify how prefabrication exposure to moisture affects the matrix hardening process and to establish the relative importance between prefabrication moisture exposure and post-fabrication moisture exposure. The relative humidity (RH) in composite manufacture had a severe impact on the IFSS of the resulting composites. The IFSS achieved at normal room conditions (20 °C, 60% RH) was only a half of what was achieved in the dry condition. Composites produced at high RH (80 and 90%) had a negligible interfacial strength. They suggested that the high moisture content of henequen fibres would evaporate during processing, resulting in micro-voids around the fibres, which would thus reduce the interface area and decrease the IFSS and, consequently, the composite properties.[75,76] The IFSS decreased almost linearly with the increase in bamboo moisture content from 0 to about 12%. The rate of decrease suddenly accelerated after this point. A further increase of 2% in bamboo moisture content caused a reduction of more than 80% in the IFSS. Reported optical photographs showed that the specimen produced at dry condition showed no visible voids at the interface between bamboo fibres and the matrix whereas the sample produced at 80% RH exhibited thin void lines along the fibrous region and wide void spaces along the ground matrix regions, which was in accordance with the IFSS.

Several studies[77–79] reported that the presence of lignocellulosic fibres could lead to the development of transcrystallinity (TC). Some studies[32,60,78,79] were conducted with natural fibre–polymer composites to observe the effect of a transcrystalline region on composite interfacial properties. The results reported are fairly contradictory regarding the effect of a transcrystalline interphase on the composite interfacial properties. Zafeiropoulos *et al.*[78] used four different types of flax fibres, green flax, dew-retted flax, Duralin flax, and stearic acid-treated flax with two different iPP matrices. The effect of various processing conditions such as the crystallization temperature, time and cooling rates on TC was investigated and the effect of the TC layer upon the mechanical properties of the interface was studied using single-fibre fragmentation tests. They concluded that the presence of TC in the flax–PP system had a profound effect on the interface characteristics. The interface was strongly enhanced, as shown by fragmentation tests. However, there was no difference for different thickness or crystallization temperature and the cooling rate from the melt to the crystallization temperature did not affect the morphology of the transcrystalline layer.

Eichhorn *et al.*[62] investigated the effect of cooling rates upon the interface of flax–isotactic PP composites by single-fibre fragmentation tests. Two different flax fibres were used: dew retted flax and green flax. The results showed that there was a significant improvement in the IFSS when TC was present, as shown by the sharp decrease of the critical length. They concluded that TC improved the interfacial stress transfer and the thickness of the transcrystalline layer did not affect the interface. Felix and Gatenholm[79] employed a single-fibre fragmentation test to evaluate the effect of a transcrystalline interphase morphology on the shear stress transfer in cotton fibre–PP system. They found that the presence of a trancrystalline interphase improved the shear transfer up to ~100% depending on the thickness of the transcrystalline layer. They suggested that slow cooling favours the kinetics of the approach of PP molecules and, hence, interfacial adsorption, which yields an ordered transcrystalline PP interphase having a high density of secondary bonds with the cellulose surface. They concluded that a careful control of thermal conditions during processing may be sufficient for obtaining satisfactory interfacial and thus composite properties.

However, Garkhail *et al.*[80] studied the effect of transcrystallization on the interfacial shear strength in PP–flax fibre composites by fibre pull-out tests. They found that the IFSS of the PP–flax system was slightly decreased in the presence of a transcrystalline interphase.

George *et al.*[32] studied the effect of TC on the interfacial shear strength of flax fibre-reinforced polypropylene by pull-out tests. They observed that with increasing transcrystallinity thickness the value of IFSS decreased. However, the differences observed were very small.

Finally, some mention should be made of the effect of the type of matrix chosen for composite fabrication. Khalil *et al.*[41] observed that thermosetting polymer showed a higher IFSS than thermoplastic matrices. They suggested that the thermosetting systems had a higher wettability than thermoplastic ones owing to the fact that thermosetting plastics are cured 'in situ' from relatively low molecular weight compounds while mixing with fibres, followed by the polymerization reaction. On the other hand, thermoplastics are mixed in the molten state where the wettability is limited. Another reason for the higher IFSS of thermosets compared with thermoplastics could be the chemical components of natural fibres that could be more reactive or compatible with thermoset matrices.

5.4   Conclusions

Each method for characterization of the interfacial adhesion in the single fibre–polymer matrix is suitable for studying various surface treatments in a given fibre–matrix system. However, the results obtained by different methods are not similar because of the different loading modes and different stress distribution. Moreover the intrinsic problems of each method for measuring IFSS, and also the lack of homogeneity in these fibres makes the interpretation of the results difficult.

Usually lignocellulose fibres become more hydrophobic after treatments indicating that treated fibres can be better wetted out by hydrophobic matrix. However, wetting is not the unique requisite to form a good fibre–matrix adhesion. Each combination of fibre–matrix presents a coupling agent or an optimum treatment condition that can improve fibre–matrix adhesion. This improvement may be related to a mechanism of adhesion or be the result of a combination of different adhesion mechanisms.

Another way to obtain a better fibre–matrix adhesion is to control the processing conditions, such as the relative humidity in composite manufacture or cooling rate in semicrystalline thermoplastic matrix.

The results of the single-fibre tests in the studies that have been reviewed here are possibly not indicative of the performance of an actual composite because the composite microstructure deviates significantly from that used on single-fibre tests and usually the improvement observed in single-fibre tests is not immediately representative of the actual composite where the increase in performance is generally much lower.

5.5   References

1 Netravali, AN, Henstenburg, RB, Phoenix, SL & Schwartz, P 1989, 'Interfacial shear strength studies using the single-filament-composite test. I: Experiments on graphite fibres in epoxy', *Polymer Composites*, **10**(4), 226–241.
2 Narkis, M & Chen, EJH 1988, 'Review of methods for characterization of interfacial fibre-matrix interactions', *Polymer Composites*, **9**(4), 245–251.

3 Wimolkiatisak, AS & Bell, JP 1989 'Interfacial shear strength and failure modes of interphase-modified graphite-epoxy composites', *Polymer Composites*, **10**(3), 162–172.

4 Arbelaiz, A, Fernandez, B, Cantero, G, Llano-Ponte, R; Valea, A & Mondragon, I 2005, 'Mechanical properties of flax fibre/polypropylene composites. Influence of fibre/matrix modification and glass fibre hybridization', *Composites, Part A: Applied Science and Manufacturing*, **36A**(12), 1637–1644.

5 Valadez-Gonzalez, A, Cervantes-Uc, JM, Olayo, R & Herrera-Franco, PJ 1999, 'Effect of fibre surface treatment on the fibre-matrix bond strength of natural fibre reinforced composites', *Composites, Part B: Engineering*, **30B**(3), 309–320.

6 Gauthier, R, Joly, C, Coupas, AC, Gauthier, H & Escoubes, M 1998, 'Interfaces in polyolefin/cellulosic fibre composites: chemical coupling, morphology, correlation with adhesion and aging in moisture', *Polymer Composites*, **19**(3), 287–300.

7 Van de Velde, K & Kiekens, P 2001, 'Influence of fibre surface characteristics on the flax/polypropylene interface', *Journal of Thermoplastic Composite Materials* **14**(3), 244–260.

8 Joffe, R, Andersons, J & Wallstroem, L 2005, 'Interfacial shear strength of flax fiber/thermoset polymers estimated by fibre fragmentation tests', *Journal of Materials Science*, **40**(9/10), 2721–2722.

9 Rao, V & Drzal LT 1991, 'The dependence of interfacial shear strength on matrix and interphase properties', *Polymer Composites*, **12**(1), 48–56.

10 Felix, J 1993, 'Enhancing interactions between cellulose fibres and synthetic polymers', PhD thesis, Chalmers University of Technology, Goteborg.

11 Scarponi, C & Pizzinelli, CS 2009, 'Interface and mechanical properties of natural fibres reinforced composites: a review', *International Journal of Materials & Product Technology* **36**(1–4), 278–303.

12 Pisanova, EV, Zhandarov SF & Dovgyalo, VA 1994, 'Interfacial adhesion and failure modes in single filament thermoplastic composites', *Polymer Composites*, **15**(2), 147–155.

13 Lee SM, 1989, *International encyclopedia of composites*. Volume 1, VCH publishers, New York.

14 Kelly, A & Tyson, WR 1965, 'Tensile properties of fibre-reinforced metals. Copper/tungsten and copper/molybdenum', *Journal of the Mechanics and Physics of Solids*, **13**(6), 329–350.

15 Bian, XS, Kenny, JM, Ambrosio, L & Nicolais, L 1991, 'Development and application of the single fibre technique (SFT) for the evaluation and analysis of fibre/matrix adhesion in polymeric composites', *Chimica Oggi* **9**(6), 44–49.

16 Netravali, AN, Li, ZF; Sachse, W & Wu, HF 1991, 'Determination of fibre matrix interfacial shear strength by an acoustic emission technique', *Journal of Materials Science*, **26**(24), 6631–6638.

17 Wu, HF, Biresaw, G & Laemmle, JT 1991, 'Effect of surfactant treatments on interfacial adhesion in single graphite/epoxy composites', *Polymer Composites*, **12**(4), 281–288.

18 Tripathy, SS, Levita, G & Di Landro, L 2001, 'Interfacial adhesion in jute polyolefin composites', *Polymer Composites* **22**(6), 815–822.

19 Li, ZF, Grubb, DT & Phoenix, SL 1995, 'Fibre interactions in the multi-fibre composite fragmentation test', *Composites Science and Technology*, **54**(3), 251–266.

20 Ohsawa, T, Nakayama, A, Miwa, M & Hasegawa, A 1978, 'Temperature dependence of critical fibre length for glass fibre-reinforced thermosetting resins', *Journal of Applied Polymer Science*, **22**(11), 3203–3212.

21 Drzal, LT 1985, 'Adhesion of graphite fibres to epoxy matrices. III. The effect of hygrothermal exposure', *Journal of Adhesion*, **18**(1), 49–72.

22 Favre, JP & Jacques, D 1990, 'Stress transfer by shear in carbon fibre model composites. Part 1. Results of single-fibre fragmentation tests with thermosetting resins', *Journal of Materials Science*, **25**(2B), 1373–1380.

23 Goda, K, Park, JM & Netravali, AN 1995, 'A new theory to obtain Weibull fibre strength parameters from a single-fibre composite test', *Journal of Materials Science* **30**(10), 2722–2728.

24 Zhao, FM, Okabe, T & Takeda, N 2000, 'The estimation of statistical fibre strength by fragmentation tests of single-fibre composites', *Composites Science and Technology*, **60**(10), 1965–1974.

25 Weibull, W 1939, 'A statistical theory of the strength of materials', Ingenioersvetenskapsakademien, Handlingar Stocklholm.

26 Arbelaiz, A, Cantero, G, Fernandez, B, Ganan, P, Kenny, JM & Mondragon, I 2005, 'Flax fibre surface modifications: Effects on fibre physico mechanical and flax/polypropylene interface properties', *Polymer Composites*, **26**(3), 324–332.

27 Tripathy, SS, Di Landro, L, Fontanelli, D, Marchetti, A & Levita, G 2000, 'Mechanical properties of jute fibres and interface strength with an epoxy resin', *Journal of Applied Polymer Science*, **75**(13), 1585–1596.

28 Luo, S & Netravali, AN 2001, 'Characterization of henequen fibres and the henequen fibre/poly(hydroxybutyrate-co-hydroxyvalerate) interface', *Journal of Adhesion Science and Technology*, **15**(4), 423–437.

29 Prasad, BM & Sain, MM 2003, 'Mechanical properties of thermally treated hemp fibres in inert atmosphere for potential composite reinforcement', *Materials Research Innovations*, **7**(4), 231–238.

30 Lodha, P & Netravali, AN 2002, 'Characterization of interfacial and mechanical properties of green composites with soy protein isolate and ramie fibre', *Journal of Materials Science*, **37**(17), 3657–3665.

31 Behzad, T & Sain, M 2009, 'Surface and interface characterization of untreated and SMA imide-treated hemp fibre/acrylic composites', *Polymer Composites*, **30**(6), 681–690.

32 George, J, Garkhail, SK, Wieland, B & Peijs, T 2000, 'A study on the trans-crystallisation behaviour of flax fibre reinforced polypropylene composites and effect on mechanical properties', *Proceedings from the third international symposium on natural polymers and composites, and the workshop on progress in production and processing of cellulosic fibres and natural polymers*, Sao Pedro, Brazil, pp. 408–413.

33 Liu, FP, Wolcott, MP, Gardner, DJ & Rials, TG 1994, 'Characterization of the interface between cellulosic fibres and a thermoplastic matrix', *Composite Interfaces*, **2**(6), 419–432.

34 Stamboulis, A, Baillie, C & Schulz, E 1999, 'Interfacial characterisation of flax fibre-thermoplastic polymer composites by the pull-out test', *Angewandte Makromolekulare Chemie*, **272**, 117–120.

35 Li, Y, Pickering, KL & Farrell, RL 2009, 'Determination of interfacial shear strength of white rot fungi treated hemp fibre reinforced polypropylene', *Composites Science and Technology*, **69**(7–8), 1165–1171.

36 Herrera-Franco, PJ & Valadez-Gonzalez, A 2005, 'A study of the mechanical properties of short natural-fibre reinforced composites'. *Composites, Part B: Engineering*, **36B**(8), 597–608.

37 Pisanova, E, Zhandarov, S & Mader, E 2001, 'How can adhesion be determined from micromechanical tests?', *Composites, Part A: Applied Science and Manufacturing*, **32A**(3–4), 425–434.

38 Zhang, C & Qiu, Y 2003, 'Modified shear lag model for fibres and fillers with irregular cross-sectional shapes', *Journal of Adhesion Science and Technology*, **17**(3), 397–408.

39 Pitkethly, MJ, Favre, JP, Gaur, U, Jakubowski, J, Mudrich, SF, Caldwell, DL, Drzal, LT, Nardin, M, Wagner, HD, Di Landro, L, Hampe, A, Armistead, JP, Desaeger, M & Verpoest, I 1993, 'A round-robin program on interfacial test methods', *Composites Science and Technology*, **48**(1–4), 205–214.

40 Miller B, Muri P & Rebenfeld L 1987, 'A microbond method for determination of the shear strength of a fibre/resin interface', *Composites Science and Technology*, **28**(1), 17–32.

41 Khalil, HPSA, Ismail, H, Rozman, HD & Ahmad, MN 2001, 'The effect of acetylation on interfacial shear strength between plant fibres and various matrices', *European Polymer Journal*, **37**(5), 1037–1045.

42 Joly, C, Gauthier, R & Chabert, B 1996, 'Physical chemistry of the interface in polypropylene/cellulosic-fibre composites', *Composites Science and Technology*, **56**(7), 761–765.

43 George, J, Sreekala, MS & Thomas, S 2001, 'A review on interface modification and characterization of natural fibre reinforced plastic composites', *Polymer Engineering and Science*, **41**(9), 1471–1485.

44 Garcia-Hernandez, E, Licea-Claverie, A, Zizumbo, A, Alvarez-Castillo, A & Herrera-Franco, PJ 2004, 'Improvement of the interfacial compatibility between sugar cane bagasse fibres and polystyrene for composites', *Polymer Composites*, **25**(2), 134–145.

45 Barkakaty, BC 1976, 'Some structural aspects of sisal fibres', *Journal of Applied Polymer Science*, **20**(11) 2921–2940.

46 Mukherjee, PS & Satyanarayana, KG 1984, 'Structure and properties of some vegetable fibres. Part 1. Sisal fibre', *Journal of Materials Science*, **19**(12), 3925–3934.

47 Joly, C, Kofman, M & Gauthier, R 1996, 'Polypropylene/cellulosic fibre composites: chemical treatment of the cellulose assuming compatibilization between the two materials', *Journal of Macromolecular Science, Pure and Applied Chemistry*, **A33**(12), 1981–1996.

48 Felix, JM & Gatenholm P 1991, 'The nature of adhesion in composites of modified cellulose fibres and polypropylene', *Journal of Applied Polymer Science*, **42**(3), 609–620.

49 Bledzki, AK & Gassan, J 1999, 'Composites reinforced with cellulose based fibres', *Progress in Polymer Science* **24**(2), 221–274.

50 Mwaikambo, LY & Ansell, MP 2002, 'Chemical modification of hemp, sisal, jute, and kapok fibres by alkalization', *Journal of Applied Polymer Science*, **84**(12), 2222–2234.

51 Karlsson, JO, Blachot, JF, Peguy, A & Gatenholm, P 1996, 'Improvement of adhesion between polyethylene and regenerated cellulose fibres by surface fibrillation', *Polymer Composites*, **17**(2), 300–304.

52 Adusumali, RB, Reifferscheid, M, Weber, H, Roeder, T, Sixta, H & Gindl, W 2006, 'Mechanical properties of regenerated cellulose fibres for composites', *Macromolecular Symposia*, **244**, 119–125.

53 Lopez Manchado, MA, Arroyo, M, Biagiotti, J & Kenny, JM 2003, 'Enhancement

of mechanical properties and interfacial adhesion of PP/EPDM/flax fibre composites using maleic anhydride as a compatibilizer', *Journal of Applied Polymer Science*, **90**(8), 2170–2178.

54 Lopattananon, N, Payae, Y & Seadan, M 2008, 'Influence of fibre modification on interfacial adhesion and mechanical properties of pineapple leaf fiber-epoxy composites', *Journal of Applied Polymer Science*, **110**(1), 433–443.

55 Park, JM, Quang, ST, Hwang, BS & DeVries, K.L 2006, 'Interfacial evaluation of modified jute and hemp fibers/polypropylene (PP)–maleic anhydride polypropylene copolymers (PP–MAPP) composites using micromechanical technique and nondestructive acoustic emission', *Composites Science and Technology*, **66**(15), 2686–2699.

56 Morales, J, Olayo, MG, Cruz, GJ, Herrera-Franco, P & Olayo, R 2006, 'Plasma modification of cellulose fibers for composite materials', *Journal of Applied Polymer Science*, **101**(6), 3821–3828.

57 Li, Y, Hu, C & Yu, Y 2008, 'Interfacial studies of sisal fiber reinforced high density polyethylene (HDPE) composites', *Composites, Part A: Applied Science and Manufacturing*, **39A**(4), 570–578.

58 Joseph, S, Sreekala, MS, Oommen, Z, Koshy, P & Thomas, S 2002, 'A comparison of the mechanical properties of phenol-formaldehyde composites reinforced with banana fibres and glass fibres', *Composites Science and Technology*, **62**(14), 1857–1868.

59 Snijder, MHB & Bos, HL 2000, 'Reinforcement of polypropylene by annual plant fibers: optimisation of the coupling agent efficiency', *Composite Interfaces*, **7**(2), 69–79.

60 Felix, JM & Gatenholm, P 1991, 'The effect of compatibilizing agents on the interfacial strength in cellulose-polypropylene composites', *Polymeric Materials Science and Engineering*, **64**(3), 123–124.

61 Arbelaiz, A, Fernandez, B, Ramos, JA, Retegi, A, Llano-Ponte, R & Mondragon, I 2005, 'Mechanical properties of short flax fibre bundle/polypropylene composites: influence of matrix/fibre modification, fibre content, water uptake and recycling', *Composites Science and Technology*, **65**(10), 1582–1592.

62 Eichhorn, SJ, Baillie, CA, Zafeiropoulos, N, Mwaikambo, LY, Ansell, MP, Dufresne, A, Entwistle, KM, Herrera-Franco, PJ, Escamilla, GC, Groom, L, Hughes, M, Hill, C, Rials, TG & Wild, PM 2001, 'Current international research into cellulosic fibers and composites', *Journal of Materials Science*, **36**(9), 2107–2131.

63 Sydenstricker, THD, Mochnaz, S & Amico, C 2003, 'Pull-out and other evaluations in sisal-reinforced polyester biocomposites', *Polymer Testing*, **22**(4), 375–380.

64 Cho, D, Lee, HS, Han, SO & Drzal, LT 2007, 'Effects of e-beam treatment on the interfacial and mechanical properties of henequen/polypropylene composites', *Advanced Composite Materials*, **16**(4), 315–334.

65 Cho, D, Seo, JM, Lee, HS, Cho, CW, Han, SO, Park & WH 2007, 'Property improvement of natural fiber-reinforced green composites by water treatment', *Advanced Composite Materials*, **16**(4), 299–314.

66 Thamae, T & Baillie, C 2007, 'Influence of fibre extraction method, alkali and silane treatment on the interface of *Agave americana* waste HDPE composites as possible roof ceilings in Lesotho', *Composite Interfaces*, **14**(7–9), 821–836.

67 Huber, T & Muessig, J 2008, 'Fibre matrix adhesion of natural fibres cotton, flax and hemp in polymeric matrices analyzed with the single fibre fragmentation test', *Composite Interfaces*, **15**(2–3), 335–349.

68 Lee, HS, Cho, D & Han, SO 2008, 'Effect of natural fiber surface treatments on the

interfacial and mechanical properties of henequen/polypropylene biocomposites', *Macromolecular Research*, **16**(5), 411–417.

69 Czigany, T, Morlin, B & Mezey, Z 2007, 'Interfacial adhesion in fully and partially biodegradable polymer composites examined with microdroplet test and acoustic emission', *Composite Interfaces*, **14**(7–9), 869–878.

70 Zafeiropoulos, NE 2007, 'On the use of single fiber composites testing to characterize the interface in natural fiber composites', *Composite Interfaces*, **14**(7–9), 807–820.

71 Torres, FG & Cubillas ML 2005, 'Study of the interfacial properties of natural fibre reinforced polyethylene', *Polymer Testing*, **24**(6), 694–698

72 Wazzan, A 2006, 'The effect of surface treatment on the strength and adhesion characteristics of *Phoenix dactylifera*-L (date palm) fibers', *International Journal of Polymeric Materials*, **55**(7), 485–499.

73 Park, JM, Son, TQ, Jung, JG & Hwang, BS 2006, 'Interfacial evaluation of single ramie and kenaf fiber/epoxy resin composites using micromechanical test and nondestructive acoustic emission', *Composite Interfaces*, **13**(2–3), 105–129.

74 Joseph PV, Rabello MS, Mattoso LHC, Joseph K & Thomas S 2002, 'Environmental effects on the degradation behaviour of sisal fibre reinforced polypropylene composites', *Composites Science and Technology*, **62**(10–11), 1357–1372.

75 Chen, H, Miao, M & Ding, X 2009, 'Influence of moisture absorption on the interfacial strength of bamboo/vinyl ester composites', *Composites, Part A: Applied Science and Manufacturing*, **40**(12), 2013–2019.

76 Stark, N 2001, 'Influence of moisture absorption on mechanical properties of wood flour–polypropylene composites', *Journal of Thermoplastic Composite Materials*, **14**(5), 421–432.

77 Arbelaiz, A, Fernandez, B, Ramos, JA & Mondragon, I 2006, 'Thermal and crystallization studies of short flax fiber reinforced polypropylene matrix composites: effect of treatments', *Thermochimica Acta*, **440**(2), 111–121.

78 Zafeiropoulos, NE, Baillie, CA & Matthews, FL 2001, 'A study of transcrystallinity and its effect on the interface in flax fibre reinforced composite materials', *Composites, Part A: Applied Science and Manufacturing*, **32A**(3–4), 525–543.

79 Felix, JM & Gatenholm, P 1994, 'Effect of transcrystalline morphology on interfacial adhesion in cellulose/polypropylene composites', *Journal of Materials Science*, **29**(11), 3043–3049.

80 Garkhail, S, Wieland, B, George, J, Soykeabkaew, N & Peijs, T 2009, 'Transcrystallisation in PP/flax composites and its effect on interfacial and mechanical properties', *Journal of Materials Science*, **44**(2), 510–519.

6

Assessing fibre surface treatment to improve the mechanical properties of natural fibre composites

K. L. PICKERING, Waikato University, New Zealand

Abstract: An overview is presented of the influence of treatments conducted to improve interfacial strength on the mechanical properties of natural cellulose-based fibres. Methodology used for mechanical testing of natural fibres to obtain fibre strength and stiffness data is described, and the statistical treatment that can be applied to this type of data and the influence of treatment conducted to improve interfacial strength on the mechanical properties of natural fibre are explored.

Key words: single-fibre tensile testing, Weibull, surface treatment.

6.1 Mechanical testing of fibres

Although at first there seems to be ready availability of fibre strength and stiffness data, commonly this has been measured using fibre yarn, bundles or unspecified material and is therefore not necessarily reflective of single fibre properties. Studies conducted to assess the influence of fibre treatments on fibre properties commonly neglect mechanical assessment and give more emphasis to assessment of other aspects including crystallinity using x-ray diffraction and differential scanning calorimetry (DSC), surface topography using scanning electron microscopy (SEM) and assessment of fibre surface chemistry using inverse gas chromatography (IGC) and Fourier transform infrared (FTIR) spectroscopy (Mwaikambo and Ansell 1999; Zafeiropoulos *et al*. 2002; Tserki *et al*. 2005). Caution also needs to be exercised when comparing data based on single fibres as there can be an absence of information on gauge length and testing rate, both of which influence mechanical properties. However, there are sufficient numbers of studies where single fibres are tested to give a reasonable expectation of the potential strength and stiffness of single natural fibres including the variability that can be expected and the influence of a number of fibre treatments. In addition to fibre treatment, mechanical properties have been shown to be dependent on harvesting time and extraction procedure, details of which are discussed in 6.4 of this chapter. They are also dependent on damage incurred during processing and humidity (Davies *et al*. 1998).

186

The most common testing methodology used for testing of single fibres is based on ASTM D3379-75 (Tensile strength and Young's modulus for high modulus single-filament materials), although sometimes ASTM D3822-01 is quoted, which is similar but based on single textile fibres and includes the requirement to remove any fibre crimp and calculation of tenacity commonly used in the textile industry instead of tensile strength. Fibres are most conveniently mounted onto slotted testing cardboard windows, as shown in Fig. 6.1, with the fibre aligned along the centre of the window. They are attached using glue (commonly cyanoacrylate, epoxy or polyvinyl acetate) thus defining the gauge length as the distance between the glue attachments. The cardboard window with mounted fibre can then be placed in the grips of a universal tester, following which the side sections of the card can be severed using a heated wire, leaving the fibre freely suspended between the glued parts of the window card such that a tensile load can now be applied to the fibre. Cross-sectional area is commonly obtained by measurement of a number of fibre diameters equally spaced along the fibre length before testing using a calibrated optical microscope and taking an average of these values and assuming that fibres are circular in cross-section (which will introduce some inaccuracy as they are generally irregular tending towards polygonal). Laser diffractometry has also been attempted for measurement

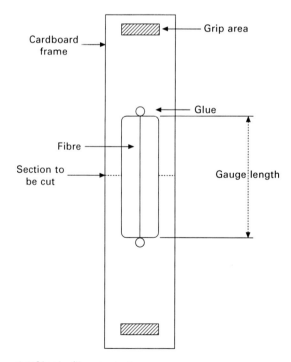

6.1 Single-fibre test piece.

of fibre diameter, but concerns have been raised of the accuracy of this technique owing to the difficulty of interpretation of the diffraction patterns produced (Romao *et al.* 2004). Researchers have also optically measured the diameter at the actual point of failure (rupture diameter) to calculate the cross-sectional area in order to compensate for variability in fibre diameter along the fibre length and improve the accuracy of strength measurements (Romao *et al.* 2004). Measurements of fibre diameter tended to be smaller from these measurements than the average fibre diameter calculated before testing. However, tensile testing is likely to have resulted in some permanent deformation that would affect fibre diameter.

For accurate measurement of Young's modulus, system compliance is required which can be determined experimentally as described in ASTM D3379-75. Another complication in terms of an accurate Young's modulus is the variability in shape of the stress–strain curve obtained for single fibres. As depicted in Fig. 6.2, observations of cellulose-based natural fibre, including flax and hemp, support that from the same fibre population behaviour can range from strain hardening (Fig. 6.2a), through linear elastic (Fig. 6.2b) to the inclusion of some plastic flow (Fig. 6.2c) (Nechwatal *et al.* 2003; Andersons *et al.* 2005; Pickering *et al.* 2007a). Departure from linearity at low strains has been explained by the orientation of fibrils along the fibre axis during loading. This has been shown to be irreversible, such that reloading of fibres produces a linear graph having a gradient similar to the linear part of the original graph (Baley 2002). A logical solution would therefore seem to be to use the linear part of the stress–strain graph to obtain Young's modulus, although, some workers have proposed a calculation based on the whole stress–strain curve (Nechwatal *et al.* 2003) simply by taking the maximum stress and dividing it by the strain at maximum stress (secant modulus) (Eichhorn and Young 2004).

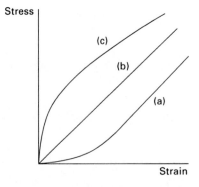

6.2 Typical stress–strain curves for natural cellulosic fibres: (a) strain hardening, (b) linear elastic, (c) plastic.

6.2   Statistical treatment of single-fibre strength

Natural cellulose-based fibres, similarly to synthetic fibres, undergo little plastic deformation and therefore can be considered to have strengths, as described by Griffith theory, determined by the flaws which occur along their length. In synthetic fibres, these could have originated in the precursor material or have been introduced during processing. For cellulose-based fibres, defects occur naturally, but are also introduced during fibre processing (Bos *et al.* 2002; Joffe *et al.* 2003). The strength of a fibre depends on the variation of flaws, or more precisely, the worst flaw that exists within it. Owing to the varying severity of flaws, the strength of fibres is found to be variable and a large number of tests are required for strength characterisation. The statistical distribution of natural fibre strengths is shown to be well approximated to the two-parameter Weibull distribution:

$$P_\mathrm{f} = 1 - \exp\left[- V\left(\frac{\sigma}{\sigma_0}\right)^w \right]$$ [6.1]

where P_f is the cumulative probability of failure of a fibre, at a stress less than or equal to σ, σ_0 is the Weibull scale parameter or characteristic strength, which is the failure stress for a probability of failure of 0.632 (and can be compared for different fibre populations similarly to average strength), w is the shape parameter or Weibull modulus, which describes the variability of strength (a higher value of w means lower variability) and V is fibre volume (Joffe *et al.* 2003). For synthetic fibres with consistent fibre diameters, fibre length (L) has replaced volume in the two-parameter Weibull distribution:

$$P_\mathrm{f} = 1 - \exp\left[-L\left(\frac{\sigma}{\sigma_0}\right)^w \right]$$ [6.2]

This equation has also been shown to give good agreement with natural fibre strength data (Biagiotti *et al.* 2004; Andersons *et al.* 2005). It should be noted that L is a relative measure of length and can be considered in multiples of the particular gauge length used.

Rearrangement of the two-parameter Weibull equation gives:

$$\ln\ln\left(\frac{1}{1 - P_\mathrm{f}}\right) = w\ln\sigma - \ln\sigma_0 + \ln L$$ [6.3]

The scale and shape parameters can therefore be obtained from a plot of $\ln\ln(1/(1 - P_\mathrm{f}))$ versus $\ln\sigma$ (commonly referred to as a Weibull plot), which, given that the data agrees with the Weibull distribution, should give a straight line of gradient w and intercept σ_0 at $\ln\ln(1/(1 - P_\mathrm{f})) = 0$. For this procedure, linear regression has been shown to give similar results to the maximum likelihood method (Zafeiropoulos and Baillie 2007).

The dependence of single-fibre tensile strength on fibre length for all fibres is well known (Nechwatal *et al.* 2003) and is also seen, but to a lesser extent, for natural fibre bundles (Mieck *et al.* 2003). The cause of this dependence can also be related to the defects within fibres; the longer the fibre length, the more likely it is that there is a defect of sufficient severity to bring about failure at a particular stress and therefore on average the strength will reduce. It can be considered that a fibre can be divided into segments or 'links', each containing a flaw of varying severity (Fig. 6.3). Hence the fibre is considered to have failed if one of its links has failed. This type of model is referred to as the 'Chain of Links' model. A longer fibre can be considered to have a larger number of links than a shorter one and therefore, there is an increased probability of the existence of a more severe flaw along the length and so longer fibres would be expected to have on average lower strength. Therefore, it is important to specify the gauge length of fibre for interpretation and comparison of fibre strength data. Embedded within the Weibull equation, is the ability to use values of strength obtained at any given gauge length to predict the strength for another length, for a similar probability of failure by means of the following 'weak-link scaling' equation:

$$\sigma_{(2)} = \sigma_{(1)} \left(\frac{L_1}{L_2} \right)^{\frac{1}{w}} \tag{6.4}$$

where $\sigma_{(1)}$ and $\sigma_{(2)}$ are the strengths for L_1 and L_2, respectively, at a particular probability. Most conveniently, characteristic strengths can be scaled for different lengths using this equation. Therefore, a plot of the logarithm of characteristic strength versus the logarithm of length should give a straight

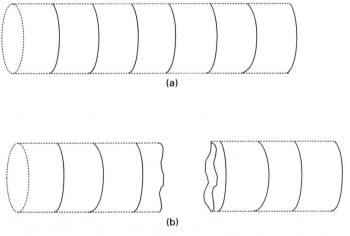

6.3 Chain of links model showing (a) fibre demonstrating chain of links and (b) fibre with a failed link.

line if weak link scaling is observed, from which the Weibull modulus can be obtained from the reciprocal of the gradient. However, considerable mismatch has been found between Weibull moduli obtained from Weibull plots at single gauge lengths and those from log log plots of strength versus length (Pickering *et al.* 2007a; Sparnins and Andersons 2009) despite good agreement with the Weibull distribution. For example, Pickering *et al.* (2007a) obtained Weibull moduli of 3.4 and 4.2 at gauge lengths of 1.5 and 10 mm, respectively, (and characteristic strengths of 876 and 745 MPa respectively) however, the weak link scaling plot from the same data gave a Weibull modulus of 11.7. A similar discrepancy found in another study (Andersons *et al.* 2005) has led to the suggestion of use of a modified Weibull distribution, which has been developed for use with synthetic fibres:

$$P_{f(L)} = 1 - \exp\left[-L^{\gamma}\left(\frac{\sigma}{\sigma_0}\right)^{w}\right] \qquad [6.5]$$

which overcomes the apparent mismatch. Here it is considered that the factor γ takes account of the inter-fibre variation of strength as would be expected owing to differing growth and processing history (Andersons *et al.* 2009a) where individual fibres could be modelled using the Weibull equation, but the Weibull parameters varied between fibres. Attempts have been made to correlate the modified Weibull equation with defects comprising of local misalignment of microfibrils occurring during growth and processing, commonly called kink bands (Andersons *et al.* 2009a). This has shown that although it can largely be accounted for by such defects, some discrepancy remained suggesting other factors such as irregularity in fibre geometry also accounts for strength distribution. Alternatively, the mismatch between Weibull modulus calculated from one gauge length and multiple gauge lengths has been explained as being caused by an artificial broadening of the distribution of tensile strength owing to variable fibre volume at a fixed length (Doan *et al.* 2006).

Another variation from the basic two-parameter Weibull distribution, suggested because of the better fit observed for hemp, jute, ramie and kenaf, is a bimodal Weibull distribution of the form given below (Park *et al.* 2006a; 2006b):

$$P_{f(L)} = 1 - \left\{ p\exp\left[-L\left(\frac{\sigma}{\sigma_{01}}\right)^{w1}\right] + q\exp\left[-L\left(\frac{\sigma}{\sigma_{02}}\right)^{w2}\right] \right\} \qquad [6.6]$$

where p and q give the relative amount of fibre influenced by two different populations of link strength (such that $p + q = 1$) and $\sigma_{01/2}$ and $w_{1/2}$ are the characteristic strengths and Weibull moduli, respectively, for those two different populations. The potential of a multimodal Weibull distribution has also been raised elsewhere owing to observation of clustering of residual values suggesting departure from linearity with a unimodal Weibull distribution (Zafeiropoulos and Baillie 2007). Although this could be justified by different types of defect (perhaps internal and external), as Park *et al.* (2006a; 2006b) state, deviations from a unimodal Weibull distribution can largely be related to fibre damage during preparation of tensile specimens leading to the lower strength population of fibre links, which can therefore be considered as merely an artefact of preparation. On this assumption the true strength distribution would be better described by applying a unimodal Weibull distribution to the higher strength fibres.

One requirement to enable use of Weibull statistics is for an estimator of P_f to be calculated from the number of fibres that have failed from a population of tested fibres for which there are a number of alternatives as given below (Zafeiropoulos and Baillie 2007):

$$P_f = \frac{i - 0.5}{n} \qquad [6.7]$$

$$P_f = \frac{i}{n + 1} \qquad [6.8]$$

$$P_f = \frac{i - 0.3}{n + 0.4} \qquad [6.9]$$

$$P_f = \frac{i - 0.375}{n + 0.25} \qquad [6.10]$$

where i is the ith fibre to fail and n is the total number of fibres tested. The first of these is recommended for sample sizes larger than 20 as it is considered to give the least biased results (Pickering *et al.* 2007a). However, this has been shown to give the least linearity for Weibull plots of flax fibre, with the second estimator giving the greatest linearity (Zafeiropoulos and Baillie 2007).

Although, the vast majority of data available is based on the single-fibre test, single-fibre fragmentation has also been explored for characterisation of fibre strength distribution (Andersons *et al.* 2005). Compared with data obtained from single-fibre testing, variability was found to be underestimated, although the influence of fibre length was well described, leading to the conclusion that fibre could be fully characterised using a limited number of single-fibre tensile tests at a fixed gauge length along with single-fibre fragmentation tests to characterise influence of length.

6.3   Mechanical properties of untreated single fibres

Cellulose-based fibre properties have been shown to be highly variable for different plant varieties and also for the same variety. For example, one study has shown the strength of hemp, nettle, milkweed, flax and linseed fibres of length approximately 1 mm to be 1080, 918, 728, 649 and 371 MPa with Young's moduli 9.2, 9.7, 9.6, 19.4 and 10.4 GPa, respectively (Snell *et al.* 1997), whereas another study (Robson and Hague 1996) states average strengths of approximately 2000, 1450, 1250, 900, 850 and 450 MPa and stiffnesses of approximately 75, 68, 58, 25, 23 and 17 GPa for flax, linseed, kenaf, hemp, softwood and wheat, respectively, with unspecified gauge length extracted by a number of unstated chemical methods. Part of the variability observed for fibre from the same plant species can be explained by the different procedures used to extract the fibre from the plant. Initially, removal of the outer bast material from plant stalks as well as separation of fibre from the bast material may be eased by retting. This is a biological process and could be regarded as a biological treatment, but for the purposes of this chapter is included in this section as it is commonly part of the extraction process that can influence the mechanical properties. Retting involves the action of bacteria within an aqueous environment and can be conducted after harvesting in the field or away from the field at a processing plant (Van de Velde and Baetens 2001). Alternatively, steam explosion has been used for fibre separation and has produced hemp fibre with almost 70% higher strength than that produced biologically (Keller 2001). Mechanical removal of bast material from stalks commonly involves pulling them through rollers to break the inside part of the stalk and beating with metal blades in a process known as scutching. The bast material can then be hackled (combed) to separate long and short fibres using increasingly fine sets of pins leading to removal of non bast material as well as alignment. Chemical agents may also be used to assist fibre extraction, however, this is commonly directed at influencing interfacial strength and is mainly covered within this chapter under the chemical treatment section (6.4.1).

Work conducted to assess the influence of harvesting period on single-fibre properties for hemp (Pickering *et al.* 2007a) involved monitoring average single fibre strength of retted material from the initiation of pollen release (99 days after planting) until harvesting. The average strength of 10 mm gauge length fibre was found to increase from 510 up to 670 MPa for fibre lengths of 10 mm during the 10 days following pollen release, with a subsequent decline in average fibre strength. This pattern of strength variation can be correlated with a change in fibre structure as seen in the literature (Mediavilla *et al.* 2001) such that initially long, empty primary fibres are created which fill between vegetative growth and flower formation, followed by greater production of weaker secondary fibres. Mechanical separation has been highlighted as reducing mechanical strength (Bos *et al.* 2002); the average tensile strength of single flax fibres (gauge length 3 mm) reduced from 1800 MPa for hand separated to 1500 MPa for mechanised separation. Comparison of retted fibre with unretted also conducted by Pickering *et al.*

(2007) showed unretted hemp fibre strength to be 28% higher (857 MPa for a gauge length of 10 mm) than the retted fibre, suggesting that degradation may be occurring during retting, although issues of selectivity regarding ease of extraction of primary relative to secondary fibres was quite marked for the unretted material. Alternatively, the tensile strength of flax fibre has been found to be unaffected by retting (Van de Velde and Baetens 2001), although chemical extraction had been employed that could be overriding the influence of retting as supported by lower average flax fibre strength of between 700–950 MPa for a gauge length of 10 mm. Also in this study, the influence of hackling was found to depend on the degree of retting, such that the reduction in strength was found for hackled material subjected to a longer retting procedure, but not with shorter retting times. Young's modulus for flax was found to vary between 26 and 36 GPa depending on the degree of retting and mechanical processing, although no trend was observed relating to processing.

Bag retting has been assessed as a more controlled and environmentally friendly (regarding water pollution) technique than field retting (Li *et al.* 2009). Despite an approximate halving of fibre strength seen for one such treatment, composite strength has been increased by approximately 23% using this method.

Drying of retted, scutched and hackled flax fibre has been found to result in a 42–46% reduction in strength from 1500 and 1320 MPa at a gauge length of 10 mm for fibres of diameter 20–22.5 μm and 22.5–25 μm, respectively, (Baley *et al.* 2005). Subsequent water absorption resulted in only about half the strength loss being recovered, which raised the problem of permanent fibre damage and modification of chemical fibre components during drying.

Most of the data available from Weibull analysis of natural fibres is based on flax fibre. Biagiotti *et al.* (2004) have applied Weibull statistics to single-fibre strength and stiffness data. The characteristic strength and characteristic Young's modulus (E_0) have been reported to be 1054 MPa and 33.2 GPa, respectively, for untreated flax fibres with Weibull moduli within the range 2.6–3.6 describing high variability. However, sadly, there is no mention of gauge length or testing rate thus preventing proper comparison (Biagiotti *et al.* 2004). Based on the modified Weibull distribution, Andersons *et al.* (2005) obtained parameters of $\sigma_0 = 1400$ MPa at a scaled length of 1 mm, $w = 2.8$ and $\gamma = 0.46$ for enzyme retted flax. A later study quoted values for flax of $w = 2$–4.3 and $0 \leq \gamma \leq 1$ (Andersons *et al.* 2009a).

Lower values of characteristic strength 577–743 MPa for lengths of 5–20 mm (characteristic strength reducing with increased length as expected) with values of Weibull modulus of 2.6–3.8 were obtained for single fibres taken from textile flax yarn (Doan *et al.* 2006). Similar values of $\sigma_0 = 812$ MPa (average strength ~720 MPa) at a length of 10 mm and $w = 2.7$ were obtained elsewhere (Andersons *et al.* 2009b).

Weibull analysis of other cellulosic fibres is quite limited. One study has obtained a value for characteristic stress of 320 MPa and Weibull modulus of 2.5 for 40 mm gauge length hemp fibre (Schlddjewski *et al.* 2006). This compares well using weak link scaling with other work where the characteristic strength and Weibull modulus were found to be 876 MPa and 3.4, respectively, at a gauge length of 1.5 mm, and 745 MPa and 4.2, respectively, at a gauge length of 10 mm for alkali-treated fibre (Pickering *et al.* 2007a). Weibull analysis for jute fibres resulted in calculation of characteristic strengths of 378–624 MPa (average strengths of 336–558 MPa) for jute fibres of length of 6–50 mm with Weibull moduli between 2.84 and 3.31 (Virk *et al.* 2009). The characteristic strength and Weibull modulus for kenaf fibre have been found to be 876 MPa and 3.4 at a gauge length of 40 mm (Schlddjewski *et al.* 2006).

Exposure to elevated temperature has been found to reduce tensile strength quite dramatically; a reduction in strength of 36% was seen for flax fibre exposed to 180 °C for 2 h, although little effect on tensile strength was seen up to 120 °C (despite a significant reduction of failure strain) (Van de Velde and Baetens 2001). A reduction in tensile strength was also observed for hemp and jute of the order of 16–25% when subjected to 140 °C for 2 h (Park *et al.* 2006a; Liu and Dai 2007).

It has been observed for wood, hemp and sisal that generally fibre strength decreases as fibre diameter increases, which is consistent with strength reducing with increased volume owing to the increased likelihood of a critical flaw (Joffe *et al.* 2003; Pickering *et al.* 2003; Pickering *et al.* 2005). Young's modulus of flax has also been seen to decrease with increased fibre diameter (Joffe *et al.* 2003). However, the contrary relationship has been seen for jute fibre strength with increased diameter, highlighting the likelihood that structural changes within the fibre that occur during its growth, and not taken account of by Weibull statistics, such as microfibril angle, are likely to affect mechanical performance (Doan *et al.* 2006).

6.4   Influence of fibre treatment on mechanical properties of natural fibres

6.4.1   Influence of chemical treatment

Alkali treatment with sodium hydroxide is one of the most common treatments utilised to improve interfacial strength. The primary aims of this treatment are to remove weak non-cellulosic components, allow increased exposure to cellulose and roughen the fibre to enhance mechanical interlocking. Its reported effect on fibre strength has been mixed. Although some studies showed a reduction in hemp and flax fibre strength owing to treatment with sodium hydroxide, improvement has been seen for jute and sisal as well as hemp in other studies (Romao *et al.* 2004; Arbelaiz *et al.* 2005; Park *et al.* 2006a; Liu and Dai 2007). Gassan and Bledski (1999) demonstrated an increase in jute yarn strength and stiffness using sodium hydroxide and attributed this to increased degree of polymerisation and crystallinity. A comprehensive study by Pickering *et al.* (2007a) has shown that fibre strength depended on sodium hydroxide treatment severity. Optimised treatment in this study increased the average single-fibre tensile strength by over 11%, although harsher treatments could result in a dramatic reduction of strength. Removal of non-cellulosics including lignin, pectin and hemicelluloses was supported by less cluttered surfaces with etched striations observed by SEM as shown in Fig. 6.4, as well as by chemical analysis. Crystallinity was also observed to increase, thought to be the result of better packing of cellulose chains with the removal of amorphous material from the fibre. Flash hydrolysis of flax using sodium hydroxide at very high pressure to separate fibre has, therefore, not surprisingly been shown to drastically reduce single fibre strength (Ansari *et al.* 2001).

Acetylation was applied to natural fibres to increase their compatibility with polymer matrices and has commonly been seen to reduce fibre strength. Acetylation has been shown to reduce average flax fibre strength by 9–23% with a greater reduction generally observed with increased exposure. Possible explanations include fibre damage owing to the acid used as part of the treatments leading to cleavage of cellulose and reduced crystallinity, although this has been shown to depend on degree of fibre retting, to the extent that increased strength has been observed with unretted fibre (Biagiotti *et al.* 2004; Arbelaiz *et al.* 2005; Zafeiropoulos and Baillie 2007).

Silane treatment using vinyltrimethoxysilane (VTMO) used as a coupling agent to compatibilise the fibre and matrix has also been seen to reduce fibre strength by up to 40%, again suggested to be the result of fibre damage by an acid carrier (Biagiotti *et al.* 2004; Arbelaiz *et al.* 2005). However, the average hemp and jute fibre strength have been seen to increase with silane treatment, (Park *et al.* 2006a). This was attributed to removal of flaws using a silane coupling agent.

Grafting of flax with maleic anhydride polypropylene (MAPP) copolymer to enhance the fibre/matrix compatibility has been observed to reduce strength by 6% as well as increase strength by 9%, the increase being explained by MAPP leading to a smoother surface and decreasing stress concentration (Biagiotti *et al.* 2004; Arbelaiz *et al.* 2005).

Although not statistically significant, the average strength of flax fibre reduced with stearic acid treatment, which was also used to increase fibre/matrix compatibility up to a loss of 35% (Zafeiropoulos and Baillie 2007).

Variability and characteristic stiffness have been observed to be unchanged by treatment with maleic anhydride, MAPP and VTMO (Biagiotti *et al.* 2004).

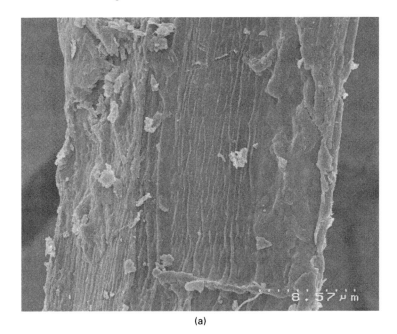

(a)

(b)

6.4 Scanning electron micrographs of hemp fibre surface (a) before treatment in sodium hydroxide and (b) after treatment in sodium hydroxide.

EDTMPA (ethylenediamine tetra(methylene phosphonic acid)) has been used because of its chelating properties for treating natural fibres; chelating agents that are generally capable of reacting with metals, have been identified as able to remove calcium ions from pectin in plant cell walls, suggesting the potential for fibre separation and better exposure to cellulose (Li and Pickering 2008). Composite strength has been found to be increased, despite fibre strength reduction of the order of 20% which was attributed to interfacial strength.

6.4.2   Influence of biological treatment

In addition to retting, which is a biological process applied to fibre for extraction purposes, other forms of biological treatment have been applied as a potentially environmentally friendly technique to improve interfacial bonding. White rot fungi from the *Basidomycetes* group, have been employed for this purpose (Pickering *et al.* 2007b) because of their ability to degrade lignin and expose cellulose. Although treatment with white rot fungi has been seen to reduce fibre strength by up to 50% with an unoptimised treatment, an increase in composite strength of 22% compared with untreated fibre composites has been observed. This appears to result from exposure of cellulose and the production of pits in the fibre walls leading to improved chemical bonding and mechanical interlocking, caused by the presence of fungal hyphae. Further improvement in strength has been observed by combined alkali and fungal treatment (Pickering *et al.* 2007b).

Fibres modified using bacteria through fermentation resulting in deposition of nano-cellulose and increased interfacial bonding has also been examined mechanically (Pommet *et al.* 2008). It has been found that treatment can be conducted without significantly affecting the strength of sisal fibres, although hemp fibres were found to be drastically reduced in strength.

6.5   Conclusion

Overall, there is a growing number of studies on the mechanical properties of single cellulose-based natural fibre largely obtained by single-fibre tensile testing. The two-parameter Weibull distribution is most commonly used to describe strength variation, but modified and bimodal Weibull distributions have been demonstrated by some workers to provide a better fit. However, it is unclear as to whether this is a true reflection of strength distribution or related to testing inadequacies owing to factors such as fibre damage during handling and inaccuracies in measurement of fibre diameter. Weibull moduli from single gauge lengths have been obtained in the range of 2–4.3 describing high variability, although at the upper end, this is close to that obtained for carbon fibre used in advanced engineering structures. Much focus has been on flax fibre which has strength and stiffness measured at up to 2000 MPa and 75 GPa, respectively, although there is increasing interest in other cellulose-based fibres including hemp, which also has high strength and stiffness and a potential advantage relating to reduced herbicide and pesticide requirements. Mechanical properties are seen to be dependent on many factors including harvest time, extraction procedure and treatments for improvement of interfacial bonding. Despite large reductions of fibre strength with some treatments, large improvements with composite strength can be achieved by their use, demonstrating interfacial strength to be the major limitation for untreated fibre and the importance of optimisation of these treatments.

6.6   Acknowledgements

The author would like to express gratitude to collaborators, contributors and supporters of their research work included within this chapter including Drs. Gareth Beckermann and Maggie (Yan) Li, Professor Roberta Farrell as well as Hemptech and the Foundation for Research in Science and Technology, New Zealand.

6.7   References

Andersons, J., E. Porike and E. Sparnins (2009a). 'The effect of mechanical defects on the strength distribution of elementary flax fibres.' *Composites Science and Technology* **69**(13): 2152–2157.

Andersons, J., E. Sparnins and R. Joffe (2009b). 'Uniformity of filament strength within a flax fiber batch.' *Journal of Materials Science* **44**(2): 685–687.

Andersons, J., E. Sparnins, R. Joffe and L. Wallström (2005). 'Strength distribution of elementary flax fibres.' *Composites Science and Technology* **65**(3–4): 693–702.

Ansari, I. A., G. C. East and D. J. Johnson (2001). 'Structure-property relationships in natural cellulosic fibres. Part II: Fine structure and tensile strength.' *Journal of the Textile Institute* **92**(4): 331–348.

Arbelaiz, A., G. Cantero, B. Fernandez, I. Mondragon, P. Ganan and J. M. Kerry (2005). 'Flax fiber surface modifications: effects on fiber physico mechanical and flax/polypropylene interface properties.' *Polymer Composites* **26**(3): 324–332.

Baley, C. (2002). 'Analysis of the flax fibres tensile behaviour and analysis of the tensile stiffness increase.' *Composites Part A–Applied Science and Manufacturing* **33**(7): 939–948.

Baley, C., C. Morvan and Y. Grohens (2005). 'Influence of the absorbed water on the tensile strength of flax fibers.' *Macromolecular symposia* **222**(1): 195–202.

Biagiotti, J., D. Puglia, L. Torre, J. M. Kerry, A. Arbelaiz, G. Cantero, C. Marieta, R. Llano-Ponte and I. Mondragon (2004). 'A systematic investigation on the influence of the chemical treatment of natural fibers on the properties of their polymer matrix composites.' *Polymer Composites* **25**(5): 470–479.

Bos, H. L., M. J. A. Van den Oever and O. C. J. J. Peters (2002). 'Tensile and compressive properties of flax fibres for natural fibre reinforced composites.' *Journal of Materials Science* **37**(8): 1683–1692.

Davies, G. C., D. M. Bruce, *et al.* (1998). 'Effect of environmental relative humidity and damage on the tensile properties of flax and nettle fibers.' *Textile Research Journal* **68**: 623–629.

Doan, T. T. L., S. L. Gao and E. Mader (2006). 'Jute/polypropylene composites I. Effect of matrix modification.' *Composites Science and Technology* **66**(7–8): 952–963.

Eichhorn, S. J. and R. J. Young (2004). 'Composite micromechanics of hemp fibres and epoxy resin microdroplets.' *Composites Science and Technology* **64**(5): 767–772.

Gassan, J. and A. K. Bledzki (1999). 'Alkali treatment of jute fibers: relationship between structure and mechanical properties.' *Journal of Applied Polymer Science* **71**(4): 623–629.

Joffe, R., J. Andersons and L. Wallstrom (2003). 'Strength and adhesion characteristics of elementary flax fibres with different surface treatments.' *Composites Part A–Applied Science and Manufacturing* **34**(7): 603–612.

Keller, A. (2001). 'Compounding and mechanical properties of biodegradable hemp fibre composites.' *1st International EcoComp Conference*, London, England.

Li, Y. and K. L. Pickering (2008). 'Hemp fibre reinforced composites using chelator and enzyme treatments.' *Composites Science and Technology* **68**(15–16): 3293–3298.

Li, Y., K. L. Pickering and R. L. Farrell (2009). 'Analysis of green hemp fibre reinforced composites using bag retting and white rot fungal treatments.' *Industrial Crops and Products* **29**(2–3): 420–426.

Liu, X. Y. and G. C. Dai (2007). 'Surface modification and micromechanical properties of jute fiber mat reinforced polypropylene composites.' *Express Polymer Letters* **1**(5): 299–307.

Mediavilla, V., M. Leupin and A. Keller (2001). 'Influence of the growth stage of industrial hemp on the yield formation in relation to certain fibre quality traits.' *Industrial Crops and Products* **13**(1): 49–56.

Mieck, K. P., T. Reussmann and A. Nechwatal (2003). 'About the characterization of the mechanical properties of natural fibres.' *Materialwissenschaft Und Werkstofftechnik* **34**(3): 285–289.

Mwaikambo, L. Y. and M. P. Ansell (1999). 'The effect of chemical treatment on the properties of hemp, sisal, jute and kapok for composite reinforcement.' *Angewandte Makromolekulare Chemie* **272**: 108–116.

Nechwatal, A., K. P. Mieck and T. Reussmann (2003). 'Developments in the characterization of natural fibre properties and in the use of natural fibres for composites.' *Composites Science and Technology* **63**: 1273–1279.

Park, J. M., S. T. Quang, B.-S. Hwang and K. L. DeVries (2006a). 'Interfacial evaluation of modified Jute and Hemp fibers/polypropylene (PP)-maleic anhydride polypropylene copolymers (PP-MAPP) composites using micromechanical technique and nondestructive acoustic emission.' *Composites Science and Technology* **66**(15): 2686–2699.

Park, J. M., T. Q. Son, J. G. Jung and B. S. Hwang (2006b). 'Interfacial evaluation of single ramie and kenaf fiber/epoxy resin composites using micromechanical test and nondestructive acoustic emission.' *Composite Interfaces* **13**(2–3): 105–129.

Pickering, K. L., A. Abdalla, C. Ji, A. G. McDonald and R. A. Franich (2003). 'The effect of silane coupling agents on radiata pine fibre for use in thermoplastic matrix composites.' *Composites Part A-Applied Science and Manufacturing* **34**(10): 915–926.

Pickering, K. L., G. W. Beckermann, S. N. Alam and N. J. Foreman (2007a). 'Optimising industrial hemp fibre for composites.' *Composites Part A–Applied Science and Manufacturing* **38**(2): 461–468.

Pickering, K. L., M. Priest, T. Watts, G. Beckermann and S. N. Alam (2005). 'Feasibility study for NZ hemp fibre composites.' *Journal of Advanced Materials* **37**(3): 15–20.

Pickering, K. L., Y. Li, R. L. Farrell and M. C. Lay (2007b). 'Interfacial modification of hemp fiber reinforced composites using fungal and alkali treatment.' *Journal of Biobased Materials and Bioenergy* **1**(1): 109–117.

Pommet, M., J. Juntaro, J. Y. Y. Heng, A. Mantalaris, A. F. Lee, K. Wilson, G. Kalinka, M. S. P. Shafer and A. Bismarck (2008). 'Surface modification of natural fibers using bacteria: depositing bacterial cellulose onto natural fibers to create hierarchical fiber reinforced nanocomposites.' *Biomacromolecules* **9**(6): 1643–1651.

Robson, D. and J. Hague (1996). 'A comparison of wood and plant fiber properties. *Proceedings of woodfiber-plastic composites conference*. Madison, WI, Forest Products Society, pp. 41–46.

Romao, C., P. Vieira, F. Pieto, A. T. Marques and J. L. Esteves (2004). 'Single filament mechanical characterisation of hemp fibres for reinforcing composite materials.' *Molecular Crystals and Liquid Crystals* **418**: 87–99.

Schlddjewski, R., L. Medina and A. K. Schlarb (2006). 'Mechanical and morphological characterization of selected natural fibres.' *Advanced Composites Letters* **15**(2): 55–61.

Snell, R., J. Hague and L. Groom (1997). 'Characterizing agrofibers for use in composite materials.' *4th International Conference on Woodfiber-Plastic Composites*, Madison, Wisconsin.

Sparnins, E. and J. Andersons (2009). 'Diameter variability and strength scatter of elementary flax fibers.' *Journal of Materials Science* **44**(20): 5697–5699.

Tserki, V., N. E. Zafeiropoulos, F. Simon and C. Panayiotou (2005). 'A study of the effect of acetylation and propionylation surface treatments on natural fibres.' *Composites Part A – Applied Science and Manufacturing* **36**(8): 1110–1118.

Van de Velde, K. and E. Baetens (2001). 'Thermal and mechanical properties of flax fibres as potential composite reinforcement.' *Macromolecular Materials and Engineering* **286**(6): 342–349.

Virk, A. S., W. Hall and J Summerscales (2009). 'Tensile properties of jute fibres.' *Materials Science and Technology* **25**(10): 1289–1295.

Zafeiropoulos, N. E. and C. A. Baillie (2007). 'A study of the effect of surface treatments on the tensile strength of flax fibres: Part II. Application of Weibull statistics.' *Composites Part A – Applied Science and Manufacturing* **38**(2): 629–638.

Zafeiropoulos, N. E., D. R. Williams, C. A. Baillie and F. L. Matthews (2002). 'Engineering and characterisation of the interface in flax fibre/polypropylene composite materials. Part I. Development and investigation of surface treatments.' *Composites Part A – Applied Science and Manufacturing* **33**(8): 1083–1093.

Part II
Testing interfacial properties in natural fibre composites

7

Electrokinetic characterisation of interfacial properties of natural fibres

K. K. C. HO and A. BISMARCK, Imperial College
London, UK

Abstract: The benefits of electrokinetic measurements in the study of electrostatic interface forces and electrokinetic surface properties for many industrial, biological and medical applications are reviewed. Measuring ζ-potentials is a useful tool for the characterisation of natural fibres, including acid/base surface characterisation, and evaluating the relative acid/base strength as well as the swelling behaviour of natural fibres.

Key words: natural fibres, electrochemical double layer, ζ-potential, electrokinetic analysis, surface analysis, interfaces.

7.1 Introduction

The qualitative and quantitative characterisation of the surface properties and the reactivity of solid surfaces is of crucial importance for many liquid as well as solid formulations, such as for the stability of dispersions, the adhesion between components in, for instance, composites (Jacobasch *et al.* 1992; Jacobasch 1993; Mäder *et al.* 1994), but also for the determination of materials properties, such as biocompatibility of materials (Werner *et al.* 1995). Measurements of ζ-potential have been used to characterise the influence of various processing parameters and surface treatments on the surface properties of textile (Flath and Saleh 1980; Grosse and Jacobasch 1994), synthetic (Jacobasch and Grosse 1991; Rybicki and Jacobasch 1992) and (sized) glass fibres (Bledzki *et al.* 1997). However, ζ-potential measurements are particularly useful for the characterisation of rather hydrophilic materials, such as cellulosic materials, because ζ-potential (as well as contact angle) measurements enable the characterisation of surfaces in the wet state (Jacobasch *et al.* 1995; Ribitsch and Stana-Kleinscheck 1997), which for a number of applications of cellulosic materials, for instance during papermaking (Beck *et al.* 1980) as well as for membrane separations, is of crucial importance. Moreover, it was found that the adhesion in fibre reinforced polymers correlates with the results obtained from ζ-potential measurements. For instance, it was found that the minimum value of the ζ-potential measured as a function of the KCl concentration ($\zeta = f([\text{KCl}])$) for glass fibres correlates with the adhesion behaviour of these fibres in

205

unsaturated polyester resins (Jacobasch 1984). This correlation depends on the KCl concentration, which is governed by the dispersion forces present on a particular surface.

Dispersion forces determine the adhesion if no specific (direct chemical) interactions occur between the two phases. It was shown that the difference in the adsorption-free energies of the electrolyte ions at the solid surface corresponds to the dispersive forces occurring at the solid/electrolyte interface (Jacobasch *et al.* 1995). Moreover, ζ-potential measurement is a straightforward and reliable tool for predicting adhesive properties (Häßler and Jacobasch 1994); in this study for an increasing difference in the ζ-potential plateau values determined by measuring the ζ-potential as function of pH between adherent (substrate) and adhesive (matrix), the measured adhesive strength increases. For polymer composites, it was reported that the trend of changing adhesion could be predicted by ζ-potential measurements (Häßler and Jacobasch 1994; Chiu and Wang 1998; Bismarck *et al.* 1999b; Campagne *et al.* 2002).

Natural cellulose-based fibres have many advantages but also several disadvantages (see preceding chapters on moisture uptake of natural fibres). Natural fibres contain a high portion of hydroxyl groups, which makes them highly polar; these groups readily form hydrogen bonds at the interfaces in composites (Bellmann *et al.* 2005). However, natural fibres are usually covered with pectin and waxy substances (Stana-Kleinschek and Ribitsch 1998), which hinder the hydroxyl groups from interacting with polar matrices and, therefore, affect adhesion. Owing to the intrinsic hydrophilic behaviour of natural fibres (highly dependent on the purity of the composition of the fibres and the degree of crystallinity of the cellulose), these fibres swell severely and so take up a significant amount of water (see also chapter 10). Nevertheless, natural fibres have become a favourable constituent for use as reinforcement in green composites. Many natural fibres with very different compositions and properties are available (ranging from bast to leaf and seed fibres). The physical properties of natural fibres, such as their density, mechanical properties and surface area can be easily measured, however, the surface properties, such as surface chemistry and wetting behaviour are more difficult to determine because of the intrinsic hydrophilicity of the fibres and the fact that the fibres are essentially all made up of carbohydrates and lignin (carbon, hydrogen and oxygen-based organic compounds). In order to characterise the surface properties of various natural fibres and to understand the underlying mechanisms that may lead to better natural fibre composites, it is beneficial to explore the electrokinetic properties of these fibres.

The measurement of ζ-potential is an efficient means of characterising natural fibres. To aid understanding we start by introducing electrokinetic phenomena and ζ-potential. If a solid in chemical and thermal equilibrium is in contact with an aqueous electrolyte solution, an electrochemical double layer (EDL) (Fig. 7.1) forms at the phase boundary, i.e. the interface. This EDL forms because of the dissociation of functional surface groups (should they be present) and/or the preferential adsorption of ions, the adsorption of ionic surface active agents or because of the isomorphic substitution of cations and anions at the interface (Lyklema 1981), which results in an asymmetric distribution of charges in the interface. Any excess charges at the surface are compensated for by counter ions, some of which are partially/strongly adsorbed at the solid surface whereas the remainder (caused by Brownian movement) of the counter ions required for complete charge neutrality is situated at a larger distance away from the surface. The phenomenon is described by the generally accepted Gouy–Chapman–Stern–Grahame (GCSG) model (Stern 1924; Hunter 1981; Börner and Jacobasch 1985) of the electrochemical double layer (Grahame 1947). Although the widely used model was proposed by Grahame in 1947, the EDL concept itself dates back to Helmholtz (1853). The GCSG model (Fig. 7.1) for a solid with negative surface potential ψ, which is experimentally inaccessible, assumes that the EDL consists of two layers of immobile adsorbed charges, the inner (IHP) and outer Helmholtz plane (OHP), and a diffuse layer, called the Gouy or diffuse layer (Fisher 1996), in which the ions are trapped at a large distance away from the solid surface by a balance of attractive electrostatic forces and thermal (Brownian) motion (Werner et al. 1995). The adsorption of anions from the electrolyte phase is driven by attractive interactions on the solid surface by partially losing their hydration shell (specific adsorption), whereas hydrated cations adsorb only by electrostatic forces on the interface. The potential drop across the Helmholtz plane is linear and depends only on the size of the ions as well as the size of the hydration shell of the

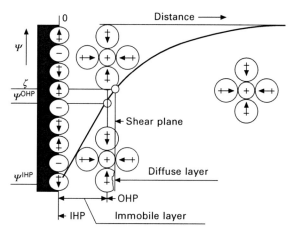

7.1 A schematic of the electrochemical double layer: Gouy–Chapman–Stern–Grahame model (Jacobasch et al. 1998).

counter ions (Bergmann and Voinov 1976). However, in the diffuse part of the EDL, the potential drop is exponential and can be described by a Boltzmann distribution. By creating a relative movement between the solid and the liquid, the diffusive part of the EDL is sheared off of the surface creating an electrical potential across the surface, which can be measured. The measured potential is called the ζ-potential, which is the potential at the plane of shear between the immobile and diffusive part of the EDL. The position of the plane of shear is physically unambiguously defined by the equilibrium between the attractive surface forces and the force field causing the relative movement between the solid and liquid phase (Jacobasch *et al.* 1996a). The relative movement between the solid and the electrolyte phase causes a number of effects on electrokinetic phenomena; which are named, depending on the type of relative movement: electrophoresis, electro-osmosis, streaming potential and sedimentation potential.

These electrokinetic phenomena can be quantified by either measuring the speed at which a particle or the electrolyte moves as function of an applied electric field or by measuring the potential difference across a sample when the electrolyte is forced through the sample or particles sediment under gravity. Although electrophoresis is the most frequently used method for measuring the ζ-potential, it is impractical to characterise fibres or any macroscopic surface (Delgado *et al.* 2007). Electrochemical characterisation (ζ-potential) does not only offer the possibility of estimating the state, type and amount of dissociable surface functional groups on fibre surfaces but also allows dispersion forces proceeding from solid surfaces to be characterised (Uchida and Ikada 2002).

The composition of the EDL can be assessed by measuring the ζ-potentials as function of the composition of the electrolyte phase. Measuring the ζ-potentials as a function of the electrolyte concentration and pH ($\zeta = f(c, \text{pH})$) allows information to be extracted about the mechanisms that lead to the formation of the EDL. Moreover, thermodynamic quantities such as aerial charge densities, molar free adsorption enthalpies and equilibrium constants for the specific adsorption of ions and the dissociation of functional surface groups can be determined (Jacobasch *et al.* 1996a).

7.2   Streaming potential measurements

The ζ-potentials of macroscopic materials, such as bundles of fibres, powder beds, films or porous media, which can be arranged in a capillary system, are best characterised using the streaming potential measurement method (Jacobasch *et al.* 1996b). For streaming potential measurements the relative movement between the solid and the electrolyte phase is induced by forcing the electrolyte through the capillary system using a pump, which causes a pressure drop Δp across the sample. The complete or partial shear-off of the ions in the diffusive part of the double layer results in a potential difference, the streaming potential U_p across the sample (i.e. the capillary system). The potential at the plane of shear, i.e. the ζ-potential, can be determined from streaming potential measurements using the Smoluchowski equation:

$$\frac{\Delta U}{\Delta p} = -\frac{\varepsilon_r \varepsilon_0 \zeta}{\eta \kappa} \qquad [7.1]$$

where ε_r is the dielectric constant of the electrolyte, ε_0 = permittivity of vacuum, η = dynamic viscosity of electrolyte and κ the conductivity of the electrolyte. Following equation [7.1] it is possible to determine the ζ-potential by measuring the streaming potential as a function of the pressure drop across the sample $p = f(U)$. However, equation [7.1] is only valid for single capillaries. Moreover, it is usually impossible to characterise or describe the geometrically irregular capillary bundles, such as packed fibre beds. Nevertheless, it is possible to determine the ζ-potential of macroscopic solids by correcting for surface conductivity of the solid. This can be achieved by a relative simple means (Fairbrother and Mastin 1924) but is not the topic of this chapter. However, if ζ-potential measurements are used for the qualitative comparison between samples, one does not necessarily have to correct for the surface conductivity, which would, at a given pH and electrolyte concentration only change the amplitude of the ζ-potentials but does not affect the position of the isoelectric point and the shape of $\zeta = f(\text{pH})$ (Jacobasch *et al.* 1996a).

The ζ-potential of macroscopic surfaces, including (natural) fibres, can easily be measured using an electrokinetic analyser (EKA) (Fig. 7.2), made by Anton Paar KG (Graz, Austria). The measuring system (Jacobasch *et al.* 1986) consists of a chassis for holding the measuring cells and also contains piping for the electrolyte transport, a valve to change the flow direction of the electrolyte, a computer controlled pump, and a differential pressure sensor situated between the inlet and outlet of the measuring cell, as well as a conductivity meter and a pH electrode. The electrolyte is held in a large beaker. A computer is used to control the EKA and to measure the pH, conductivity, pressure drop and streaming potential. The streaming potential is usually measured using two perforated Ag/AgCl electrodes, which sandwich the sample of interest.

The ζ-potential as property of the EDL is influenced by a number of factors (Jacobasch 1984) the surface potential ψ_0, pH, the swelling behaviour of the sample in the electrolyte, temperature, the valence and concentration of the ions in the electrolyte as well as the surface composition of the solid sample. The ζ-potential measurements allow the determination of a number of properties, such as nature (i.e. acid/base (Brønsted) properties) and concentration of dissociable functional groups, the adsorption of water

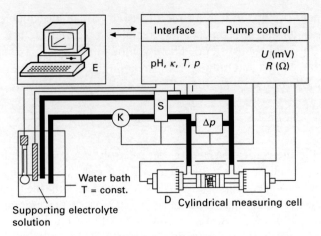

Water bath
T = const.

Supporting electrolyte
solution

D Cylindrical measuring cell

7.2 Schematic illustration of the EKA equipped with a cylindrical glass cell used to measure the ζ-potential of (natural) fibres (with kind permission of Dr R. Tahhan). (E = internal conductivity meter, T = thermostat containing the electrolyte, D = fibre measuring cell, K = pump, Δp = pressure sensor, S = valve to control flow direction of electrolyte.

and ions from the electrolyte, as well as the interaction forces that originate from a surface (Rätzsch *et al.* 1990).

As mentioned earlier, natural fibres are highly polar owing to the large amount of hydroxyl groups present and, therefore, they take up, i.e. swell readily in, water. According to Kanamaru (1960), the change of the ζ-potential of a solid swelling in an aqueous electrolyte can be described as follows:

$$-\frac{d\zeta}{dt} = k(\zeta - \zeta_{\infty})$$

[7.2]

which leads to:

$$-\ln\frac{\zeta - \zeta_{\infty}}{\zeta_0 - \zeta_{\infty}} = kt$$

[7.3]

where ζ is the measured ζ-potential value at a certain time, ζ_{∞} is the potential value which the function $\zeta = f(t)$ reaches asymptotically and k is a relative constant, which depends on the structure of the solid, and of the swelling process (Jacobasch 1984; Bismarck *et al.* 2001). It was found that the quotient $\Delta\zeta = (\zeta_0 - \zeta_{\infty})/\zeta_0$ is proportional to the water uptake at 100% relative humidity (the sorption capacity) of the natural fibres (Fig. 7.3) (Baltazar-y-Jimenez and Bismarck 2007b).

Time dependent ζ-potential measurements are a relatively fast way to characterise the swelling behaviour of solids, however, it should be noted that investigating the kinetics of water uptake by natural fibres using streaming

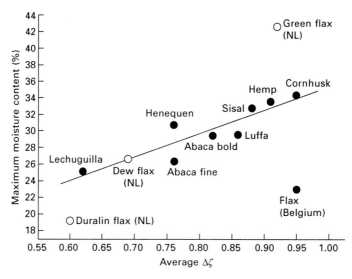

7.3 Maximum moisture content (MC) of natural fibres as a function of the $\Delta\zeta$ quotient, which correlates almost linearly with the MC of the fibres (adapted from Baltazar-y-Jimenez and Bismarck 2007b, Bismarck, *et al.* 2002).

potential measurements can be problematic. This is because the first step in the measurement of streaming potential involves rinsing the fibre plug to remove all the trapped air. Only after that can the time dependent ζ-potential measurement be started. This, however, does not take into account that water uptake of natural fibres occurs instantly as soon as the fibres are in contact with the water-based electrolyte (1 mM KCl, pH = 5.6). Therefore, the precision of locating the initial starting point ζ_0 is difficult. Nevertheless, it has been confirmed that good agreement on determining ζ_0 can be obtained by fitting an exponential first order decay function to the measured ζ-potential-time graph and subtracting ζ_∞ (Bismarck *et al.* 2002).

It can be seen from Fig. 7.4 that the value of the ζ-potential decreases rapidly to approach a much smaller value asymptotically, and that the decay of the ζ-potential differs in magnitude for the different fibres. For most natural fibres, with the exception of flax and cornhusk (Fig. 7.4b and 7.4c), the ζ-potential of the fibres increases from a value ζ_0 on different time scales asymptotically to a constant but larger value ζ_∞. For flax fibres and cornhusk the ζ-potential was initially positive and decreased with time to a much smaller ζ-potential. Moreover, as the measurements advanced and the plateau developed, the flax fibres produced a negative ζ_∞ value (Fig. 7.4b), whereas the cornhusk produced a positive one (Fig. 7.4c).

Once the water uptake behaviour of natural fibres has reached saturation,

ζ-potential as a function of pH can also be measured. Measurements of ζ-potential as a function of the pH of the electrolyte solutions ($\zeta = f(\text{pH})$) provide information about the presence of Brønsted acidic and/or basic surface functional groups as well as their relative acid/base strength provided that the formation of the EDL is mainly dominated by the dissociation of functional surface groups according to their pK_a (Jacobasch 1989). Figure 7.5 shows schematically the $\zeta = f(\text{pH})$ graph. If an investigated material contains only acidic functional groups (i.e. surface–COOH), a plateau region is observed

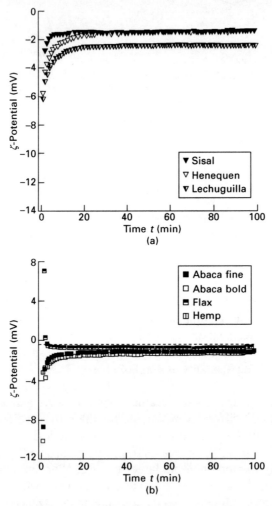

7.4 ζ-Potential as a function of time for (a) leaf fibres (henequen, lechuguilla and sisal; (b) bast fibres (abaca fine, abaca bold, flax and hemp; and (c) fruit fibres (cornhusk and luffa) (Baltazar-y-Jimenez and Bismarck 2007b).

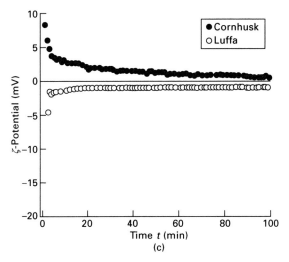

7.4 Continued

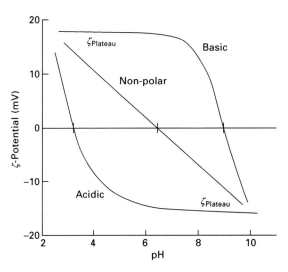

7.5 Schematic illustration of pH dependent ζ-potential curves measured on acidic, basic or non-polar samples.

at pH values ($\mathrm{d}\zeta/\mathrm{d(pH)} = 0$) indicating the increasing dissociation of acidic surface groups and, therefore, increased number of negative surface charges on the surface. Once all surface groups are dissociated, the ζ-potential (ζ_{Plateau}) stays the same for increasing pH. However, with decreasing pH, the dissociation of acidic surface groups is reduced causing an increase in the ζ-potential until complete protonation of the acid groups, which results in an isoelectric point (iep) at which $\zeta = 0$. A ζ-potential reversal is sometimes observed because of H_3O^+ adsorption on the surface. If, however, a solid only contains basic functional groups (i.e. surface–NH_2), at high pH, the dissociation of these groups is suppressed while OH^- ions are adsorbed onto the surface. When the pH decreases, the number of positively charged groups increases eventually resulting in a plateau in the $\zeta = f(pH)$ curve. The $\zeta = f(pH)$ of solids that do not contain dissociable groups has no plateau. In this instance, the ζ-potential is determined by the increasing adsorption of H_3O^+/ OH^- ions, respectively (Bismarck et al. 1999a). One should note that even in this instance the iep is located in the acid pH range as OH^- ions are smaller and so polarise the surface more effectively (Zimmermann et al. 2009). However, the point at which $\zeta = 0$ depends on the concentration of ions in the electrolyte solution; thus, if iep depends on the concentration of ions in the electrolyte, the specific adsorption of ions from the electrolyte occurs on the materials surface (Jacobasch et al. 1996b). Moreover, if iep is independent of the concentration of the electrolyte, no specific ion adsorption takes place and it is called the point of zero charge (pzc).

The sign of the ζ-potential (negative or positive) and the position of the iep provide information about the presence of the acidic or basic surface groups. Amphoteric materials, which contain both acid and basic dissociable functional groups, for instance proteins, wool or polyamides, have a sigmoidal shape of $\zeta = f(pH)$.

When the ζ-potential is zero, which is known as the isoelectric point (iep), the surface carries no net electrical charge because the number of negative charges equals the number of positive ones. The position of the iep is determined by the function of the pKa concentration of all functional groups at the surface; if the iep is situated in the low pH range the solid surface has an acidic character whereas an iep at high pH signifies that the surface contains dissociable (Brønsted) basic surface groups (Mader et al. 1994). According to Ottewill and Shaw (Ottewill and Shaw 1967; Healy and White 1978) the pK_a und pK_b-values of functional groups present can be determined for small ζ-potentials using the following equation:

$$pK_a = pH_{\frac{\zeta_{Plateau}}{2}} + \frac{0.4343 F \zeta_{Plateau}}{2RT} \qquad [7.4]$$

where $pH_{\zeta Plateau/2}$ is the pH at $2\zeta = \zeta_{Plateau}$, F = Faraday constant, R = gas constant, and T = absolute temperature. The pK_a of a solid surface is calculated using the measured quantities $pH_{\zeta Plateau/2}$ and $\zeta_{Plateau}$. However, since $\zeta_{Plateau}$ depends on the ionic strength of the electrolyte, it is necessary to determine $\zeta = f(pH)$ for various electrolyte concentrations and to extrapolate the calculated pK_a to zero ionic strength. A full derivation of the thermodynamic approach to quantifying the pK_a of acidic and basic function groups is available (Hunter 1981; Börner and Jacobasch 1985; Jacobasch 1989).

7.3   Electrokinetic properties of natural fibres

Numerous investigations have been conducted to investigate various natural fibres using electrokinetic measurements. However, it should be noted that all properties of natural fibres, such as composition, fibre geometry, mechanical properties, even if extracted from the same crop species, are very dependent on the location the plants were grown at, time of harvest, location within the plant that the fibre came from and the extraction method used. Furthermore, natural fibres are often treated or modified to improve the fibre's properties, which will also influence its surface properties. In conclusion, all natural fibre properties are intrinsically rather variable.

It can be seen from Table 7.1 that electrokinetic properties between different natural fibres, but also the same natural fibres from different regions in the world, do vary widely. This variation is expected as different fibres have different chemical compositions, which determines their surface properties. For example coir and hemp fibres exhibit very different $(\zeta_0 - \zeta_\infty)/\zeta_0$; this is because coir fibres have a high lignin content (up to 45 wt%) and a low cellulose content (as low as 32 wt%), Whereas hemp fibres have a very low lignin content (as low as 3.7 wt%) and a high cellulose content (up to 74 wt%) (Mohanty et al. 2005). Apart from lignin and cellulose content, waxes also make up to 1.7 wt% of the composition of natural fibres and can be mainly found at the surface of the natural fibres. This, in turn, affects the water uptake of natural fibres. The effect that dewaxing of jute, sisal and coir fibres had on the surface properties was studied by Bismarck et al. (Table 7.2) using ζ-potential measurements (Bismarck et al. 2000). It was found that dewaxed jute fibres swell much faster and adsorb more water as shown

Table 7.1 ζ-Potential measurements determined water uptake at 100% relative humidity $(\zeta_0 - \zeta_\infty)/\zeta_0$) and dissociations of functional groups ($\zeta_{Plateau}$) properties of some commonly found natural fibres (adapted from Bismarck et al. 2000; 2001; Baltazar-y-Jimenez and Bismarck, 2007b)

Fibre	$(\zeta_0 - \zeta_\infty)/\zeta_0$	$\zeta_{Plateau}$ (mV)
Abaca bold (Philippines)	0.82	0
Abaca fine (Philippines)	0.76	−0.6
Coir (India)	0.22	−3.8
Flax (Belgium)	0.95	−1.1
Hemp (Central Asia)	0.91	−0.1
Henequen (Mexico)	0.76	−0.6
Lechuguilla (Mexcio)	0.62	−2.5
Lyocell (regenerated wood cellulose)	0.55	−3.2
Jute (India)	0.18	−2.6
Sisal (India)	0.76	−1.7
Sisal (Mexico)	0.88	−0.4

Table 7.2 Influence of dewaxing of jute, sisal and coir fibres on electrokinetic properties

Treatment	Fibre	ζ_0 (mV)	ζ_∞ (mV)	$(\zeta_0 - \zeta_\infty)/\zeta_0$	Moisture content (wt%) (RH = 65%)
None	Jute	−15	−2.8	0.18	–
	Sisal	−3.4	−2.6	0.76	5.8
	Coir	−6.9	−5.4	0.22	8.6
Dewaxed	Jute	0.2	−2.6	14	–
	Sisal	−2.4	−1.5	0.38	6.5
	Coir	−7.0	−5.0	0.28	6.5

by $\zeta = f(t)$ measurements. The natural wax layer of the original jute fibres, consisting of non-water soluble alcohols as well as palmic, oleaginous and stearic acids (Bledzki and Gassan 1997), slows down the water absorption considerably.

The variation of the magnitude of $\zeta_{Plateau}$, obtained from $\zeta = f(pH)$ reflects the hydrophilic nature of the fibres (Ribitsch and Stana-Kleinscheck 1998). The swelling of natural fibres caused by water uptake resulted in a reduction of the measured ζ-potential because the electrical double layer is withdrawn from mechanical interactions with the electrolyte. This, in turn, leads to a small negative plateau value. The negative $\zeta_{Plateau}$ of natural fibres is caused by the presence of dissociable carboxyl and hydroxyl groups. For instance, strong acids such as uronic acids are present in hemicellulose whereas weaker acids can be found in residual lignin (Laine 1994).

The increasing utilisation of natural fibres as reinforcement for polymers caused by the huge cost saving potential, but also the increasing environmental awareness, growing global waste problems etc., requires a good understanding of the fibre surface properties that can lead to green composites with improved mechanical properties. Measurement of ζ-potential is a reliable method for the characterisation of natural fibres. The interface between untreated natural fibres and polymer matrices is usually quite poor (Heng *et al.* 2007). The poor bonding between natural fibres and hydrophobic polymers is usually attributed to the extreme hydrophilic nature of natural fibres and, therefore, extensive research has been focused on the modification of natural fibres to promote better adhesion to intrinsically hydrophobic matrices (Mohanty *et al.* 2005). It should be noted that x-ray photoelectron spectroscopy (XPS) measurements on natural fibres are difficult to perform because of the complex fibre composition and the fact that almost all components present in natural fibres contain carbon, oxygen and hydrogen. Measurements of ζ-potential have been used to characterise alkali, atmospheric air pressure plasma, and fungal treated natural fibres as well as polymethylmethacrylate (PMMA) grafted natural fibres and fibres modified during novel extraction processes, such as Duralin fibres (Bismarck *et al.* 2001; Stamboulis *et al.* 2001; Baltazar-y-Jimenez and Bismarck 2007a; Pickering *et al.* 2007).

Alkali treatment removes some, if not all, of the wax components from the fibre surface via saponification (Ribitsch *et al.* 1996; Bledzki and Gassan 1997). Such treatment increases the active fibre surface simultaneously with accessibility of the dissociable surface groups. For sisal fibres, the $\zeta_{Plateau}$ of the fibres after treatment remains at -1.7 ± 1 mV, which suggests that the surface properties were not affected by alkali treatment.

Atmospheric air pressure plasma treatment of natural fibres (flax, hemp and sisal) led to an increase in positively charged surface functional groups, such as carbonyl and carboxyl groups, at the fibre/electrolyte interface, as shown by the increased $\zeta_{Plateau}$ (Baltazar-y-Jimenez and Bismarck 2007a). This is attributed to the air plasma functionalisation and/or the plasma-induced change in the surface morphology, which resulted in an increased exposure and accessibility of zwitterionic functional compounds (Homola and James 1977; Rendall and Smith 1978).

Stamboulis *et al.* (2001) studied the $\zeta = f(pH)$ of Duralin treated flax fibres. The process to obtain Duralin flax was developed by CERES B.V., which uses simply deseeded flax straw as starting material (Stamboulis *et al.* 2000). More importantly, no retting process is required to extract natural fibres from the straw. A description of the Duralin process can be found in the literature (Stamboulis *et al.* 2000; Stamboulis *et al.* 2001; Bismarck *et al.* 2002). During the Duralin fibre extraction process, hemicellulose and lignin are depolymerised into lower molecular weight aldehyde and phenolic compounds, which recombine into a water-resistant resin, which affects the composition of the flax fibres and thus the water uptake behaviour as well as the $\zeta = f(pH)$ curve. It was observed that as a result of the Duralin fibre extraction process, the surface contained a large number of acidic surface groups. Furthermore, as compared with the control, it was reported that the $\zeta_{Plateau}$ became more negative as the fibres are more moisture sensitive. This upgrading treatment of the flax fibres does not affect the mechanical properties of the fibres.

Pickering *et al.* (2007) applied a fungal treatment to hemp fibres and observed that the $\zeta_{Plateau}$ decreased compared with the original fibres, suggesting that the removal of non-cellulosic materials from the fibre's surface leads to the exposure of more reactive hydroxyl sites. In the meantime it was also reported that fungal treatment led to surface roughening of the fibres, owing to the action of the fungal hyphae, which can produce fine holes in the hemp fibre cell wall, supposedly resulting in more negative $\zeta_{Plateau}$. The removal of non-cellulosic materials also leads to a drop in the tensile strength of the fibres (Li *et al.* 2009).

Methylmethacrylate (MMA) grafting on waxed and dewaxed jute fibres was studied by Bismarck *et al.* (2000). A small decrease in the $\zeta_{Plateau}$ was found when the original fibres were grafted with a large amount of PMMA. However, if the grafting took place after the fibres were dewaxed, a higher grafting density could be achieved. This would lead to a less negative $\zeta_{Plateau}$, implying that a higher pH would be required to protonate all the available ether functions on the surface. A mechanism was proposed (Bismarck *et al.* 2000).

7.4   Conclusion

During the last couple of decades, there has been a renewed interest in using natural fibres as a substitute for synthetic fibres. This motivation is driven by potential advantages of weight savings, low costs, advantages in recycling and ecological aspects. Natural fibres are largely made up of cellulose, hemicelluloses, lignin, pectin, and wax, and much research has been undertaken with the aim of optimising the properties of natural fibres in order to ultilise them more widely as reinforcement for polymers. Surface treatments applicable to natural fibres have been developed and characterised using streaming ζ-potential measurements so that surface properties could be better understood. Natural fibres exhibit a very hydrophilic surface owing to the presence of hydroxyl groups, which is responsible for their unique swelling behaviour. Most natural fibres contain acidic surface functional groups which result in pronounced $\zeta_{Plateau}$ in the alkaline region of $\zeta = f(pH)$. By employing ζ-potential measurements for the characterisation of natural fibres together with other characterisation methods, valuable information about the surface properties of natural fibres can be gathered. The most important outcome is probably the fact that the results of measuring $\zeta = f(t)$ correlate well with the maximum moisture uptake of natural fibres, which makes this type of measurement a fast and reliable method for characterising the swelling behaviour of natural fibres.

7.5   References

Baltazar-y-Jimenez, A. and A. Bismarck (2007a). 'Surface modification of lignocellulosic fibres in atmospheric air pressure plasma.' *Green Chemistry* **9**: 1057–1066.

Baltazar-y-Jimenez, A. and A. Bismarck (2007b). 'Wetting behaviour, moisture up-take and electrokinetic properties of lignocellulosic fibres.' *Cellulose* **14**(2): 115–127.

Beck, U., F. Müller and E. Rohloff (1980). 'Zetapotentialmessungen in der Papierindustrie: Neuere Meßmethoden und Anwendungen.' *Acta Polymerica* **31**: 504–509.

Bellmann, C., A. Caspari, V. Albrecht, T. T. Loan Doan, E. Mäder, T. Luxbacher and R. Kohl (2005). 'Electrokinetic properties of natural fibres.' *Colloids and Surfaces A: Physicochemical and Engineering Aspects* **267**(1–3): 19–23.

Bergmann, E. and M. Voinov (1976). 'Extension of Gouy–Chapman double-layer theory to interface between a liquid and a solid electrolyte'. *Journal of Electroanalytical Chemistry* **67**(2): 145–154.

Bismarck, A., C. Wuertz and J. Springer (1999a). 'Basic surface oxides on carbon fibers.' *Carbon* **37**(7): 1019–1027.

Bismarck, A., D. Richter, C. Wuertz and J. Springer (1999b). 'Basic and acidic surface oxides on carbon fiber and their influence on the expected adhesion to polyamide.' *Colloids and Surfaces A: Physicochemical and Engineering Aspects* **159**(2–3): 341–350.

Bismarck, A., I. Aranberri-Askargorta, J. Springer, A. K. Mohanty, M. Misra, G. Hinrichsen and S. Czapla (2001). 'Surface characterization of natural fibers; surface properties and the water up-take behavior of modified sisal and coir fibers.' *Green Chemistry* **3**(2): 100–107.

Bismarck, A., I. Aranberri-Askargorta, J. Springer, T. Lampke, B. Wielage, A. Stamboulis, I. Shenderovich and H.-H. Limbach (2002). 'Surface characterization of flax, hemp and cellulose fibers: surface properties and the water uptake behavior.' *Polymer Composites* **23**(5): 872–894.

Bismarck, A., J. Springer, A. K. Mohanty, G. Hinrichsen and M. A. Khan (2000). 'Characterization of several modified jute fibers using zeta-potential measurements.' *Colloid and Polymer Science* **278**(3): 229–235.

Bledzki, A. and J. Gassan (1997). *Handbook of engineering polymeric materials*. N. Cheremisinoff. New York, Dekker.

Bledzki, A. K., J. Lieser, G. Wacker and H. Frenzel (1997). 'Characterization of the surfaces of treated glass fibres with different methods of investigation.' *Composite Interfaces* **5**(1): 41–53.

Börner, M. and H.-J. Jacobasch (1985). Elektrokinetische Erscheinungen, Dresden.

Campagne, C., E. Devaux, A. Perwuelz and C. Cazé (2002). 'Electrokinetic approach of adhesion between polyester fibres and latex matrices.' *Polymer* **43**(25): 6669–6676.

Chiu, H. T. and J. H. Wang (1998). 'The relationship between zeta-potential and pull-out shear strength on modified UHMWPE fiber reinforced epoxy composites.' *Polymer Composites* **19**(4): 347–351.

Delgado, A. V., F. Gonzalez-Caballero, R. J. Hunter, L. K. Koopal and J. Lyklema (2007). 'Measurement and interpretation of electrokinetic phenomena.' *Journal of Colloid and Interface Science* **309**(2): 194–224.

Fairbrother, F. and H. Mastin (1924). 'Studies in electro-endosmosis.' *Journal of Chemical Society* **125**: 2319–2330.

Fisher, A. C. (1996). *Electrode dynamics*, Oxford Science Publication.

Flath, H. J. and N. Saleh (1980). 'Zur Bedeutung des Zetapotentials von Faserstoffen für den Färbeprozeß.' *Acta Polymerica* **31**: 510–517.

Grahame, D. (1947). 'The electrical double layer and the theory of electrocapillarity.' *Chemical Reviews* **41**: 441–501.

Grosse, I. and H. J. Jacobasch (1994). 'Grenzflächenchemische Grundlagen der Präparation von Textilfaserstoffen.' *Tenside Surfactants Detergents* **31**: 377–384.

Häßler, R. and H.-J. Jacobasch (1994). 'Wird Adhäsion endlich meßbar?' *Kleben & Dichten: Adhäsion* **38**: 36–38.

Healy, T. W. and L. R. White (1978). 'Ionizable surface group models of aqueous interfaces.' *Advances in Colloid and Interface Science* **9**(4): 303–345.

Helmholtz, H. L. F. v. (1853). 'Ueber einige Gesetze der Vertheilung elektrischer Ströme in körperlichen Leitern mit Anwendung auf die thierisch-elektrischen Versuche.' *Annalen der Physik und Chemie* **89**(222–233): 354–377.

Heng, J. Y. Y., D. F. Pearse, F. Thielmann, T. Lampke and A. Bismarck (2007). 'Methods to determine surface energies of natural fibres: a review.' *Composite Interfaces* **14**(7–9): 581–604.

Homola, A. and R. O. James (1977). 'Preparation and characterisation of amphoteric polystyrene lattices.' *Journal of Colloid and Interface Science* **59**(1): 123–134.

Hunter, R. (1981). *Zeta potential in colloid science.* New York, Academic Press.

Jacobasch, H.-J. (1984). *Oberflächenchemie faserbildender Polymerer.* Berlin, Akademie Verlag.

Jacobasch, H.-J. (1989). 'Characterization of solid-surfaces by electrokinetic measurements.' *Progress in Organic Coatings* **17**(2): 115–133.

Jacobasch, H.-J. (1993). 'Surface phenomena at polymers.' *Makromolekulare Chemie: Macromolecular Symposia* **75**: 99–113.

Jacobasch, H.-J. and I. Grosse (1991). 'Chemische Modifizierung und Charakterisierung von Chemiefaseroberflächen.' *Chemiefasern und Textilindustrie* **93**: 1294–1300.

Jacobasch, H.-J., F. Simon, C. Warner and C. Bellmann (1996a). 'Determination of the zeta potential from streaming potential and streaming current measurements.' *Technisches Messen* **63**(12): 447–452.

Jacobasch, H.-J., F. Simon, C. Werner and C. Bellmann (1996b). 'Electrokinetic measuring methods: Principles and applications.' *Technisches Messen* **63**(12): 439–446.

Jacobasch, H. J., F. Simon and P. Weidenhammer (1998). 'Adsorption of ions onto polymer surfaces and its influence on zeta potential and adhesion phenomena.' *Colloid and Polymer Science* **276**(5): 434–442.

Jacobasch, H.-J., G. Bauböck and J. Schur (1986). 'Comparative measurements of zeta-potential of fibers by electroosmosis and streaming current streaming potential.' *Monatshefte für Chemie* **117**(10): 1133–1144.

Jacobasch, H.-J., K. Grundke and C. Werner (1995). 'Surface characterization of cellulose materials.' *Papier* **49**(12): 740–745.

Jacobasch, H.-J., K. Grundke, E. Mäder, K.-H. Freitag and U. Panzer (1992). 'Application of surface free-energy concept in polymer processing.' *Journal of Adhesion Science and Technology* **6**(12): 1381–1396.

Kanamaru, K. (1960). 'Wasseraufnahme in ihrer Beziehung zur zeitlichen Erniedrigung des ζ-Potentials von Fasern in Wasser.' *Colloid & Polymer Science* **168**(2): 115–121.

Laine, J. (1994). Surface properties of unbleached kraft pulp fibres, determined by different methods, *Laboratory of Forest Products Chemistry Report.*

Li, Y., K. L. Pickering and R. L. Farrell (2009). 'Analysis of green hemp fibre reinforced composites using bag retting and white rot fungal treatments.' *Industrial Crops and Products* **29**(2–3): 420–426.

Lyklema, H. (1981). *Colloidal Dispersions.* London, Royal Society of Chemistry.

Mäder, E., K. Grundke, H.-J. Jacobasch and G. Wachinger (1994). 'Surface, interphase and composite property relations in fibre-reinforced polymers.' *Composites* **25**(7): 739–744.

Mohanty, A. K., M. Misra and L. T. Drzal Eds. (2005). *Natural Fibers, Biopolymers, and Biocomposites*, CRC Press.

Ottewill, R. H. and J. N. Shaw (1967). 'Studies on the preparation and characterisation of monodisperse polystyrene lattices.' *Colloid & Polymer Science* **218**: 34-40.

Pickering, K. L., Y. Li, R. L. Farrell and M. Lay (2007). 'Interfacial modification of hemp fiber reinforced composites using fungal and alkali treatment.' *Journal of Biobased Materials and Bioenergy* **1**(1): 109–117.

Rätzsch, M., H.-J. Jacobasch and K.-H. Freitag (1990). 'Origin, characterisation and technological effects of interfaction forces proceeding from polymer surfaces.' *Advances in Colloid and Interface Science* **31**(3–4): 225–320.

Rendall, H. M. and A. L. Smith (1978). 'Surface and electrokinetic potentials of interfaces containing 2 types of ionizing group.' *Journal of the Chemical Society-Faraday Transactions 1* **74**: 1179–1187.

Ribitsch, V. and K. Stana-Kleinscheck (1997). 'Characterization of textile fiber surfaces by streaming potential measurements.' *Beitrag zum Zeta Potential Workshop*, Graz, Austria.

Ribitsch, V. and K. Stana-Kleinscheck (1998). 'Characterizing textile fiber surfaces with streaming potential measurements.' *Textile Research Journal* **68**(10): 701–707.

Ribitsch, V., K. Stana Kleinschek and S. Jeler (1996). 'The influence of classical and enzymatic treatment on the surface charge of cellulose fibres.' *Colloid and Polymer Science* **274**(4): 388–394.

Rybicki, E. and H.-J. Jacobasch (1992). 'Electrokinetic studies on synthetic fibres in surfactant solutions.' *Tenside Surfactants Detergents* **29**: 311–314.

Stamboulis, A., C. A. Baillie and T. Peijs (2001). 'Effects of environmental conditions on mechanical and physical properties of flax fibers.' *Composites Part A–Applied Science and Manufacturing* **32**(8): 1105–1115.

Stamboulis, A., C. A. Baillie, S. K. Garkhail, H. G. H. van Melick and T. Peijs (2000). 'Environmental durability of flax fibres and their composites based on polypropylene matrix.' *Applied Composite Materials* **7**(5–6): 273–294.

Stana-Kleinschek, K. and V. Ribitsch (1998). 'Electrokinetic properties of processed cellulose fibers.' *Colloids and Surfaces A: Physicochemical and Engineering Aspects* **140**(1–3): 127–138.

Stern, O. (1924). 'The theory of the electrolytic double-layer' *Zeitschrift für Elektrochemie und Angewandte Physikalische Chemie* **30**: 508–516.

Uchida, E. and Y. Ikada (2002). 'Zeta-potential of polymer surfaces'. *Encyclopedia of surface and colloid science*. A. Hubbard. New York, Marcel Dekker: 5657.

Werner, C., H.-J. Jacobasch and G. Reichelt (1995). 'Surface characterisation of hemodialysis membranes based on streaming potential measurements.' *Journal of Biomaterials Science: Polymer Edition* **7**(1): 61–76.

Zimmermann, R., N. Rein and C. Werner (2009). 'Water ion adsorption dominates charging at nonpolar polymer surfaces in multivalent electrolytes.' *Physical Chemistry Chemical Physics* **11**(21): 4360–4364.

8
Mechanical assessment of natural fiber composites

P. J. HERRERA-FRANCO and
A. VALADEZ-GONZÁLEZ, Yucatan
Center for Scientific Research, Mexico

Abstract: The effect of the interfacial fiber–matrix interactions on the mechanical performance of a natural fiber reinforced thermoplastic–polymer matrix composite is reviewed. Two static testing modes are used, namely, the tensile and the shear loading modes. A well characterized fiber–matrix model system consisting of high-density polyethylene reinforced with henequen fibers (*Agave fourcroydes*) was selected. Special emphasis is given to the material behaviour and the state of stress for each loading, in order to better assess the fiber–matrix interface effects on the mechanical properties and failure modes.

Key words: natural fibers, composites, interfacial interactions, mechanical properties, tensile properties, shear properties.

8.1   Introduction

New government environmental regulations and a growth in society's environmental awareness and concern about the greenhouse effect have stimulated the search for sustainable materials[1–3] in order to replace conventional synthetic polymeric fibers in the construction, automotive, and packing industries. Natural fibers are being increasingly used as reinforcements for the production of low-cost and lightweight polymer composites. Natural fibers are cheap, readily available and renewable and can be extracted from plant leaves and the stems of fruits at very low costs. Also, some fibers are residues of agroindustrial processes such as the residues from the processing of pineapple,[4] banana,[5] rice,[6] coconut[7] and sugar cane.[8] Natural fibers offer other advantages as reinforcing materials in polymer matrix composites such as: a non-abrasive nature, high specific mechanical properties, and biodegradability. However, their limitations include: moisture absorption, poor wettability by polymeric resins, large scattering in mechanical properties, and poor understanding of the mechanisms that control their mechanical behavior and failure modes. Because of these limitations, the use of natural fiber reinforced composites is still confined to non-structural applications.[9]

The experimental characterization of composite materials, specifically,

222

the characterization of their mechanical behavior has been an elusive topic, because it has been a continuously evolving one. The wide use of man-made fiber composite materials in highly specialized fields, such as the aeronautical and aerospace applications has forced a continuous revision of the subject of mechanical testing. As a result, new test methods have been developed and the existing techniques have been continuously verified and reexamined. The results found in the literature for the characterization of natural fiber reinforced composites (NFRC) have mainly relied on test protocols developed for metals, wood, polymers and advanced fibrous composites. Issues such as the non-homogeneity and anisotropy of unidirectional and multiply oriented NFRC laminates are also present. Also, the inherent variability of the geometrical, physical, and mechanical properties of natural fibers makes the testing and interpretation of test results for NFRC more difficult.

In recent years, the composites scientific community has devoted considerable efforts to improve the interfacial compatibility between natural fibers and several traditional polymeric matrices (thermoplastics or thermosets) and biodegradable polymer matrices. The mechanical and physical properties of the composite are generally dependent on the fiber content. Several physical and chemical treatment methods that improve the fiber matrix adhesion have been proposed. These different treatments change amongst others, the hydrophilic character of the natural fibers, so that the mechanical properties are improved and the moisture absorption effects in the composite are reduced.

In this chapter, a review of the effect of the interfacial fiber–matrix interactions on the mechanical performance of a natural fiber reinforced thermoplastic–polymer matrix composite under two static testing modes, namely, the tensile and the shear loading modes is performed. Special emphasis on the material behavior and the state of stress is made for each loading mode, in order to better assess the fiber–matrix interface role.

8.2   Materials and experimental procedures

A fiber–matrix model system consisting of high density polyethylene (HDPE) reinforced with henequen fibers (*Agave fourcroydes*) was selected. Details of several studies that have been conducted in the last few years for this fiber–matrix system, including a systematic physicochemical, micromechanical and mechanical experimental program have been reported.[10,11] HDPE (Petrothene) extrusion grade from Quantum Chemical Inc. with a melt flow index of 0.33 g (10 min)$^{-1}$ and a density of 0.96 g cm^{-3} (determined using the ASTM standards D-1238-79 and D-792-86, respectively) and a melting point of 135 °C determined in a DSC-7 Perkin Elmer calorimeter was used as thermoplastic matrix. Henequen fibers with an average diameter of 180 μm and an average length of 6 mm, (Desfibradora Yucateca, S.A., DESFIYUSA Co. of Mérida,

Yucatán, México were used as reinforcement. For the various surface treatments, sodium hydroxide and xylene, reagent grade from Técnica Química S.A., and vinyltris (2-methoxyethoxy) silane (Silane A-172) from Union Carbide as a coupling agent were used. Dicumyl peroxide from Polyscience was used as a catalyst in the reaction between the silane coupling agent and the HDPE.

8.2.1   Physicochemical modification of the fiber surface

The fiber surface treatments, aimed at increasing the fiber–matrix interactions, were denoted as follows:

Physical surface treatment

Increased mechanical interlocking:

(a) henequen fibers without any surface treatment (FIB);
(b) henequen fibers treated with an aqueous NaOH solution (FIBNA);
(c) henequen fibers treated with an aqueous NaOH solution and then pre-impregnated with dissolved HDPE in xylene (FIBNAPRE).

Chemical surface treatment

Chemical bonding between fiber and matrix:

(a) henequen fibers treated with an aqueous NaOH solution and then with a silane coupling agent (FIBNASIL);
(b) FIBNASIL plus the pre-impregnation with dissolved HDPE (FIBNASILPRE).

8.2.2   Composites processing

A 20% v/v fiber content HDPE–henequen fiber composite was chosen in order to determine the effect of the different fiber surface treatments on its mechanical properties. The fibers were incorporated into the HDPE matrix at 180 °C using a Brabender plasticorder intensive mixer, model PL330. The resulting material was compression molded at a pressure of 1 tonne using a Carver laboratory press at a temperature of 180 °C. The specimens for the mechanical test were obtained from these laminates according to the ASTM standards.

8.2.3   Specimen conditioning and test conditions

The behavior of a fiber-reinforced composite specimen in testing can be determined by its past history to a considerable extent. For NFRC, the specimen not only preserves a 'memory' of the production method and the storage time, but of the conditions directly before the tests. Therefore, the reproducibility of the manufacturing conditions, the environmental conditioning before tests and during testing must be ensured. In brief, time, temperature and relative humidity should be recorded before and during the tests.

8.3   Mechanical testing

8.3.1   Tensile behavior

Experimental procedures

The tensile properties of the untreated and treated NFRC are usually determined as a function of fiber content. Also, the tensile strength and Young's modulus are reported as a function of the fiber surface treatment, that is, as a function the effect of the fiber-surface modification on the adhesion between the fiber and the matrix.[12–42]

The uni-axial tensile test is the most widely used and studied mechanical test for composites. Its popularity as a test method is mainly attributed to its simplicity of procedure and ease of processing and analysis of the test results. The characteristics obtained from uni-axial tension are for both material specification and estimation of load-carrying capacity. However, despite its apparent simplicity, the tension test is prone to a series of problems owing to the microstructure and properties of the fibrous polymeric composites. The main difficulty in achieving an acceptable tensile test typically increases as the anisotropy of the material increases, i.e., as the ratio of the axial stiffness (or strength) to the transverse stiffness (or strength) increases.

On the experimental side, the proper introduction of the applied force to the test specimen is very important to ensure a successful test. Pneumatically actuated or mechanical clamping grips are perhaps the most commonly used so that the tensile force is introduced to the specimen via a shear force at the clamp–specimen interface. This shear force is equal to the clamping force times the effective coefficient of friction at the interface. Care should be taken to avoid penetration of the rough grip surface in the sample, to prevent premature specimen failure. Standard typical test-sample dimension for this type of test from ASTM D638 and ASTM D3039[43,44] are shown in Fig. 8.1. The cross-head speed is selected according to the stiffness of the material to stay in a quasi-static loading mode. The elastic properties are determined from the stress–strain curve. The initial slope is equal to the elastic stiffness or modulus of elasticity (Young's Modulus) in the direction of the applied load

$$E_1 = \frac{\sigma_1}{\varepsilon_1}$$ [8.1]

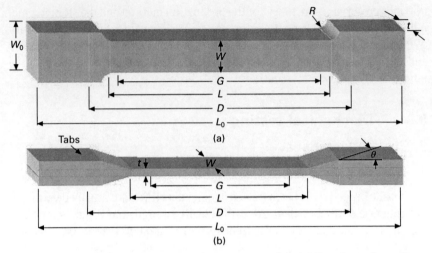

8.1 Typical tensile test specimen geometries. (a) Dog-boned tensile specimen; (b) straight-sided specimen with tapered tabs. Dimensions: *W* width of narrow section; *L* length of narrow section; W_0 width overall; L_0 length overall; *G* gage length; *D* distance between grips; *t* thickness of specimen; θ tab bevel angle; *R* radius of fillet.

For anisotropic materials, a second test should be performed with specimens fabricated at 90° with respect to the first material direction of material symmetry to obtain the second stiffness value E_2. From the measured longitudinal and transverse strains ε_1 and ε_2, either by means of a biaxial extensometer or with strain gage rosettes, the main Poisson's ratio v_{12} can be determined from the negative ratio $(\varepsilon_1-\varepsilon_2)$. Also, the longitudinal and transverse tensile strengths, σ_1 and σ_2 are defined as the ultimate values of stress for the 0° and 90° tensile tests, respectively. The ultimate strains ε_1 and ε_2 are the strains corresponding to σ_1 and σ_2.

Tensile test results

Typical stress–strain curves from the tensile tests for the short fiber henequen–HDPE composite are shown in Fig. 8.2. As expected, there is no noticeable effect on the elastic modulus as a result of the different interface interactions between the HDPE and the henequen fibers (Table 8.1). It is well known that the fiber reinforcing effect is most efficient along the fiber axis orientation. However, the processing technique used to produce the composite, will dictate the final fiber-orientation distribution, which is one of the most important characteristics that determines the effective mechanical properties. The anisotropy expected from an injection-molded part is different from that found in extrusion- or compression-molded parts. Furthermore, the mathematical prediction of the stiffness in a composite stiffness is well

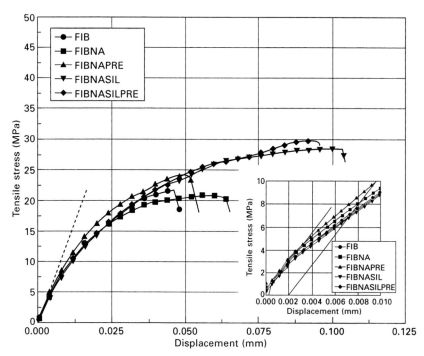

8.2 Representative curves of tensile stress versus strain for henequen–HDPE composites for various fiber surface treatments. Insert shows the slope of line used to calculate the yield stress at $\varepsilon = 0.2\%$.

Table 8.1 Tensile properties of a short fiber henequen–HDPE composite for various different surface treatments

Type of fiber surface treatment	Tensile strength (MPa)	Elastic modulus (MPa)	Strain to failure (%)
FIB	22.63 ± 1.06	653.77 ± 17.06	5.29 ± 0.76
FIBNA	21.11 ± 1.15	666.09 ± 13.35	5.11 ± 0.68
FIBNAPRE	24.40 ± 1.9	679.82 ± 14.18	5.66 ± 0.64
FIBNASIL	29.14 ± 1.72	671.85 ± 10.11	9.01 ± 1.17
FIBNASILPRE	29.89 ± 1.81	674.33 ± 14.42	7.89 ± 1.43

established for common fiber orientation distributions, i.e., parallel to the test direction, in a transverse direction to the test direction and when the fibers are randomly oriented.

As shown in Fig. 8.2, beyond the linear elastic region, each fiber treatment results in a different composite behavior and, therefore, the fiber–matrix compatibilization has an effect on the strength and elongation of the composite. The type of treatment of the natural fiber, modifies the reinforcing capability of the natural fibers and, consequently, the behavior of the composite in the nonlinear region. At this point, it is convenient to separate the different parameters that govern the tensile strength behavior of the composite. First, shear loading of the matrix is one of the coupling mechanisms that transfers loads between the individual fibers and thus is a general way of distributing the load through the material. For unidirectional composites, the shear behavior is dictated by the resin and in the case of randomly oriented fibers, the fiber–matrix adhesion also comes into play in the load transfer mechanism. For short-fiber reinforced polymers, the existence of stress concentrations at the fiber ends also contributes to the strength of composite, because they are points of possible failure initiation, either by crack propagation along the fiber–matrix interface or through the matrix or failure by matrix and/or interface yielding.

It is known that alkaline treatments result in an improvement in the interfacial bonding by giving rise to additional sites of mechanical interlocking, hence promoting more resin/fiber interpenetration at the interface but with due precaution to avoid fiber degradation. In this instance, this treatment is reflected as a slightly higher elongation capability of FIBNA as compared to FIB. When the fiber surface was modified with an aqueous NaOH solution followed by a resin pre-impregnation process (FIBNAPRE), a marginal improvement of the tensile strength of the composite of approximately 11% with respect to the untreated-fiber composite was observed. In these three cases, the interfacial interactions are mainly physical, i.e., these are frictional and/or weak interfacial physical and chemical interactions. Therefore, any stress concentration at the fiber ends should be withstood by such fiber–matrix interactions. However, when the same alkaline surface modified fiber was treated with a silane coupling agent (FIBNASIL and FIBNASILPRE), a 30% increase in the relative tensile strength and a 100% increase in the elongation capability was observed. This increment in the tensile strength and elongation is attributed to chemical interactions between cellulose fibers–silane coupling agent–HDPE matrix. In this instance, even when the HDPE has been loaded well beyond its yield strength, the covalent bonds between the coupling agent and the polymer resin are more difficult to break and, thus, even when the matrix has yielded and its molecular chains are oriented, matrix failure occurs at high tensile strain. It should be pointed out that the matrix shear yield stress, is below the composite tensile strength which is approximately equal to 7.9 MPa at 0.2% strain.

Stress analysis

From the traditional approach of engineering mechanics of solids used to determine the internal stress on arbitrary inclined sections in axially loaded bars (Fig. 8.3), the state of stress is given by equation [8.2] as follows:

$$\sigma_{ij} = \begin{pmatrix} \sigma\cos^2\theta & \sigma\sin\theta\cos\theta \\ \sigma\sin\theta\cos\theta & \sigma\cos^2\theta \end{pmatrix}$$ [8.2]

where σ is the normal tensile stress applied to the bar. Strictly speaking, upon application of an axial load, on planes which are inclined with respect to the load axis, there are also shear stress components. When θ is equal to zero, σ_{ij} = σ = P/A and for a non-zero value of θ, there is a shear stress component ($\sigma\sin\theta\cos\theta$) whose maximum value occurs in planes located at ±45°.

Effect of the interface on the tensile failure mode

Even when an axial tensile load is being applied to the composite sample, it should be remembered that some materials are weaker to shearing loads than to tensile loads. Thus, upon a load increase, yielding may occur along planes where the shear stress attains a maximum value and this type of failure is referred to as 'a ductile failure'. Failure of the tensile test sample is assumed to occur in a plane transverse to the applied load. However, observation of the failure surfaces and the micro photographs indicates that for an increasing degree of adhesion between fiber and matrix, the failure mode changes from fiber pull-out and 'brittle like' failure to a combination of matrix yielding and/or tearing to a shear-type failure as depicted in Fig. 8.3a and 8.3b. Also, for low stiffness materials, yielding of the matrix together with fiber splitting are indicative that the shear stress component is also responsible for failure. As depicted in Fig. 8.3b and 8.3c, the equivalent stress of a tensile stress for a plane rotated 45° with respect to the direction of loading is represented by shear and normal stress components. Shear-induced failure is very strongly influenced by the matrix and then it can be said that shear induced failure is matrix dominated, with some possible contribution from fiber failures by splitting (Fig. 8.4). Then, it can be stated in general that the increase in the tensile strength of the composite is an effect of the fiber–matrix interface. Also, for low fiber–matrix adhesion, the failure mode is dominated by fiber pull-out and matrix failure and for higher fiber–matrix adhesion, the failure mode is more like matrix tearing and flow and fiber tearing. Therefore, the feeble resistance of short fiber composites to stresses, especially those oriented transverse to the fibers is mainly the result of the relatively low strength of the matrix. In this case, for HDPE, with a tensile strength of 25.0 MPa, the effect of the interface strength is seen to have an effect and that an upper bound for the composite tensile strength is imposed by the properties of the matrix. Again, the fiber–matrix interfacial failure modes observed at the micro-mechanical level can be translated to the macro-mechanical behavior of the composite. The role of the covalent bonds formed because of the coupling agent, in particular at high stress and strain values is important

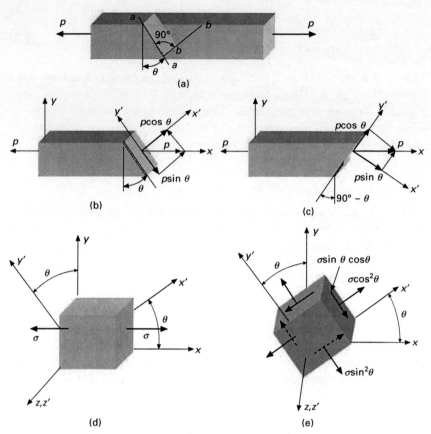

8.3 (a) Sectioning of a prismatic bar loaded in uniaxial tension; (b) normal and shear load components on two arbitrary orthogonal planes; (c) shear and normal components of tensile stress of a plane rotated 45°; (d) normal axial stresses acting on the prismatic bar; (e) normal and shear stress components acting on arbitrary orthogonal planes.

considering that the tensile and shear strength properties of the matrix are a limiting value and that the composite is able to withstand larger strains than the physically-modified henequen fiber HDPE composites.

8.3.2   Shear behavior

Iosipescu shear testing

In 1967, Iosipescu proposed an in-plane shear testing method for metal with a double-notched beam shape.[41] Since then, this method has been widely used to measure the shear stress/shear strain relation of such materials as fiber-reinforced plastics (FRP). The configuration and specimens are shown

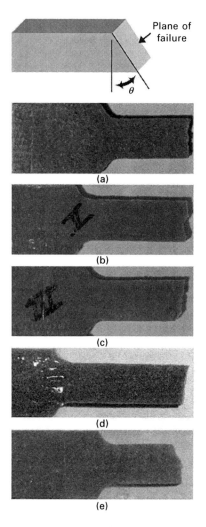

8.4 Photographs of fracture surfaces of a HDPE/henequen fiber (80:20 v/v) composite subjected to tensile stresses. (a) FIB, (b) FIBNA, (c) FIBNAPRE, (d) FIBNASIL and (e) FIBNASIL. Notice the change of angle of failure with respect to load direction, that is with increasing fiber surface treatment.

in Fig. 8.5. This test method is described in ASTM D-5379 standard[42] using a Wyoming shear test fixture adapted to a universal testing machine. One of the conditions offered by the Iosipescu shear test has been the uniform shear load in the double notch specimen cross-section. This, in turn, results in a uniform shear stress condition under an in-plane condition. The average shear stress across the notched section of the specimen is calculated using the simple formula:

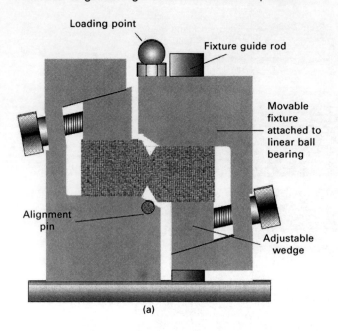

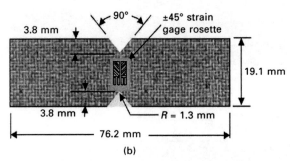

(b)

8.5 Configuration and specimen for the shear test method as described in ASTM D-5379 standard using a Wyoming shear test fixture.

$$\tau_{12} = \frac{P}{A} \qquad\qquad [8.3]$$

where P is the applied force and A is the cross-sectional area of the specimen between the notches.

Experimental procedures

In this instance, the tests were performed after conditioning the samples at 25 °C. The Iosipescu shear test specimens were cut from the laminates previously obtained, and the dimensions of specimens were 76 mm of length, 19 mm of wide, 2 mm of thickness and the distance between the two 90° notches was 12 mm. The cross-speed used on the universal testing machine (Shimadzu Model AG1) equipped with a 5 kN load cell was 0.5 mm min^{-1}. The average shear stress across the notched section of the specimen is calculated using equation [8.3].

Stress analysis and failure modes

Fiber reinforced thermoplastic polymer composites have a typical shortcoming: their low shear resistance, especially in planes where the properties are determined by the matrix and the interfacial shear strength. A low shear resistance refers to both low shear modulus and low shear strength. The main difficulty in determining shear properties of a material is to provide a pure shear load in the specimen and thereby assure sufficient precision of the method and the processing of experimental results.

Again, using a simple approach of engineering mechanics, when a body is subjected to pure shear loads, as in the Iosipescu shear specimen, other planes, inclined with respect to the plane of applied load, also contain tensile stresses, as follows:

$$\sigma_{ij} = \begin{pmatrix} 2\tau_{12}\sin\theta\cos\theta & \tau_{12}(\cos^2\theta - \sin^2\theta) \\ \tau_{12}(\cos^2\theta - \sin^2\theta) & -2\tau_{12}\sin\theta\cos\theta \end{pmatrix} \qquad [8.4]$$

As shown in Fig. 8.6, there is a normal tensile stress component $(2\tau_{12}\sin\theta\cos\theta)$ acting in one plane and a compressive stress $(-2\tau_{12}\sin\theta\cos\theta)$

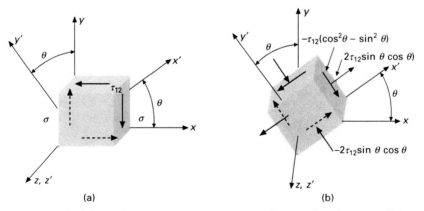

8.6 (a) Pure shear stress components acting on the plane parallel to the double notch specimen cross-section; (b) normal and shear stress components acting on a plane oriented at an angle θ to the double notch specimen cross-section.

at 90° from the first. The shear stress component is equal to $\tau_{12}(\cos^2\theta - \sin^2\theta)$ on both planes.

Effect of the interface on the shear failure mode

Typical force–displacement curves from the Iosipescu shear tests for the short fiber henequen–HDPE composites are shown in Fig. 8.7. The shear strength of the HDPE–henequen fiber composites for the various fiber surface treatments is shown in Table 8.2. Similar observations made earlier for the tensile strength on the effect of fiber–matrix adhesion are also evident

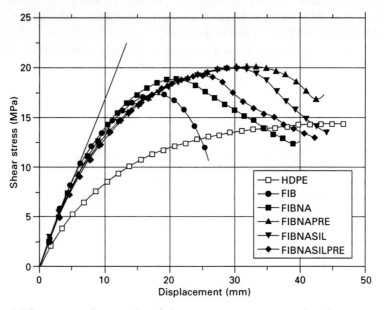

8.7 Representative graphs of shear stress versus cross-head displacement for henequen–HDPE composites for various fiber surface treatments. The graph for neat HDPE is included for comparison with composites.

Table 8.2 Shear properties of a short fiber henequen–HDPE composite for various different surface treatments

Type of fiber surface treatment	Shear strength (MPa)
FIB	17.17
FIBNA	18.90
FIBNAPRE	19.19
FIBNASIL	19.90
FIBNASILPRE	20.15

here. The stress–displacement curves also show a linear–elastic behavior only applicable in a very small range of strain and for larger strains, the stress–strain graph shows a nonlinear elastic behavior. Within a well-defined stress–displacement region, the curves level off markedly towards the displacement.

The nonlinear behavior is noticed in all fiber–surface treatments. This behavior is nonreversible and thus the material is nonlinear viscoelastic. When compared with the pure HDPE resin, the addition of the fibers changes the rigidity of the composite, although it still exhibits the same linear anelastic and the nonlinear viscoelastic behavior. The linear region of the material behavior is observed for stresses close to 7.5 MPa. The fibers with no surface treatment (FIB) show a higher strength than the neat HDPE but the maximum shear stress failure and ductility are considerably reduced. The effect of increasing the fiber–surface area of contact with the matrix seems to improve the shear strength and the ductility of the material. Fiber preimpregnation by itself has no great contribution to the shear strength. The largest increase in shear strength is observed for the fiber treated with both the aqueous alkaline solution and the silane coupling agent. From previous observations of the tensile properties for this fiber surface treatment combination, increments of approximately 20 and 30%, respectively, were obtained. For shear strength, the increase is of the order of 25%.

As mentioned earlier, the inclusion of natural fibers in the polymeric matrix changes the stiffness or rigidity of the material. It is also known that the brittle or ductile character of a material is relevant to the mechanism of failure. Therefore, it is expected that the inclusion of natural fibers in the polymer will result in a decrease of the ductility of the material. However, a clear distinction between a ductile and brittle character or the transition from one to the other is not as simple as might be inferred. The nature of the stress field, testing temperature and the material inhomogeneity itself all play a role in defining the boundary between ductility and brittleness.

As observed in Fig. 8.8a, from the Iosipescu shear test, yielding could occur in the plane joining the two notches. From a finite element model, the shear stress component is not uniform as might be thought from the applied loads (Fig. 8.8b). At the notch roots there is a slightly higher shear stress level than that observed at points located along the center line joining the v-notches of the test specimen. Thus, if yielding is possible, the notch root is a point of possible yield failure initiation. In some instances, a crack was initiated at the notch root, at a point located slightly off the centerline between notches (Fig. 8.8c). This crack was initiated after yielding along the center line between notches and at points located slightly off. Also, according to the finite element model, this point of crack initiation corresponds to a point of maximum normal stress in the sample (Fig. 8.8d). It should be added that the crack starts growing when there is high stress–strain on the specimen.

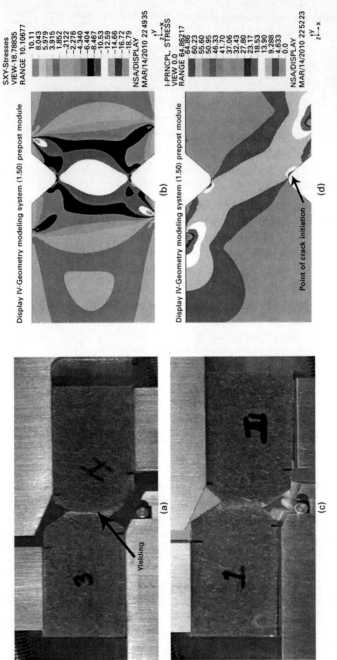

SXY-Stresses
VIEW-18.78835
RANGE 10.10677

10.11
8.043
5.979
3.915
1.852
-.2122
-2.276
-4.340
-6.404
-8.467
-10.53
-12.59
-14.66
-16.72
-18.79

NSA/DISPLAY
MAR/14/2010 22.4935

Display IV-Geometry modeling system (1.50) prepost module

(b)

I-PRNCPL, STRESS
VIEW 0.0
RANGE 64.86217

64.86
60.23
55.60
50.95
46.33
41.70
37.06
32.43
27.80
23.17
18.53
13.90
9.288
4.633
0.0

NSA/DISPLAY
MAR/14/2010 22.5223

Point of crack initiation

Display IV-Geometry modeling system (1.50) prepost module

(d)

Yielding

(a)

(c)

8.8 Failure modes in Iosipescu test specimens: (a) sample showing yielding along line joining the v-notches; (b) finite element model of Iosipescu shear specimen showing the stress band contours of shear stress, the white area at the center is the possible site of shear failure; (c) sample showing yielding along line joining the v-notches and incipient crack growth on both sides of v-notches; (d) finite element model of Iosipescu shear specimen showing the stress band contours of principal normal stress, the possible site of crack opening is shown.

As for tensile specimens, the interfacial conditions play a very important role in the overall properties of the composites. The physical modifications of the fiber surface contributed to the reinforcing capability of the natural fibers. However, the larger improvement of strength and the elongation capability of the composites were related to the chemical fiber surface treatment. The covalent bonds obtained from the use of a coupling agent result in better mechanical properties of the composites.

8.4   Conclusions

A study of the effect of the fiber–matrix interfacial behavior on the tensile and shear behavior of a model system consisting of HDPE reinforced with henequen fibers (*Agave fourcroydes*) was performed. Some of the aspects of the mechanical testing program that included loading modes in tension and shear were reviewed and the most important aspects of each test mode were revised. It has become evident that the difficulty in establishing a pure loading mode to obtain a particular answer is enhanced by the geometry of the sample, the complexity of the material's microstructure, failure modes of both fiber and matrix and most important, the interfacial fiber–matrix conditions.

The mechanics of each test is revised and related to the events observed on each test and on the stiffness and failure modes. Also, the different fiber surface conditions and the ensuing fiber–matrix interface were related to the failure modes. It was observed that the physical modifications of the fiber contributed to the reinforcing capability of the natural fibers. However, the improvement of strength and the elongation capability of the composites were related to the chemical fiber surface treatment. The covalent bonds formed from the use of a coupling agent result in a greater improvement in the mechanical properties of the composites. Therefore, from these test methods it is possible to establish a structure–property relationship between the fiber–matrix adhesion levels and the strength and failure modes for each loading mode.

8.5   References

1. Rowell, R. M., The state of art and future development of bio-based composite science and technology towards the 21st century, *Proceedings, the fourth Pacific Rim bio-based composites symposium*, November 2–5, 1998, Indonesia. p. 1–18.
2. Reddy N. and Yang Y., Biofibers from agricultural byproducts for industrial applications, *Trends in Biotechnology*, 23(1), 22–27, 2005.
3. Li, X., Tabil, L., Panigrahi, S., Chemical treatments of natural fiber for use in natural fiber-reinforced composite: a review. *Journal of Polymers and the Environment*, **15**(1), 25–33, 2007.
4. Samal, R. and Bhuyan, L., Chemical modification of lignocellulosic fibers. I.

Functionality changes and graft copolymerization of acrylonitrile onto pineapple leaf fibers; their characterization and behavior, *Journal of Applied Polymer Science*, **52**, 1675–1685, 1994.

5. Joseph S., Oommen Z., Thomas S., Environmental durability of banana-fiber-reinforced phenol formaldehyde composites, *Journal of Applied Polymer Science* **100**(3): 2521–2531, 2006.

6. Kamel S., Preparation and properties of composites made from rice straw and poly(vinyl chloride) (PVC), *Polymers For Advanced Technologies* **15**(10): 612–616, 2004.

7. Brahmakumar M., Pavithran C., Pillali R. M., Coconut fiber reinforced polyethylene composites: effect of natural waxy surface layer of the fiber on fiber–matrix interfacial bonding, *Composites Science and Technology*, **65**, 563–569, 2005.

8. García-Hernández E., Licea-Claverie A., Zizumbo A., Herrera-Franco, P., Improvement of the interfacial compatibility between sugar cane bagasse fibers and polystyrene for composites, *Polymer Composites* **25**(2): 134–145, 2004.

9. Marsh, G., Next step for automotive materials, *Materials Today*, April, 36–43, 2003.

10. Valadez-Gonzalez, A., Cervantes-Uc, J. M., Olayo, R., Herrera-Franco, P.J., Effect of fiber surface treatment on the fiber–matrix bond strength of natural fiber reinforced composites, *Composites Part B*, **39**(3): 309–320, 1999.

11. Valadez-Gonzalez, A., Cervantes-Uc, J. M., Olayo, R., Herrera-Franco, P.J., Chemical modification of henequen fibers with an organosilane coupling agent, *Composites Part B*, **39**(3), 321–331, 1999.

12. Gu, H., Tensile behaviours of the coir fiber and related composites after NaOH treatment, *Materials & Design*, **30**(9), 3931–3934, October 2009.

13. Park, J.-M., Quang, S. T., Hwang, B.-S., DeVries, K. L., Interfacial evaluation of modified jute and hemp fibers/polypropylene (PP)–maleic anhydride polypropylene copolymers (PP–MAPP) composites using micromechanical technique and nondestructive acoustic emission, *Composites Science and Technology*, **66**(15), 2686–2699, December 2006.

14. Facca, A. G., Kortschot, M. T., Yan, N., Predicting the tensile strength of natural fiber reinforced thermoplastics, *Composites Science and Technology*, **67**(11–12), 2454–2466, September 2007.

15. Brahmakumar, M., Pavithran, C., Pillai, R. M., Coconut fiber reinforced polyethylene composites: effect of natural waxy surface layer of the fiber on fiber/matrix interfacial bonding and strength of composites, *Composites, Science and Technology*, **65**(3–4), 563–569, March 2005.

16. Gomes, A., Matsuo, T., Goda, K., Ohgi, J., Development and effect of alkali treatment on tensile properties of curaua fiber green composites, *Composites Part A: Applied Science and Manufacturing*, **38**(8), 1811–1820, August 2007.

17. Bachtiar, D., Sapuan, S. M., Hamdan, M. M., The effect of alkaline treatment on tensile properties of sugar palm fiber reinforced epoxy composites, *Materials & Design*, **29**(7), 1285–1290, 2008.

18. Demir, H., Atikler, U., Balköse, D., Tıhmınlıoğlu, F., The effect of fiber surface treatments on the tensile and water sorption properties of polypropylene–luffa fiber composites, *Composites Part A: Applied Science and Manufacturing*, **37**(3), 447–456, 2006.

19. Beckermann, G. W., and Pickering, K. L., Engineering and evaluation of hemp fiber reinforced polypropylene composites: fiber treatment and matrix modification,

Composites Part A: Applied Science and Manufacturing, **39**(6), 979–988, 2008.

20. Bax, B., and Müssig, J., Impact and tensile properties of PLA/Cordenka and PLA/ flax composites, *Composites Science and Technology*, **68**(7–8), 1601–1607, 2008.

21. Graupner, N., Herrmann, A. S., Müssig, J., Natural and man-made cellulose fiber-reinforced poly(lactic acid) (PLA) composites: an overview about mechanical characteristics and application areas, *Composites Part A: Applied Science and Manufacturing*, **40**(6–7), 810–821, 2009.

22. Arbelaiz, A., Fernández, B., Cantero, G., Llano-Ponte, R., Valea, A., Mondragon, I., Mechanical properties of flax fiber/polypropylene composites. Influence of fiber/ matrix modification and glass fiber hybridization, *Composites Part A: Applied Science and Manufacturing*, **36**(12), 1637–1644, 2005.

23. Kim, S.-J., Moon, J.-B., Kim, G.-H., Ha, C.-S., Mechanical properties of polypropylene/ natural fiber composites: comparison of wood fiber and cotton fiber, *Polymer Testing*, **27**(7), 801–806, 2008.

24. Ochi, S., Mechanical properties of kenaf fibers and kenaf/PLA composites, *Mechanics of Materials*, **40**(4–5), 446–452, 2008.

25. Ismail, H., Shuhelmy, S., Edyham, M. R., The effects of a silane coupling agent on curing characteristics and mechanical properties of bamboo fiber filled natural rubber composites, *European Polymer Journal*, **38**(1), 39–47, 2002.

26. Abdelmouleh, M., Boufi, S., Belgacem, M. N., Dufresne, A., Short natural-fiber reinforced polyethylene and natural rubber composites: effect of silane coupling agents and fibers loading, *Composites Science and Technology*, **67**(7–8), 1627–1639, 2007.

27. Morsyleide F. R., Chiou, B.-S., Medeiros, E. S., Wood, D. F., Williams, T. G., Mattoso, L. H. C., Orts, W. J., Imam, S. H. Effect of fiber treatments on tensile and thermal properties of starch/ethylene vinyl alcohol copolymers/coir biocomposites, *Bioresource Technology*, **100**(21), 5196–5202, 2009.

28. Singleton, A. C. N., Baillie, C. A., Beaumont, P. W. R., Peijs, T., On the mechanical properties, deformation and fracture of a natural fiber/recycled polymer composite, *Composites Part B: Engineering*, **34**(6), 519–526, 2003.

29. Ruksakulpiwat, Y., Sridee, J., Suppakarn, N., Sutapun, W., Improvement of impact property of natural fiber–polypropylene composite by using natural rubber and EPDM rubber, *Composites Part B: Engineering*, **40**(7), 619–622, 2009.

30. Ragoubi, M., Bienaimé, D., Molina, S., George, B., Merlin, A., Impact of corona treated hemp fibers onto mechanical properties of polypropylene composites made thereof, *Industrial Crops and Products*, **31**(2), 344–349, 2010.

31. Ismail, H., Edyham, M. R., Wirjosentono, B., Bamboo fiber filled natural rubber composites: the effects of filler loading and bonding agent, *Polymer Testing*, **21**(2), 139–144, 2002.

32. Mulinari, D. R., Voorwald, H. J. C., Cioffi, M. O. H., da Silva, M. L. C. P., da Cruz, T. G., Saron, C., Sugarcane bagasse cellulose/HDPE composites obtained by extrusion, *Composites Science and Technology*, **69**(2), 214–219, 2009.

33. Kim, S. W., Oh, S., Lee, K., Variation of mechanical and thermal properties of the thermoplastics reinforced with natural fibers by electron beam processing, *Radiation Physics and Chemistry*, **76**(11–12), 1711–1714, 2007.

34. Vilaseca, F., Valadez-Gonzalez, A., Herrera-Franco, P. J., Pèlach, M. A., López, J. P., Mutjé, P., Biocomposites from abaca strands and polypropylene. Part I: Evaluation of the tensile properties, *Bioresource Technology*, **101**(1), 387–395, 2010.

35. Joseph, K., Thomas, S., Pavithran, C., Effect of chemical treatment on the tensile

properties of short sisal fiber–reinforced polyethylene composites, *Polymer*, **37**(23), 5139–5149, 1996.

36. Saha, P., Manna, S., Chowdhury, S. R., Sen, R., Roy, D., Adhikari, B., Enhancement of tensile strength of lignocellulosic jute fibers by alkali-steam treatment, *Bioresource Technology*, **101**(9), 3182–3187, 2010.

37. Mohanty, A. K., Wibowo, A., Misra, M., Drzal, L. T., Effect of process engineering on the performance of natural fiber reinforced cellulose acetate biocomposites, *Composites Part A: Applied Science and Manufacturing*, **35**(3), 363–370, 2004.

38. Threepopnatkul, P., Kaerkitcha, N., Athipongarporn, N., Effect of surface treatment on performance of pineapple leaf fiber–polycarbonate composites, *Composites Part B: Engineering*, **40**(7), 628–632, 2009.

39. John, M. J., and Anandjiwala, R. D., Chemical modification of flax reinforced polypropylene composites, *Composites Part A: Applied Science and Manufacturing*, **40**(4), 442–448, 2009.

40. Sreekumar, P. A., Thomas, S. P., Saiter, J. M., Joseph, K., Unnikrishnan, G., Thomas, S., Effect of fiber surface modification on the mechanical and water absorption characteristics of sisal/polyester composites fabricated by resin transfer molding, *Composites Part A: Applied Science and Manufacturing*, **40**(11), 1777–1784, 2009.

41. Iosipescu, N., New accurate procedure for single shear testing of metals. *Journal of Materials*, **2**(3), 537–566, 1967.

42. ASTM Standard D5379–05 Standard Test Method for Shear Properties of Composite Materials by the V-Notched Beam Method, *Annual book of ASTM standards.* American Society for Testing and Materials, West Conshohocken, 2000.

43. ASTM Standard D 638-00 Standard Test Method for Tensile Properties of Plastics, American Society for Testing and Materials, West Conshohocken, 2001.

44. ASTM Standard D 3039/D 3039M-00, Standard Test Method for Tensile Properties of Polymer Matrix Composite Materials, *Annual book of ASTM standards*, American Society for Testing and Materials, West Conshohocken, 2000.

9

Thermomechanical and spectroscopic characterization of natural fibre composites

K. R. RAJISHA, CMS College, India, B. DEEPA and
L. A. POTHAN, Bishop Moore College, India and
S. THOMAS, Mahatma Gandhi University, India

Abstract: A critical review is presented of the thermomechanical and spectroscopic characterization of the interface of natural fibre composites. These natural fibres offer a number of advantages over traditional synthetic fibres and several have been found to serve as efficient reinforcement in various polymers. The quality of the fibre–matrix interface is significant for the application of natural fibres as reinforcement for any polymer.

Key words: natural fibre reinforced composites, fibre–matrix interface, thermomechanical analysis, spectroscopic techniques.

9.1   Introduction

Current environmental awareness coupled with economic factors has triggered a renewed interest in natural fibres for developing new lightweight materials that are environmentally friendly. It has been proved recently, beyond doubt, that natural fibres can be effective reinforcement in polymeric matrices. Various fibres obtained from the stem, seed and leaves of various plants have been found to be an effective reinforcement in polymeric matrices. However, a few important issues that have to be addressed in preparing composites based on these fibres are their hydrophilic nature as well as their incompatibility with polymeric matrices (Kohler and Wedler, 1994; Mieck *et al.*, 1993). The main advantages of natural plant fibres compared with traditional glass fibres are economical viability, low density, reduced tool wear, enhanced energy recovery, reduced dermal irritation, and good biodegradability (Bledzki and Gassan, 1999). Moreover, these reinforcements can reach mechanical properties such as specific strength and modulus comparable with glass fibres (Leao *et al.*, 2010a ; Leao *et al.*, 2010b). Recently, natural fibres have been used to reinforce traditional thermoplastic polymers, especially polypropylene in automotive applications.

As for fibre-reinforced composites, the interfacial zone plays a leading role in load transfer between fibre and matrix and, consequently, in the mechanical properties. In the past, many attempts have been made to modify the surface properties of cellulose fibres in order to enhance adhesion with the

241

matrix. Various methods such as corona treatment (Belgacem *et al.*, 1994), plasma treatment (Felix *et al.*, 1994), mercerization (Bisanda and Ansell, 1991), heat treatment (Sapieha *et al.*, 1989), graft copolymerization (Felix and Gatenholm,1991; Maldas *et al.*,1989), silane treatment (Pothan *et al.*, 2007) and treatments with other chemicals have been reported to affect the compatibility in natural fibre composites. However, these methods possess many disadvantages, the most important of which are the use of expensive equipment or the use of expensive chemicals.

9.2   Natural fibre composites

Natural fibre composites combine plant-derived fibres with a polymeric matrix. The natural fibre component may be wood, sisal, hemp, coconut, cotton, kenaf, flax, jute, abaca, banana leaf fibres, bamboo, wheat straw or other fibrous material, and the matrix can be a polymeric material. Advantages of natural fibre composites include light weight, low-energy production and sequestration of carbon dioxide (reducing the 'greenhouse effect'). A political/ social advantage is that some production can be 'farmed out' to semi-skilled indigenous workers. Replacement of fibreglass with natural fibre removes concern about the potential of lung disease caused by the former and is a move toward sustainable development.

9.3   Interfaces in natural fibre composites and their characterization

The term interface is defined as a two-dimensional region between the fibre and the matrix having properties intermediate between those of the fibre and the matrix. Matrix molecules can be anchored to the fibre surface by chemical reaction or adsorption, thus determining the extent of interfacial adhesion. In certain instances, the interfaces may be composed of an additional constituent such as an interlayer between the two components of the composite. The adhesion between the fibre and the matrix is a major factor in determining the response of the interface and its integrity under stress.

In natural fibre reinforced composites, the strength and type of interface plays an important role in the thermomechanical properties. Therefore, the interface characterization is very important for assessing the properties of the composites. The characterization of the interface gives the chemical composition as well as information on interaction between the fibre and the matrix (Thomas *et al.*, 2009). Various methods are available for interface characterization.

9.4   Microscopic techniques

Microscopic studies such as optical microscopy, scanning electron microscopy (SEM), transmission electron microscopy (TEM) and atomic force microscopy (AFM) can be used to study the morphological changes on the surfaces and can predict the extent of mechanical bonding at the interface. The adhesive strength of fibre to various matrices can be measured by AFM.

9.4.1   Scanning electron microscopy

Scanning electron microscopy (SEM) is generally employed for the more extensive morphological inspection and interfacial characterization. It consists of the observation of fractured surface films at liquid nitrogen temperature. This technique allows for conclusions about the homogeneity of the composite, presence of voids, dispersion level of the continuous matrix, presence of aggregates, sedimentation, and possible orientation of fibres. Comparison of the surface fractures can provide information about the interfacial adhesion prevailing in the composite.

Martins *et al.* (2004) have reported on the mercerization and acetylation treatments applied to sisal fibres to enhance adhesion with polymer matrices in composites. The structures of the untreated and treated fibres were assessed with SEM (Fig. 9.1). The waste from sisal fibre decortication consisted of mechanical, ribbon, and xylem fibres, and their ultimate cells

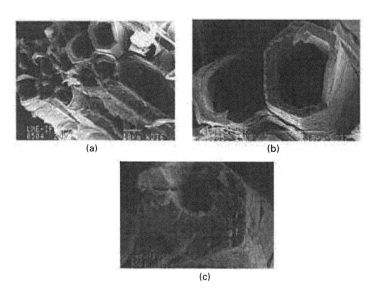

(a)   (b)

(c)

9.1 SEM micrographs of the fracture surfaces of untreated sisal fibres: (a) and (b) are ultimate cells of hexagonal shape, (c) is an ultimate cell of circular shape. (Source: Martins *et al.* 2004.)

varied considerably in size and shape. After mercerization and acetylation, the fibres and conductive vessel surfaces were successfully changed (Fig. 9.2). The parenchyma cells were partially removed, and the fibrils started to split, because of the alkali action. This increased the effective surface area available for contact with the matrix. The mercerized and acetylated fibres were coated with cellulose acetate by the grafting of the acetyl group in the fibrils. The treatment used to remove lignin and hemicellulose caused changes in the fibre surface but did not damage the fibre structure because the fibrils remained joined in a bundle.

Tserki *et al.* (2005) reported on two fibre pretreatment methods, acetylation and propionylation, applied on flax, hemp and wood fibres. SEM was used to characterize the crystallinity and the surface morphology of the untreated and esterified fibres. The highest extent of the esterification reaction was achieved for wood fibres owing to their high lignin/hemicelluloses content. The SEM results revealed that both treatments resulted in a removal of non-crystalline constituents of the fibres. Scanning electron micrographs of untreated and esterified wood fibres are shown in Fig. 9.3. The micrographs in Fig. 9.3a show that the untreated fibres are covered with a layer, whose composition is probably mainly waxy substances, as has also been reported previously (Zafeiropoulos *et al.*, 2002). It can be seen that the layer is not evenly distributed along the fibre surface, but its thickness varies from point to point. As seen in Fig. 9.3b and 9.3c the surface of the esterified materials

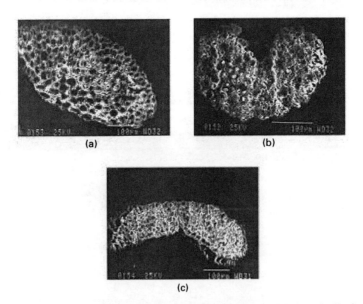

(a)   (b)

(c)

9.2 SEM micrographs of the transverse sections of chemically fixed, dehydrated, and embedded untreated sisal fibres: (a) mechanical, (b) ribbon, and (c) xylem. (Source: Martins *et al.* 2004.)

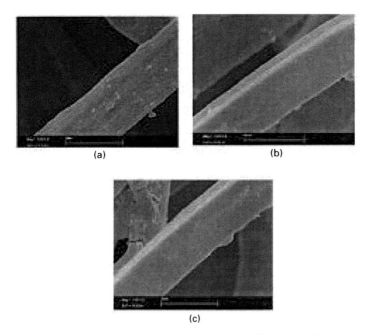

9.3 SEM micrographs of untreated and esterified wood fibre: (a) untreated, (b) acetylated and (c) propionylated. (Source: Tserki *et al.* 2005.)

became smoother than that of untreated materials. Removal of the waxy substances from the surface of lignocellulosic materials (Cherian *et al.*, 2008) and replacement of surface hydroxyl groups by acetyl and propionyl groups (Rana *et al.*, 1997) could explain smoothening of the fibre surface after esterification.

Effects of lysine-based di-isocyanate (LDI) as a coupling agent on the properties of biocomposite from poly(lactic acid) (PLA), poly(butylene succinate) (PBS) and bamboo fibre (BF) were investigated by Lee and Wang (2006). Many properties in composite materials are affected by their morphology. Figure 9.4 shows SEM micrographs of the tensile fractured surface of PLA and PBS/BF composite with or without LDI. In the PLA/BF composite without LDI, two phases can be clearly seen and many bamboo fibres were pulled out from the matrix in the fracture process, with large voids thereby being created. Also, gaps between PBS and BF in the PBS/BF composite without LDI were visible. These findings suggest that the interaction between matrix and filler was very weak, resulting in less interfacial adhesion. These features are typical of incompatible polymer composites. On the other hand, the micrographs of both composites after compounding with LDI of 0.65% showed that BF appeared to be coated with matrix polymer. This improved interfacial adhesion may be caused

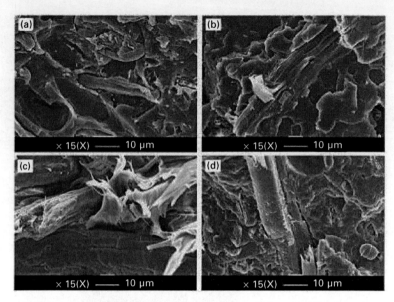

9.4 SEM micrographs of interface between matrix and BF in PLA/BF (70/30) composite without (a) or with LDI (NCO content, 0.65%) (b), and PBS/BF(70/30) composite without (c) or with LDI (NCO content, 0.65%) (d). (Source: Lee and Wang 2006.)

by the compatible effect of graft copolymer with LDI intermediates, which could be produced through a chemical reaction between the hydroxyl groups of polymer and BF under kneading conditions of higher temperatures and pressure (Wu, 2003).

Pothan *et al.* (2003) used SEM as a technique to characterize the degree of adhesion in banana fibre reinforced polyester composites. Figure 9.5 shows the composites with various banana fibre loadings. In another fascinating study done by Souza *et al.* (2010), highly crystalline nanofibrils were extracted from curava fibres and these fibres were used for the development of transparent polyvinyl alcohol (PVA) films which find application in packing, biomedical and optical devices. Environmental SEM (ESEM) was used to characterize the isolated nanofibrils; the extracted fibrils depict an aspect ratio of approximately 55 nm.

Mwaikambo and Ansell (2002) subjected hemp, sisal, jute, and kapok fibres to alkalization by using sodium hydroxide. The surface morphology of untreated and chemically modified fibres has been studied using SEM. SEM showed a relatively smooth surface for all the untreated fibres; however, after alkalization, all the fibres showed uneven surfaces. These results show that alkalization modifies plant fibres promoting the development of fibre–resin adhesion, which then results in increased interfacial energy and, hence, improvement in the mechanical and thermal stability of the composites.

9.5 (a) SEM of the composite with 10% fibre loading showing fibre–matrix debonding and fibre pull out, (b) SEM of the composite with 20% fibre loading showing fibre–matrix debonding and matrix cracking, (c) SEM of the composite with 40% fibre loading showing good fibre–matrix adhesion. (Source: Pothan *et al.* 2003.)

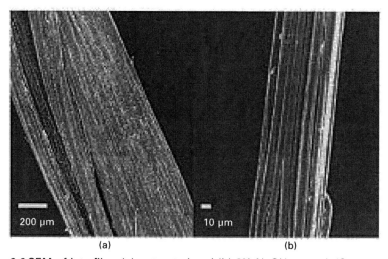

9.6 SEM of jute fibre (a) untreated and (b) 8% NaOH treated. (Source: Mwaikambo and Ansell 2002.)

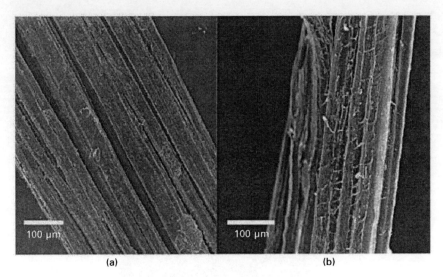

(a)   (b)

9.7 SEM of hemp fibre (a) untreated and (b) 8% NaOH treated.
(Source: Mwaikambo and Ansell 2002.)

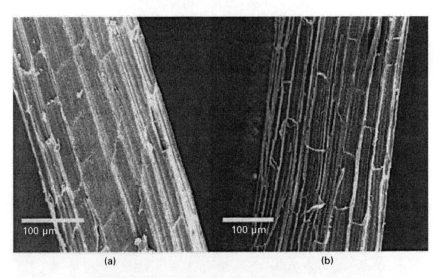

(a)   (b)

9.8 SEM of sisal fibre (a) untreated and (b) 8% NaOH treated.
(Source: Mwaikambo and Ansell 2002.)

Figures 9.6–9.8 show SEM images of the surfaces of alkali treated jute, hemp and sisal fibres.

Jacob *et al.* (2005) studied the water sorption of composites based on natural rubber and hybrid fibres of sisal and oil palm. Sisal and oil palm fibres were subjected to various treatments such as mercerization and silanation.

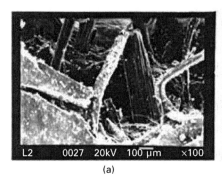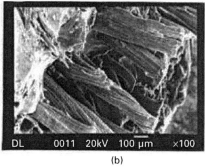

(a)   (b)

9.9 Tensile fracture surface of (a) untreated and (b) treated composite at 30 phr loading of hybrid fibres of sisal and oil palm. (Source: Jacob *et al.* 2005.)

The effect of chemical modification on moisture uptake was also analyzed. Chemical modification was seen to decrease the water uptake in the composites. The better adhesion of composites containing alkali-treated fibre is also evident from SEM. Figure 9.9a and 9.9b shows the tensile fracture surface of untreated and treated composite at 30 phr loading. The presence of holes is clearly visible in Fig. 9.9a. This indicates that the level of adhesion between the fibres and the matrix is poor, and when stress is applied, it causes the fibres to be easily pulled out from the rubber matrix, leaving behind gaping holes. In Fig. 9.9b, we can see the presence of a number of short, broken fibres projecting out of the rubber matrix. This indicates that the extent of adhesion between the fibres and the rubber matrix is greatly improved, and when stress is applied, the fibres break and do not wholly come out of the matrix.

9.4.2   Optical microscopy

Nishino and Arimoto (2007), in their studies on all-cellulose composite prepared by selectively dissolving fibre surface, used optical microscopy as a tool for assessing the interface and thereby its optical properties. They had reported that the all-cellulose composite possesses a good interface between the remaining fibres and the surrounding matrix from the selectively dissolved/resolidified fibre, resulting in an excellent bonding, a high mechanical performance, and an optical transparency. Figure 9.10 shows optical photographs of the filter paper (a) and the all-cellulose composites (immersion times: (b) 6 h and (c) 12 h). The filter paper is white and opaque, whereas the all-cellulose composite with a longer immersion time in the solvent became transparent. Recently, Nogi *et al.* (2005) reported optically transparent composites by combining bacterial cellulose with polymeric matrices. This transparency

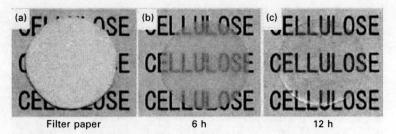

Filter paper   6 h   12 h

9.10 Optical photographs of filter paper (a) and all-cellulose composites (immersion times: (b) 6 h and (c) 12 h). (Source: Nishino and Arimoto 2007.)

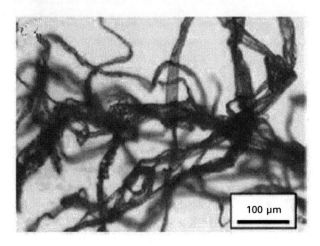

9.11 Optical micrograph of original sulfite cellulose macroscopic fibres of sizes of several tens of micrometres. (Source: Pääkko *et al.* 2007.)

was attained because of the nanodimensional effect of the fibres. Thus, the good interface, and the additional effect of the closure of the internal cell wall porosity by compression are thought to bring optical transparency to this composite.

Pääkko *et al.* (2007) in their studies on nanoscale cellulose fibrils and strong gels, made an attempt to prepare microfibrillated cellulose only by extensive mechanical shearing using the homogenizer. Figure 9.11 shows the optical micrograph of original sulfite cellulose macroscopic fibres of sizes of several tens of micrometres.

Cassano *et al.* (2007) prepared a new set of natural fibre composites with derivatized broom fibres. The fibres containing polymerizable groups were co-polymerized with dimethylacrylamide and styrene and investigated by optical polarizing microscopy (OPM). The surface roughness characterization was initially effected through investigation with high resolution optical microscope (Fig. 9.12).

Brahamakumar *et al.* (2005) studied the effect of the natural waxy surface layer of coir fibre-reinforced polyethylene composites. To assess the effectiveness of the waxy layer in interfacial bonding, coconut fibres with a wax-free surface and surface-modified wax-free fibres obtained by grafting of an isocyanate derivative of cardanol (CTDIC) were also used. The natural waxy surface layer of coconut fibre was found to provide a strong interfacial bonding between the fibre and the polyethylene matrix. Removal of the waxy layer resulted in a weak interfacial bonding which increased the critical fibre length by 100% and decreased the composite tensile strength and modulus by 40% and 60%, respectively. The waxy layer, because of its polymeric nature, showed a stronger effect on fibre–matrix bonding than a grafted layer of a C15 long alkyl chain molecule on the wax-free fibre. Figure 9.13a and 9.13b shows the optical fractographs of the as-received and wax-free fibre

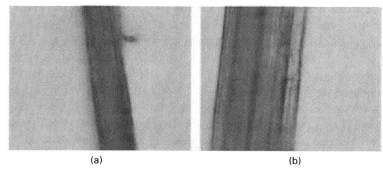

(a)   (b)

9.12 Optical micrographs of acrylated broom fibres copolymerized with (a) a stoichiometric amount of DMAA and (b) an excess of DMAA. (Source: Cassano *et al.* 2007.)

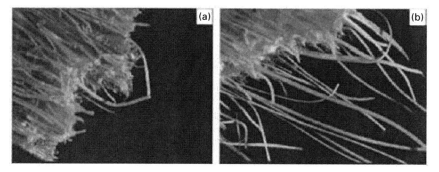

9.13 Optical fractographs of oriented discontinuous fibre composites (ODFC) of (a) as-received and (b) wax-free coconut fibres showing large fibre pullout for the latter owing to poor interfacial bonding. (Source: Brahmakumar *et al.* 2005.)

composites containing 20-mm-long coir fibres. Large fibre pullout can be seen for the wax-free fibre-reinforced composites owing to poor interfacial bonding, and this does not occur for the composites made from as-received coir fibres.

9.4.3   Atomic force microscopy

Atomic force microscopy (AFM) is a useful technique for measuring the surface roughness of fibres. Its advantages such as high resolution and non-destructivity offer a unique possibility for repetitive examinations. The force modulation mode gives a qualitative statement about the local sample surface elasticity using an oscillating cantilever tip, which indents into the sample surface. The amplitude of this deflection is measured as a function of tip position when the cantilever tip bends cyclically into the surface. AFM utilizes much smaller forces between the tip and specimen and the smaller radius of curvature of the tip gives better spatial resolution. Mai *et al.* (1998) used this technique to determine the physical properties of the interfaces. The microscopic image produced is that of a surface, representing the locus of points of constant force between the tip and the specimen. An important feature of the AFM is its ability to image non-conducting surfaces. The information collected during a scan of the surface is quantitative in three dimensions. When presented as a topographic image, the elevation of each point in the picture is encoded according to a grey scale or false colour scale. This variation is clearly understood from the three-dimensional picture of untreated and silane-treated fibres. Figure 9.14a and 9.14b shows three-dimensional images of untreated and silane-treated flax fibres (George *et al.*, 1999). After fibre treatment, the surface roughness of the fibres increased considerably compared with untreated fibre.

In an innovative study, Pickering *et al.* (2007) reported the characterization of chemically treated hemp fibre using AFM. The modifications employed were steam, alkali, a steam–alkali combination, enzymatic treatments and acetylation. The effects of various treatments on the microstructure are shown from AFM images in contact mode in Fig. 9.15a. Image from steam treatment in Fig. 9.15a suggests changes to, but ineffective removal of the primary cell wall. A combination of steam–alkali treatment in Fig. 9.15b and 9.15c shows smooth, highly aligned, and well-separated fibrils indicative of the secondary cell wall. Topography of the enzyme-treated fibres in Fig. 9.15d also shows sections, which indicate exposure of the secondary cell wall, with cross-banding of the fibrils that has been observed by other researchers using enzyme treatments.

Pääkko *et al.* (2007) presented a simple method of preparing nanoscale cellulose fibrils, where enzymatic hydrolysis is used in combination with mechanical shearing and high-pressure homogenization to promote

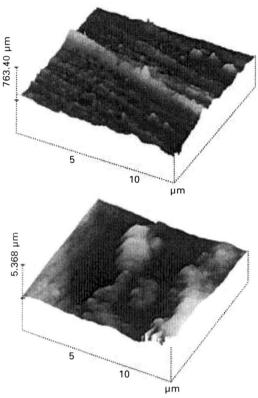

9.14 AFM three-dimensional images of flax fibres. (Source: George *et al.* 1999.)

delamination of the fibre wall. They concluded from the AFM studies that the cellulose fibrils obtained are interconnected and coiled and form an inherently entangled network (Fleming *et al.*, 2001; Lima and Borsali, 2004), in contrast to nonentangled acid hydrolyzed and sonicated rod-like cellulose crystallites or less entangled microfibrils (Lowys *et al.*, 2001), shown in Fig. 9.16.

9.4.4   Transmission electron microscopy

Transmission electron microscopy (TEM) is an excellent tool for analyzing the morphology of the composite. Wu *et al.* (2007) prepared a high-strength composite with microcrystalline cellulose and polyurethane. The morphology of the nanocomposite was studied by TEM (Fig. 9.17). Cellulose nanofibrils, approximately 20–40 nm in diameter and with considerable aspect ratio, are dispersed in the polyurethane matrix. They reported that the dispersion of the fibre in the matrix was good and thus high-strength polymer composites were obtained.

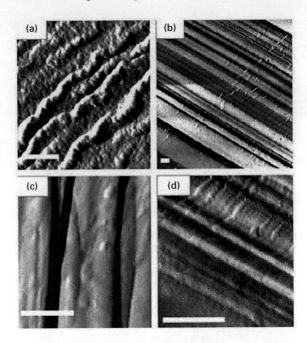

9.15 Morphology of fibre surfaces after treatment obtained from deflection AFM image data. Scale bars show 400 nm: (a) steam treatment, (b) low magnification steam–alkaline, (c) at higher magnification, and (d) after enzyme treatment (Source: Pickering *et al.* 2007.)

TEM images can also be used to confirm the nanostructure of the prepared filler. Wågberg *et al.* (2008) used TEM to analyse the structure of the filler as well as the composite. Capadona *et al.* (2009) prepared polymer nanocomposites with cellulose nanowhiskers. They also chose the same technique to confirm the nanostructure of the filler as well as the composite dispersion (Fig. 9.18).

In an interesting study, Cherian *et al.* (2008) extracted cellulose nano-fibres from the pseudo stem of the banana plant by using acid treatment coupled with high-pressure defibrillation. Characterization of the fibres by TEM showed that there is reduction in the size of banana fibres to the nanometer range (below 40 nm) and the resultant nanofibrils have a needle-like structure. The average length and diameter of the developed nanofibrils were found to be 200–250 nm and 4–5 nm, respectively. Uranyl acetate staining gave reasonable contrast between the fibrils and the carbon film. The presence of the heavy uranium in close vicinity of each whisker gave enough contrast for imaging. A tendency to agglomerate could be observed from TEM.

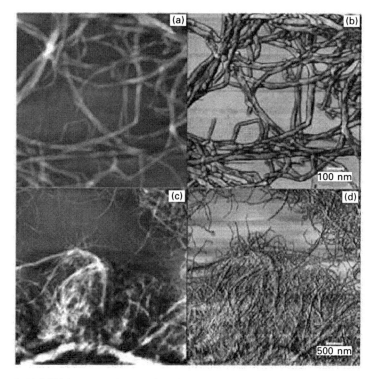

9.16 AFM images of microfibrillated cellulose on mica after drying: topographical images (a and c), and phase-contrast images (b and d). The scan size was 1 μm × 1 μm (a and b) and 5 × 5 μm (c and d). (Source: Pääkko *et al.* 2007.)

9.5   Spectroscopic techniques

Nuclear magnetic resonance (NMR), electron spin resonance (ESR), and photoacoustic spectroscopy were shown to be successful in polymer surface and interface characterization.

9.5.1   NMR analysis

Fan (2009) investigated 'A new route for the preparation of cellulose triacetate (CTA) optical films from the biomass of ramie fibre' and this was found to have environmental benefits. The structure and the thermomechanical properties of CTA and its film were investigated by different techniques, including NMR analysis. The ^{1}H NMR analysis of CTA was shown in Fig. 9.19. In the linear molecular structure of CTA, there are three kinds of hydrogen molecules which show two clusters of peak signals as shown in Fig. 9.19. The proton resonance of the glucose ring (δ) 3.30–5.20 ppm and

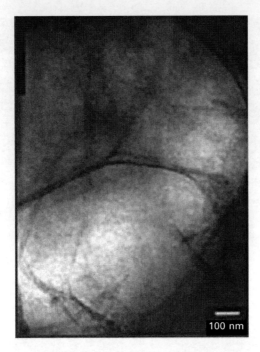

9.17 TEM image of PU/cellulose. (Source: Wu *et al.* 2007.)

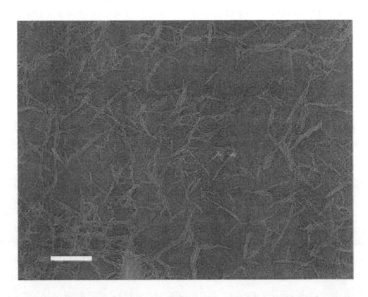

9.18 Transmission electron microscopy (TEM) images of cellulose nanowhiskers prepared by dispersion of microcrystalline cellulose. (Source: Capadona *et al.* 2009.)

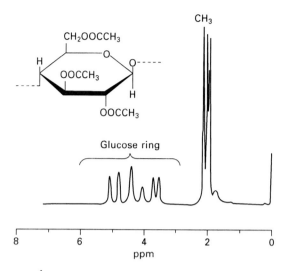

9.19 ¹H NMR spectrum of the CTA derived from ramie fibre. (Source: Fan 2009.)

the corresponding resonance for the methyl protons of the acetate group (δ) 1.50–2.20 ppm are shown in Fig. 9.19.

Megiatto *et al.* (2007) worked on the surface chemical modification of sisal fibre using reagents obtained from renewable sources. Sisal fibres have been modified to improve their compatibility with phenolic polymer matrices using furfuryl alcohol (FA) and polyfurfuryl alcohols (PFA) that can be obtained from renewable sources. The modification corresponded first to oxidation with ClO_2, which reacts mainly with guaiacyl and syringyl units of lignin, generating *o*- and *p*-quinones and muconic derivatives, followed by reaction with FA or PFA. The FA and PFA modified fibres presented a thin similar layer, indicating the polymer character of the coating. The untreated and treated sisal fibres were characterized by ¹³C CP-MAS NMR. The ¹H NMR spectrum of isolated lignin (Fig. 9.20) shows the presence of typical lignin signals (Megiatto, 2007). Signals 3 (3.8 ppm) and 1 (6.8 ppm) were assigned to methoxy and aromatic protons, respectively; signals 4 (3.4 ppm) and 5 (2.6 ppm) to residual water and DMSO, respectively; signal 2, between 5.0 and 5.5 ppm, to H_α, H_β, and H_γ of arylether units; and the signal 8 between 0.8 and 1.4 ppm to methyl and methylene protons in saturated aliphatic side chains. These were more intense than signals 6 and 7 between 1.9 and 2.1 ppm assigned to methyl or methylene protons adjacent to either a double bond or carbonyl group. This suggests that a large part of the side chains in the C9 units was saturated in the lignin sample, extracted from sisal fibres.

Solid-state ¹³C NMR using cross polarization and magic angle spinning

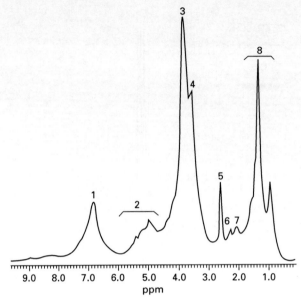

9.20 ¹H NMR spectrum of acidolysis lignin isolated from sisal fibres (solvent DMSO). (Source: Megiatto *et al*. 2007.)

(CP/MAS) is useful for characterizing wood–polymer composites (WPCs) because detailed information can be obtained from solid samples. In general, solid-state NMR involves proton–carbon cross polarization to enhance the ¹³C signal, high-power decoupling to eliminate dipolar line broadening owing to protons and spinning of the sample about the magic angle of 54.74° with respect to the static field to reduce chemical shift anisotropy effects. A spectrum of southern pinewood obtained using CP/MAS NMR is shown in Fig. 9.21 (Wright and Mathias, 1993).The peaks in the spectral region, from 160 ppm to 110 ppm, are assigned to the aromatic ring carbons of lignin, whereas those between 160 ppm and 143 ppm are assigned to oxygen substituted aromatic carbons. The peak at 56 ppm corresponds to lignin methoxy carbons, and the peaks at 122 and 135 ppm correspond to unsubstituted and alkylated aromatic carbons, respectively. The acetyl groups from hemicellulose components yield peaks at 21 ppm (methyls) and 172 ppm (carbonyls). The single peak at 105 ppm is assigned to the C1 carbon of the cellulose anhydroglucose repeat unit. Thus, chemical shift data allow qualitative and quantitative identification of the three major components of wood.

9.5.2   Acoustic emission spectrometry

Acoustic emission real-time monitoring has been frequently used during mechanical tests on traditional polymer composites i.e., reinforced with carbon

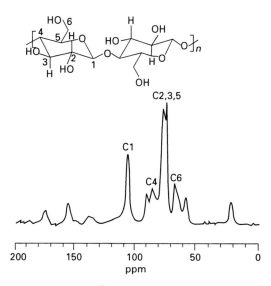

9.21 Solid state ^{13}C NMR spectrum of southern pinewood and structure of the cellulose unit. (Source: Wright and Mathias 1993.)

(Mizutani *et al.*, 2000), glass (Barre and Benzeggagh, 1994; Benevolenski and Karger-Kocsis, 2001; Margueres *et al.*, 2000) or aramid fibres (Davidovitz *et al.*, 1984). These studies involved different and progressively sounder levels of analysis, from the classical non-destructive testing (NDT) application aimed at acceptance/rejection of the material under stress to the measurement and localization of accumulated damage. In some instances, acoustic emission (AE) signal features were also correlated with specific damage phenomena occurring in composites, such as fibre failure and delamination.

The mechanical properties of fibre-reinforced polymeric composites strongly influence their final industrial application. Mechanical properties of composites depend on several factors, such as the properties of the constituent reinforcement and matrix, their relative volume fraction, and the shape, size and architecture of the reinforcement phase, but to an even greater extent on the reinforcement/matrix interfacial shear strength (IFSS). A relatively strong interfacial bond is needed for an effective transfer of the applied load, because a weak interface probably leads to a premature failure of the composite. The IFSS is a critical factor affecting the toughness, transverse mechanical properties and interlaminar shear strength of composites; hence, a detailed knowledge of the characteristics of the fibre–matrix interface is necessary when tailoring performance to applications (Luo and Netravali, 1999). Improving IFSS results in increasing the tensile and flexural strength of the composite whilst lowering the impact strength and toughness. The quality of the interface region is a significant concern for traditional man-made fibre reinforced composites and an even more worrying aspect for natural fibre-reinforced composites. In fact, what has prevented a more widespread use of natural fibres is the lack of good adhesion to most polymeric matrices. The hydrophilic nature of natural fibres adversely influences adhesion to the hydrophobic matrix, resulting in low compatibility and strength (Mohanty *et al.*, 2000). Furthermore, a strong interfacial bond represents a key aspect for the durability of composites. To investigate the interfacial behaviour of natural fibre composites, AE has been extensively used for the detection and localization of fibre breakage during single-fibre composite (SFC) tests. It has been demonstrated that almost all fibre breakages were detected and associated with a single acoustic emission event according to a one-to-one correspondence. Furthermore, the correspondence between the AE events detected and the observed fibre breakages is generally well established.

These techniques were also successfully applied to interfacial studies in natural fibre reinforced composites, thus allowing a sounder knowledge of failure modes and of the level of adhesion between reinforcement and matrix. Park *et al.* (1999) used dog-bone shaped epoxy composites reinforced with dual basalt (two basalt fibres of different diameter embedded at half-depth in epoxy resin) and SiC fibres. These specimens were tested in tension and were online monitored by AE in order to characterize the failure modes. The AE event amplitude, energy and duration was monitored using cross plots to highlight the differences during the loading of dual-fibre composites. The signals were well separated in three distinct ranges, two ranges for breakages of each basalt fibre of different diameter (15 and 97 µm) and one for matrix cracking. This was true for both AE amplitude and energy. The analysis also showed that the thicker fibre broke before the thinner one. It was reported that characteristic frequency peaks detected using fast Fourier transform (FFT) analysis could provide information for discriminating failure sources. In addition, for mineral fibres, a one-to-one correspondence was found between AE events and fibre breakages (Fig. 9.22). Therefore, AE could be used to analyze interfacial modes for semi- or non-transparent composites, which cannot be investigated by optical microscopy.

Acha *et al.* (2006) used the AE technique to study the fracture and failure behaviour of biodegradable jute fabric reinforced thermoplastic polyester composites. During the tensile test, they observed that AE amplitude increased in accordance to the following ranking from lowest to highest: debonding, fibre pull-out, and fibre fracture.

Sreekala *et al.* (2003) used AE monitoring to investigate the failure modes of oil palm and pineapple reinforced phenol formaldehyde composites. They highlighted that a correlation between the cumulative AE events and average amplitude versus elapsed time exists during the application of loading. At the beginning of the loading, the distribution of events showed small amplitude signals owing to matrix deformation and debonding. In contrast, in proximity of the maximum load value, the AE events distribution was

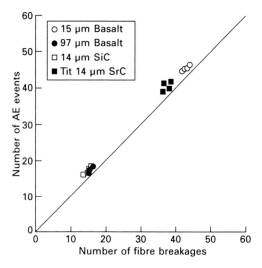

9.22 One-to-one correspondence between the number of various typed-fibres breakage and the number of AE events. (Source: Park *et al.* 1999.)

characterized by the presence of higher amplitude signals owing to pull-out and fibre fracture.

Acoustic emission may constitute a useful tool for mechanical behaviour monitoring of natural fibre composites, thus contributing to the future transition towards fully biodegradable materials, of which natural fibre composites constitute one possible path, that is being increasingly followed.

9.6   Thermomechanical methods

A detailed understanding of the degradation characteristics of polymers on heating is essential for selecting materials with improved properties for a specific application. The change in weight of the material can be measured when it is heated in an inert atmosphere or in the presence of air or oxygen. It was reported that the chemical composition, heating rate, temperature, and inorganic substances are the major factors that affect the thermal behaviour of biomass (Shafizadeh and DeGroote, 1976). The kinetics of exothermic reactions is important in assessing the potential of both materials and systems for thermal exposure (Doh *et al.*, 2004). Different treatments cause a variation in the degree of impurities removed as well as the degree of structural disruption. The effect of the difference in non-cellulosic composition and degree of structural disruption on the thermal stability is an important aspect to be investigated (Ouajai and Shanks, 2005). The assessment of the thermal stability of polymeric materials can provide valuable technical information. The demand for polymers, that could be used in high-temperature applications, stimulated the investigation to unravel the relationship between thermal properties and chemical structure (Still 1997). Most polymers degrade if exposed to elevated temperatures, especially in air, and during manufacturing, they are often exposed to thermal stress. Therefore, it is important to know the effects of the processing temperature associated with the processing duration. Various thermomechanical methods can be used to characterize the interfaces in polymer nanocomposites.

9.6.1   Thermogravimetric analysis

Thermogravimetric analysis (TGA) is an analytical technique used to determine a material's thermal stability and its fraction of volatile components by monitoring the weight change that occurs as a sample is heated at a constant rate. The measurement is normally carried out in air or in an inert atmosphere, such as helium or argon, and the weight is recorded as a function of increasing temperature. There are several reports on the interfacial characterization of natural fibre reinforced polymer composites using thermogravimetric analysis (TGA).

Alemdar and Sain (2008) studied the thermal stability of biocomposites obtained from wheat straw nanofibres using TGA. The TGA thermograms of the thermoplastic starch (TPS) and the nanocomposite filled with 5 wt% nanofibre showed that the starch starts to degrade at around 275 °C. The degradation temperature for the nanofibres was around 296 °C. The lower weight loss of the TPS and the nanocomposites is the result of evaporation of glycerol. Results also showed that the degradation temperature of the polymer matrix and the nanocomposites are close to each other and smaller than that of each component.

Manfredi *et al.* (2006) studied the thermal properties of natural fibre reinforced unsaturated polyester (UP) and modified acrylic resins (Modar) composites using TGA. They found that the Modar matrix composites were more resistant to temperature than the composites with polyester matrix owing to the presence of acrylic acid as modifier.

Mohanty *et al.* (2006) studied the thermal properties of maleic anhydride grafted polyethylene (MAPE) treated jute–high-density polyethylene (HDPE) composites. They observed that the thermal degradation of all the samples took place within the programmed temperature range of 30–900 °C. TGA and differential thermo gravimetry (DTG) thermograms displayed an increase in the thermal stability of HDPE matrix with fibre reinforcement and MAPE treatment.

Doh *et al.* (2004) studied the thermal behaviour of liquefied wood polymer composites by means of TGA. Low-density polyethylene (LDPE), HDPE, and polypropylene (PP) as polymer matrices were used with liquefied wood (LW). They observed that the HDPE exhibits higher thermal stability than PP and the thermal stability of LW decreased with the increase of LW content up to 40%. A higher heating rate provided higher thermal stability, resulting from the decelerated decomposition rate.

Gañán and Mondragon (2003) studied the thermal and degradation behaviour of fique fibre reinforced thermoplastic matrix composites. The fique fibres have been treated with various chemicals such as maleic anhydride (MA), propionic acid (PA), glycidylmethacrylate (G) or formaldehyde (F), as well as the use of a copolymer of polypropylene and maleic anhydride (MAPP) as compatibilizer. Both PP and polyoxymethylene (POM) have been used as matrices. Thermal stability has been studied by TGA, which showed that the treatments introduced an improvement in the thermal stability of fique fibres with respect to untreated fibres.

Kim et al. (2005) investigated the thermal properties of agro-flour-filled (rice husk flour, RHF and wood flour, WF) polybutylene succinate (PBS) biocomposites as a function of the agro-flour content and mesh size. The thermal decomposition, thermal stability and DTG_{max} temperature of the biocomposites were examined using TGA. The results indicated that on increasing agro-flour content, the thermal stability, degradation temperature and derivative thermogravimetric curve (DTG_{max}) temperature of the biocomposites decreased whereas the ash content increased. The thermal degradation of the biocomposites was not affected by agro-flour mesh size. It can be seen that the ash content of RHF-filled PBS biocomposites is a little higher than that of WF-filled PBS biocomposites. The thermal stability of WF-filled PBS biocomposites is higher than that of RHF-filled biocomposites owing to the higher cellulose and lignin content of WF.

Rudnik (2007) studied the thermal properties of biocomposites prepared from modified starch matrix reinforced with natural vegetable fibres. They observed that the addition of natural fibres improved thermal behaviour of biocomposites. Two kinds of natural fibres were used, i.e. flax and cellulose in the amount of 0–40 mass %. The results showed that thermal degradation of modified starch matrix and cellulose-reinforced biocomposites proceeds in three steps, whereas the degradation process of flax-reinforced biocomposites occurs in two steps. For unreinforced matrix as well as for all biocomposites, regardless of type and amount of reinforcement, the major mass loss is observed at temperatures above 300 °C. The increase in thermal stability with introduction of natural fibre is observed for both flax and cellulose-reinforced biocomposites.

Alvarez and Vázquez (2004) investigated the effect of sisal fibre content on the thermal degradation of cellulose derivatives/starch blends. Apparent kinetic parameters were determined using a variety of conventional thermogravimetric methods. The thermal degradation of cellulose derivatives/starch blends was considered to be a step-degradation process of the two different peaks, depending on the temperature range. It was found that the increase in fibre content does not significantly affect the thermal degradation of the cellulose derivatives/starch blends at the first step of the thermal degradation. A slightly higher temperature and activation energy was obtained for the first peak and a lower temperature and activation energy was obtained for the second peak when the sisal fibre content increased in the composites.

Sun-Young et al. (2009) studied the thermal properties of pure PVA and PVA films with microcrystalline cellulose (MCC) treated by 1.5M and 2.5M HBr using TGA. Thermograms of nanocellulose reinforced PVA films reveal three main weight loss regions. The TGA graphs of all the PVA films with MCC hydrolyzed by HBr follow the same pattern as observed for the neat PVA film. A marginal increase in the thermal degradation temperature was observed for PVA films with nanocellulose treated with 1.5 M HBr. The addition of 1, 3 and 5 wt% nanocellulose to PVA showed a significant effect on the thermal deformation. As the nanocellulose loading increased, thermal stability of the composite films increased. MCC with 5 wt% loading exhibited the highest thermal stability. This is similar to that of PVA nanocomposite films, where MCC was treated with 2.5M HBr. Therefore, it can be concluded that the thermal stability of the composite films increased as the nanocellulose loading increased.

Imam et al. (2005) investigated the thermal stability of biodegradable composite films prepared from blends of poly(vinyl alcohol), cornstarch, and lignocellulosic fibre. TGA indicated the suitability of formulations for melt processing, and for application as mulch films in fields at much higher temperatures. The study also revealed that both starch and lignocellulosic fibre degraded much more rapidly than PVA. The addition of fibre to the formulations was found to enhance the PVA degradation.

In another interesting study, the thermal behaviour of green composites fabricated from bagasse fibre and poly(vinyl alcohol) was investigated by (Fernandes et al., 2004).They observed an increase in thermal stability upon incorporation of bagasse fibre. The thermal properties of composites of poly(L-lactide) with hemp fibres was investigated by (Masirek et al., 2007) The thermogravimetric analysis of the composites, carried out in both nitrogen and air, showed that the degradation process of fibre-filled systems started earlier than that of plain PLA, independent of the presence of the plasticizer.

9.6.2   Differential scanning calorimetry analysis

Differential scanning calorimetry (DSC) is a technique for measuring the energy necessary to establish a nearly zero temperature difference between a substance and an inert reference material, as the two specimens are subjected to identical temperature regimes in an environment heated or cooled at a controlled rate. The basic principle underlying this technique is that, when the sample undergoes a physical transformation such as phase transitions, more (or less) heat will need to flow to it than the reference to maintain both at the same temperature. This technique has been extensively used in the measurement of various thermal parameters such as glass transition, melting and crystallization. Several studies examined the thermal properties of natural fibre reinforced polymer composites using DSC analysis.

Liu *et al.* (2010) studied the thermal behaviour of flax cellulose fibres on a PLA matrix using DSC analysis. They determined the glass transition temperature (T_g), crystallization temperature (T_c), melting temperature (T_m) and the degree of crystallinity from the DSC graphs. The results are shown in Table 9.1. The three samples show a well defined glass transition, 'cold' crystallization and melting transitions. The composite films show a broader cold crystallization temperature range and lower cold crystallization onset temperature than those of the pure PLA film. DSC results concluded that the flax cellulose nanofibre can induce crystal nucleation of the PLA polymer, which implies that this flax cellulose material can probably be used as a nucleating agent for PLA.

Huda *et al.* (2004) studied the melting and crystallization behaviour of the cellulose fibre-reinforced PLA 'green' composites using DSC. The thermal parameters obtained from the DSC results are summarized in Table 9.2. DSC thermograms of neat PLA, and the composites exhibit the glass transition temperatures and melting temperatures over almost the same temperature range. Results indicated that the T_g and T_m of PLA did not change even after reinforcement with cellulose fibres but crystallization enthalpy and melting enthalpy changed with the addition of cellulose sample.

Singh *et al.* (2008) reported the DSC analysis of bamboo fibre based poly(3-hydroxy butyrate-co-3-hydroxyvalerate) (PHBV) biocomposites.

Table 9.1 DSC results for pure PLA and PLA–flax cellulose (FC) composites for the first and second heating runs (Liu *et al.* 2010)

Composition	T_g (°C)	T_c (°C)	T_m (°C)		ΔH_m (J g^{-1})		ΔH_c (J g^{-1})	X_c (%)	
			1st	2nd	1st	2nd		1st	2nd
Neat PLA	56.0	117.3	139.9	141.8	19.6	0.16	0.14	21.1	0.0
PLA-2.5 wt% FC	55.5	112.1	140.1	142.2	19.4	1.81	1.42	21.4	0.4
PLA-5.0 wt% FC	55.3	108.7	141.2	142.4	18.9	2.81	1.90	21.4	1.0

Table 9.2 Thermal properties of PLA and PLA/cellulose composites (Huda *et al.* 2004)

PLA/cellulose (wt %)	T_g (°C)	ΔH_c (J g^{-1})	T_c (°C)	ΔH_m (J g^{-1})	χ (%)	T_m (°C)
PLA/cellulose(100/0)	59	26.74	101	43.63	46.56	172
PLA/TC 2500(70/30)	57	23.79	98	38.23	40.80	172
PLA/TC 1004(70/30)	60	25.63	90	50.22	53.60	172

The details are given in Table 9.3. The melt crystallization temperature (T_c) of PHBV was 101 °C. The addition of 30 wt % wood fibre increased this temperature to 104 °C, whereas the addition of bamboo fibre decreased it to 99 °C; indicating an increase in the rate of crystallization in the presence of wood fibre and a decrease in the presence of bamboo fibre.

Arbelaiz *et al.* (2006) investigated the effect of fibre treatments on the thermal stability of flax fibre and flax fibre/PP composites. For the thermal stability study, flax fibres were treated using MA, MAPP copolymer, vinyltrimethoxysilane and alkalization. Crystallinity studies were done with the help of DSC analysis and the details of thermal parameters are given in Table 9.4. Results showed that the crystallization rates of all systems were strongly influenced by the crystallization temperature, fibre addition, fibre surface treatment and the pressure applied. It has been observed that increasing crystallization temperature involves a decrease in the crystallization rate.

Ge *et al.* (2005) reported another study involving poly(propylene carbonate) (PPC), starch–g–poly(methyl acrylate) (S–g–PMA). S–g–PMA copolymer was used to reinforce PPC. S–g–PMA copolymer was prepared by ceric ammonium nitrate-initiated polymerization of methyl acrylate on to corn starch (CS). The results indicated the improved compatibility between grafted starch and PPC. The observation of only one T_g may be attributed to the relatively smaller content of the grafted PMA compared with that of the PPC matrix, and this led to the weak glass transition of grafted PMA.

9.6.3   Dynamic mechanical analysis

The physical and thermomechanical properties of a polymeric material are strongly dependent on its structure, relaxation processes, and morphology (Krassig, 1993). The properties of composite materials are determined by the characteristics of the polymer matrices, by the content and properties of the reinforcements, and by the fibre–matrix adhesion. Composite mechanical properties are also dependent on good fibre dispersion and minimization of voids (Sims and Broughton, 2000). The interfacial adhesion depends on the bonding strength at the interface (Shibata *et al.*, 2004). Good dispersion of

Table 9.3 Detailed information obtained from differential scanning calorimetry of PHBV and its composites (Singh *et al.* 2008)

Sample	T_m (°C)	H_f (J g^{-1})	T_c (°C)	H_c (J g^{-1})
PHBV	154	57.25	101.4	48.6
PHBV–wood fibre (70:30)	153.3	63.28	104.3	35.47
PHBV–bamboo fibre (70:30)	152.9	62.29	99.1	30.52
PHBV–bamboo fibre (60:40)	152.7	59.44	97.2	28.92

Table 9.4 Thermal parameters calculated from the fusion run after isothermal crystallization at several temperatures for neat PP, untreated and MAPP-treated flax fibre composites (Arbelaiz *et al.* 2006)

Parameter	$T_c = 134\ ^\circ\text{C}$			$T_c = 140\ ^\circ\text{C}$			$T_c = 145\ ^\circ\text{C}$		
	PP	Untreated	MAPP	PP	Untreated	MAPP	PP	Untreated	MAPP
T_m (°C)	168.4	167.7	167.8	171.5	171.0	170.5	174.6	174.4	173.9
H_f (J $g^{-1}_{composite}$)	104.3	74.8	67.4	104.9	78.6	68.0	119.9	76.8	79.9
H_f (J g^{-1}_{PP})	104.3	106.8	96.3	104.9	112.3	97.1	119.9	109.7	114.1
X_c (%)	49.9	51.1	46.1	50.2	53.7	46.5	57.4	52.5	54.9

fibres in a polymeric matrix has been reportedly difficult to achieve (Raj *et al.*, 1989).

Dynamic mechanical analysis (DMA) is a well established method in thermomechanical analysis. The DMA measurement consists of the observation of time-dependent deformation behaviour of a sample under periodic mostly sinusoidal deformation force with very small amplitudes. Thus it is possible to calculate the storage modulus E' and loss factor tan δ with varying temperature and deformation frequency.

Mishra *et al.* (2005) prepared biocomposites with recycled newspaper cellulose fibre (RNCF) and PLA and PP. They studied the thermomechanical properties of prepared composites by dynamic mechanical thermal analysis (DMTA). The mechanical and thermomechanical properties of the RNCF reinforced PLA composites compared favorably with the corresponding properties of PP composites. Compared with the neat resin, the tensile and flexural moduli of PLA composites were significantly higher as a result of reinforcement by the RNCF. From the DMA results, it is revealed that incorporation of the fibres gives rise to a considerable increase in the storage modulus (stiffness) and a decrease in the tan δ values. They showed that the PP/Talc 1004 resulted in a greater storage modulus than neat PP, owing to the reinforcement imparted by the cellulose fibres that allows stress transfer from the matrix to the cellulose fibre (Hedenberg and Gatenholm, 1995). Storage modulus values of PP matrix and its composite are not the same at low temperature, because the fibres impart stiffness to the composite (Joseph *et al.*, 2003). Figure 9.23 shows that the storage modulus of the PP/Talc 1004 composite decreased with increasing temperature. The reduction of modulus is associated with a softening of the matrix at higher temperatures.

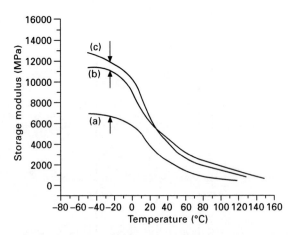

9.23 Temperature dependence of the storage modulus of PP and PP-based composites: (a) neat PP, (b) PP/TC 1004 (70/30), and (c) PP/talc (70/30). (Source: Mishra *et al.* 2005.)

These results demonstrate the reinforcing effect of RNCF on both PLA and PP matrices.

Wu *et al.* (2007) successfully prepared a high-strength elastomeric nanocomposite by dispersing microcrystalline cellulose in a polyurethane matrix. The resulting nanocomposites show increased strain-to-failure in addition to increased stiffness and strength compared with the unfilled polyurethane. They reported that in the pure polyurethane (Fig. 9.24), the modulus remains almost constant at temperatures below T_g. A rapid decrease in tensile modulus is then observed at the glass–rubber transition. In the rubbery region, the tensile modulus keeps decreasing with temperature owing to the thermoplastic nature of the material. Above T_g, a significant modulus increase is observed in the nanocomposites, although the cellulose content is only a few percent. Reinforcement effects are attributed both to micromechanical effects from stiff nanofibrils in a rubbery matrix and entropic effects from polyurethane molecules interacting physically and chemically with cellulose nanofibril surfaces. The last mechanism is important in cellulose–rubber systems with nanocellulosics as a discrete reinforcement phase rather than as a nanofibrous network. Because of the strong nanofibril–polyurethane interaction at the interface, and corresponding effects on physical network structure, the present nanocomposite has large strain-to-failure. At high strains, cellulose nanofibrils as well as polymer molecules become strongly reoriented in the loading direction, providing additional stiffening.

9.7   Conclusions

Fibre-reinforced composites were a rapidly growing part of the revolution in advanced materials development during the 1980s and 1990s. A common design feature of any composite material that must be understood, engineered,

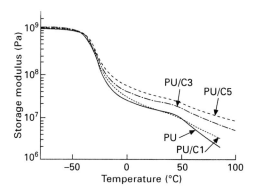

9.24 Storage modulus for polyurethane and polyurethane–cellulose composites. (Source: Wu *et al.* 2007.)

and controlled is the interfacial bond strength. Essential properties of the composite such as interlaminar shear and flexural strength can be achieved by controlling interfacial bond strengths. Adhesion at the fibre–matrix interface is, therefore, a critical property of the composite that must be measured. Often the interface can be engineered by modification of the fibre surface chemistry to optimize the adhesion between fibre and matrix. The surface energy of a reinforcing fibre is a measure of the adhesive properties of the fibre surface that can be used to predict adhesion at the fibre/matrix interface.

9.8   References

Acha B.A., Marcovich N.E., Karger-Kocsis J. (2006). Biodegradable jute cloth reinforced thermoplastic copolyester composites: fracture and failure behaviour. *Plast. Rubber Compos.* **35**(2): 73–82.

Alemdar A., Sain M. (2008). Biocomposites from wheat straw nanofibres: Morphology, thermal and mechanical properties. *Compos. Sci. Technol.* **68**: 557–565.

Alvarez V.A., Vázquez A. (2004). Thermal degradation of cellulose derivatives/starch blends and sisal fibre biocomposites. *Polym. Degrad. Stab.* **84**: 13–21.

Arbelaiz A., Fernández B., Ramos J.A., Mondragon I. (2006). Thermal and crystallization studies of short flax fibre reinforced polypropylene matrix composites: Effect of treatments. *Thermochim. Acta* **440**: 111–121.

Barre S., Benzeggagh M.L. (1994). On the use of acoustic emission to investigate damage mechanisms in glass fibre-reinforced polypropylene. *Compos. Sci. Technol.* **52**(3): 369–76.

Belgacem M.N., Bataille P., Sapeiha S. (1994). Effect of corona modification on the mechanical properties of polypropylene/cellulose composites. *J Appl. Polym. Sci.* **53**: 379–385.

Benevolenski O., Karger-Kocsis J. (2001). Fracture and failure behavior of partially consolidated discontinuous glass fibre mat-reinforced polypropylene composites (Azdel SuperLite®). *Macromol. Symp.* **170**: 165–179.

Bisanda E.T.N., Ansell M.P. (1991). The effect of silane treatment on the mechanical and physical properties of sisal–epoxy composites. *Compos. Sci. Technol.* **41**: 165–178.

Bledzki A.K., Gassan J. (1999). Composites reinforced with cellulose based fibres. *Prog. Polym. Sci.*, **24**: 221–274.

Brahmakumar M., Pavithran C., Pillai R.M. (2005). Coconut fibre reinforced polyethylene composites: effect of natural waxy surface layer of the fibre on fibre/matrix interfacial bonding and strength of composites. *Compos. Sci. Technol.*, **65**: 563–569.

Capadona J.R., Shanmuganathan K., Trittschuh S., Seidel S., Rowan S.J., Weder C. (2009). Polymer nanocomposites with nanowhiskers isolated from microcrystalline cellulose. *Biomacromolecules* **10**: 712–716.

Cassano R., Trombino S., Bloise E., Muzzalupo R., Iemma F., Chidichimo G., Picci N. (2007). New broom fibre (*Spartium junceum* L.) derivatives: preparation and characterization. *J. Agric. Food Chem.* **55**: 9489–9495.

Cherian B.M., Pothan L.A., Nguyen-Chung T., Mennig G., Kottaisamy M., Thomas S. (2008). A novel method for the synthesis of cellulose nanofibril whiskers from banana fibres and characterization. *J. Agric. Food Chem.* **56**: 5617–5627.

Davidovitz M., Mittelman A., Roman I., Marom G. (1984). Failure modes and fracture mechanisms in flexure of Kevlar–epoxy composites. *J. Mater. Sci.* **19**(2): 377–384.

Doh G.H., Lee S.Y., Kang I.A. and Kong Y.T. (2004). Thermal behavior of liquefied wood polymer composites (LWPC). *Compos. Struct.* **68**: 103–108.

Fan X. (2009). Cellulose triacetate optical film preparation from ramie fiber. *Ind. Eng. Chem. Res.* **48**(13): 6212–6215.

Felix J.M., Carlsson C.M.G., Gatenholm P. (1994). Adhesion characteristics of oxygen plasma-treated rayon fibres. *J. Adhes. Sci. Technol.* **8**(2): 163–180.

Felix J.M., Gatenholm P.J. (1991). The nature of adhesion in composites of modified cellulose fibres and polypropylene. *J. Appl. Polym. Sci.* **42**: 609–620.

Fernandes E.G., Cinelli P., Chiellini, E. (2004). Thermal behavior of composites based on poly(vinyl alcohol) and sugar cane bagasse. *Macromol. Symp.*, **21**: 231–240.

Fleming K., Gray D. G., Matthews S. (2001). *Chem. Eur. J.* **7**(9): 1831–1835.

Gañán P., Mondragon I. (2003). Thermal and degradation behavior of fique fibre reinforced thermoplastic matrix composites. *J. Therm. Anal. Calorim.*, **73**: 783–795.

Ge X.C., Xu Y., Meng Y. Z., Li R. K. Y. (2005). Thermal and mechanical properties of biodegradable composites of poly(propylene carbonate) and starch–poly(methyl acrylate) graft copolymer. *Compos. Sci. Technol.* **65**, 2219–2225.

George J., Ivens J., Verpoest I. (1999). Mechanical properties of flax fibre reinforced epoxy composites. *Angew. Macromol. Chem.*, **272**, 41–45.

Hedenberg P., Gatenholm P. (1995). Conversion of plastic/cellulose waste into composites. I. Model of the interphase. *J. Appl. Polym. Sci.* **56**: 641–651.

Huda M.S., Mohanty A.K., Drzal L.T., Misra M. (2004) Physico-mechanical properties of 'green' composites from polylactic acid (PLA) and cellulose fibres. *GPEC* 2004, abstract 11.

Imam S.H., Cinelli P., Gordon S.H., Chiellini E. (2005). Characterization of biodegradable composite films prepared from blends of poly(vinyl alcohol), cornstarch, and lignocellulosic fibre. *J. Polym. Environ.*, **13**: 47–55.

Jacob M., Varghese K.T., Thomas S. (2005). Water sorption studies of hybrid biofibre-reinforced natural rubber biocomposites. *Biomacromolecules* **6**(6): 2969–2979.

Joseph P.V., Mathew G., Joseph K., Groeninckx G., Thomas S. (2003). Dynamic mechanical properties of short sisal fibre reinforced polypropylene composites. *Compos. Part A* **34**: 275–290.

Kim H.S., Yang H.S., Kim H.J., Lee B.J., Hwang T.S. (2005). Thermal properties of agro-flour-filled biodegradable polymer bio-composites. *J. Therm. Anal. Calorim.*, **81**: 299–306.

Kohler R., Wedler M. (1994). Non-textile applications of flax fibres, *Technology of textile symposium*, lecture 331.

Krassig H.A. (1993). *Polymer monographs*, Elsevier Press: NewYork, Vol. 2.

Leao A.L., Souza S.F., Cherian B.M., Frollini E., Thomas S., Pothan L.A., Kottaisamy M. (2010a). Agro-based biocomposites for industrial applications. *Mol. Cryst. Liq. Cryst.* **522**: 18–27.

Leao A.L., Souza S.F., Cherian B.M., Frollini E., Thomas S., Pothan L.A., Kottaisamy M. (2010b). Pineapple leaf fibres for composites and cellulose. *Mol. Cryst. Liq. Cryst.* **522**: 36–41.

Lee S.H., Wang S. (2006). Biodegradable polymers/bamboo fiber biocomposite with bio-based coupling agent. *Compos. Part A* **37**, 80–91.

Lima M.M.D., Borsali R. (2004). *Macromol. Rapid Commun.* **25**(7): 771–787.

Liu D.Y., Yuan X.W., Bhattacharyya D., Easteal A.J. (2010). Characterisation of solution cast cellulose nanofibre-reinforced poly(lactic acid). *eXPRESS Polym. Lett.* **4**(1): 26–31.

Lowys M.P., Desbrieres J., Rinaudo M. (2001). *Food Hydrocolloids* **15**(1), 25–32.

Luo S., Netravali A.N. (1999). Interfacial and mechanical properties of environment-friendly 'green' composites made from pineapple fibres and poly(hydroxybutyrate-co-valerate) resin. *J. Mater. Sci.* **34**(15): 3709–3719.

Mai K., Mader E., Muhle M. (1998). Interphase characterization in composites with new non-destructive methods. *Compos. Part A* **29**(9–10): 1111–1119.

Maldas D., Kokta B.V., Daneault C. (1989). Influence of coupling agents and treatments on the mechanical properties of cellulose fibre-polystyrene composites. *J. Appl. Polym. Sci.* **37**: 751–775.

Manfredi L.B., Rodríguez E.S., Wladyka-Przybylak M., Vázquez A. (2006). Thermal degradation and fire resistance of unsaturated polyester, modified acrylic resins and their composites with natural fibres. *Polym. Degrad. Stab.* **91**: 255–261.

Margueres P., Meraghni F., Benzeggagh M.L. (2000). Comparison of stiffness measurements and damage investigation techniques for a fatigued and postimpact fatigued GFRP composite obtained by RTM process. *Compos. Part A* **31**(2): 151–163.

Martins M.A., Kiyohara P.K., Joekes I. (2004). Scanning electron microscopy study of raw and chemically modified sisal fibres. *J. Appl. Polym. Sci.* **94**: 2333–2340.

Masirek R., Kulinski Z., Chionna D., Piorkowska E., Pracella, M. (2007). Composites of poly(L-lactide) with hemp fibres: morphology and thermal and mechanical properties. *J. Appl. Polym. Sci.*, **105**, 255–268.

Megiatto J.D., Hoareau J.R., Gardrat C., Frollini E., Castellan A. (2007). Sisal fibres: surface chemical modification using reagent obtained from a renewable source; characterization of hemicellulose and lignin as model study. *J. Agric. Food. Chem.* **55**(21), 8576–8584.

Mieck K.P., Nechwatal A., Knobelsdort, C. (1993). Application of natural fibre resin composites, *Technology of textile symposium*, lecture 311.

Mishra M., Huda M.S., Drazal T.S. (2005). A study on biocomposites from recycled newspaper fibre and poly(lactic acid). *Ind. Eng. Chem. Res.* **44**: 5593–5601.

Mizutani Y., Nagashima K., Takemoto M., Ono K. (2000). Fracture mechanism characterization of cross-ply carbon-fibre composites using acoustic emission analysis. *NDT&E Int.* **33**(1): 101–110.

Mohanty A.K., Misra M., Hinrichsen G. (2000). Biofibres, biodegradable polymers and biocomposites: an overview. *Macromol. Mater. Eng.* 276–277 (1): 1–24.

Mohanty A.K., Verma S.K, Nayak SK. (2006). Dynamic mechanical and thermal properties of MAPE treated jute/HDPE composites. *Compos. Sci. Technol.* **66**: 538–547.

Mwaikambo L.Y., Ansell M.P. (2002). Chemical modification of hemp, sisal, jute, and kapok fibres by alkalization. *J. Appl. Polym. Sci.* **84**: 2222–2234.

Nishino T., Arimoto N. (2007). All-cellulose composite prepared by selective dissolving fibre surface. *Biomacromolecules* **8**: 2712–2716.

Nogi M., Handa K., Nakagaito A.N., Yano H. (2005). Optically transparent bionanofibre composites with low sensitivity to refractive index of the polymer matrix. *Appl. Phys. Lett.* **87**, 243110.

Ouajai S., Shanks R. A. (2005). Composition, structure and thermal degradation of hemp cellulose after chemical treatments. *Polym. Degrad. Stab.* **89**: 327–335.

Pääkko M., Ankerfors M., Kosonen H., Nykänen A., Ahola S., Sterberg M.O., Ruokolainen J., Laine J., Larsson P.T., Ikkala O., Lindström T. (2007). Enzymatic hydrolysis combined with mechanical shearing and high-pressure homogenization for nanoscale cellulose fibrils and strong gels. *Biomacromolecules*, **8**: 1934–1941.

Park J.M., Shin W.G., Yoon D.J. (1999). A study of interfacial aspects of epoxy-based composites reinforced with dual basalt and SiC fibres by means of the fragmentation and acoustic emission techniques. *Compos. Sci. Technol.* **59**(3): 355–70.

Pickering K.L, Beckermann G.W., Alam S.N., Foreman N.J. (2007). Optimizing industrial hemp fibre for composites. *Compos. Part A*, **38**(2): 461–468.

Pothan L. A., Oommen Z., Thomas S. (2003). Dynamic mechanical analysis of banana fiber reinforced polyester composites. *Compos. Sci. Technol.* **63**: 283–293.

Pothan L.A., George C.N., Jacob M., Thomas S. (2007). Effect of chemical modification on the mechanical and electrical properties of banana fibre polyester composites, *J. Compos. Mater.* **41**: 2371–2386.

Raj R.G., Kokta B.V., Dembele F., Sanschagrain B. (1989). Compounding of cellulose fibres with polypropylene: Effect of fibre treatment on dispersion in the polymer matrix. *J. Appl. Polym. Sci.* **38**: 1987–1996.

Rana A.K., Basak R.K., Mitra B.C., Lawther M., Banerjee A.N. (1997). Studies of acetylation of jute using simplified procedure and its characterization. *J. Appl. Polym. Sci.* **64**: 1517–1523.

Rudnik E. (2007). Thermal properties of biocomposites. *J. Therm. Anal. Calorim.*, **88**(2), 495–498.

Sapieha S., Pupo J.F., Schreiber H.P. (1989). Thermal degradation of cellulose containing composites during processing. *J. Appl. Polym. Sci.* **37**(1): 233–240.

Shafizadeh F., DeGroote W.F. (1976). Combustion characteristics of cellulosic fuels. In: Shafizadeh, F., Sarkanen, K.V., Tillman, D.A. (Eds.), *Thermal uses and properties of carbohydrates and lignins*. Academic Press Inc., New York, pp. 1–8.

Shibata M., Oyamada S., Kobayashi S., Yaginuma, D. (2004). Mechanical properties and biodegradability of green composites based on biodegradable polyesters and lyocell fabric. *J. Appl. Polym. Sci.* **92**: 3857–3863.

Sims G.D., Broughton W.R. (2000). *Comprehensive composite materials*: Kelly A., Zweben C (Eds), Elsevier Press, London, Vol. 2.

Singh S., Mohanty A.K, Sugie T., Takai Y., Hamada H. (2008). Renewable resource based biocomposites from natural fibre and polyhydroxybutyrate-co-valerate (PHBV) bioplastic. *Compos. Part A* **39**: 875–886.

Souza S. F., Leao A.L., Jenny H.C., Wu C., Sain M., Cherian B.M. (2010). Nanocellulose from curava fibres and their nanocomposites. *Mol. Cryst. Liq. Cryst.* **522**: 42–52.

Sreekala M.S., Czigany T., Romhany G., Thomas S. (2003). Investigation of oil palm and pineapple fibre reinforced phenol formaldehyde composites by acoustic emission technique. *Polym. Compos.* **11**(1): 9–18.

Still R.H. (1977). *Developments in polymer degradation* (N. Grassie ed.), Applied Science Publishers, London, vol. 1, p 1.

Sun-Young L., Mohan D.J., In-Aeh K., Geum-Hyun D., Lee S., Han S.O. (2009). Nanocellulose reinforced PVA composite films: effects of acid treatment and filler loading. *Fibr Polym.* **10**(1): 77–82.

Thomas S., Pothan L.A., Cherian B.M. (2009). Advances in natural fibre reinforced polymer composites: macro to nanoscales. *Int. J. Mater. Prod. Technol.* **36**(1–4): 317–333.

Tserki V., Zafeiropoulos N.E., Simon F., Panayiotou C. (2005). A study of the effect of acetylation and propionylation surface treatments on natural fibres. *Compos. Part A* **36**: 1110–1118.

Wågberg L., Decher G., Norgren M., Lindström T., Ankerfors M., Axnas K. (2008). The build-up of polyelectrolyte multilayers of microfibrillated cellulose and cationic polyelectrolytes. *Langmuir* **24**: 784–795.

Wright J.R., Mathias L.J. (1993). Physical characterization of wood and wood-polymer composites: An update. *J. Appl. Polym. Sci.* **48**: 2225.

Wu C.S. (2003). Performance of an acrylic acid grafted polycaprolactone/starch composites: Characterization and mechanical properties. *J. Appl. Polym. Sci.* **89**: 2888–2895.

Wu Q., Henriksson M., Liu X., Berglund L.A. (2007). High strength nanocomposite based on microcrystalline cellulose and polyurethane. *Biomacromolecules*, **8**(12): 3687–3694.

Zafeiropoulos N.E., Williams D.R., Baillie C.A., Matthews F.L. (2002). Engineering and characterisation of the interface in flax fibre/polypropylene composite materials Part I: Development and investigation of surface treatments. *Compos. Part A*, **33**: 1083–1093.

Assessing the moisture uptake behaviour of natural fibres

K.-Y. LEE and A. BISMARCK, Imperial College
London, UK

Abstract: Various methods for the determination of moisture uptake and water sorption behaviour of natural fibres are discussed; namely through simple weight gain measurement and dynamic vapour sorption (DVS), respectively. Simple weight gain measurement provides the equilibrium moisture content at a specified relative humidity. DVS, on the other hand, provides water sorption/desorption behaviour of natural fibres. Not only can the water sorption behaviour of natural fibres be determined, water sorption hysteresis can also be studied. A novel method of determining the accessible hydroxyl groups utilising heavy water using DVS apparatus was also discussed. This technique could potentially be applied to study the water sorption kinetics and mechanism of natural fibres.

Key words: natural fibres, moisture uptake, composites, dynamic vapour sorption.

10.1   Introduction

Understanding the moisture uptake behaviour of natural fibres has become increasingly important owing to their wide applications in numerous industries: the textile industry (Huda *et al.* 2007), the pulp and paper industry (Eichhorn *et al.* 2001) and more recently, the composites industry (Bledzki and Gassan 1999). The use of natural fibres is mainly driven by cost considerations and increased environmental concerns and legislations (EU 1999; EU 2000; EU 2003). Natural fibres have proven their application in reinforced polymeric materials for the production of natural fibre reinforced composites (Baltazar-y-Jimenez *et al.* 2008a; 2008b; Huda *et al.* 2008; Juntaro *et al.* 2007; Juntaro *et al.* 2008; Pommet *et al.* 2008; Sahoo *et al.* 2007; Satyanarayana *et al.* 2009). However, natural fibres' high moisture sorption properties have proven to be a major disadvantage for composites application (Dhakal *et al.* 2007; Stamboulis *et al.* 2000). The high moisture uptake affects not only the fibres' properties but also the mechanical performance of the resulting composites. High moisture sorption of natural fibres caused a reduction in the fibres' tensile strengths as a result of water molecules forcing the cellulose molecules in natural fibres apart, thereby reducing the overall rigidity of natural fibres (Stamboulis *et al.* 2000). The effect of moisture on

275

the mechanical performance of natural fibre reinforced composites was also studied extensively (Akil *et al.* 2009; Dhakal *et al.* 2007; Rashdi *et al.* 2009; Stamboulis *et al.* 2000). It was found that the mechanical performance of the composites reduces when the composites were exposed to moisture for an extended period of time. This is attributed to the high moisture uptake of natural fibres, which reduces the tensile strength of the fibre (Stamboulis *et al.* 2000) and weakens the fibre–matrix interface owing to water absorption (Dhakal *et al.* 2007).

In addition to the application of natural fibres in the textile (Huda *et al.* 2007) and composites industries, natural fibres also have the potential to be used as an environmental friendly material for building construction, utilising their moisture uptake behaviour (Collet *et al.* 2008). When used appropriately, new natural fibre based construction materials, such as lime-hemp render, hemp mortar and hemp wool have the potential to be used as an insulating material and a material that can control the humidity in an indoor environment (Hill *et al.* 2009b). Therefore, an understanding of the physical properties of natural fibres, in particular of the moisture uptake behaviour, is important. With this knowledge in hand, materials engineers will be able to design new environmentally friendly materials for different applications.

10.1.1 Background on the moisture sorption of natural fibres

The major constituents of natural fibres are cellulose, hemicellulose and lignin. Natural fibres are inherently hydrophilic in nature owing to the presence of a large number of hydroxyl (–OH) groups available, particularly in cellulose and hemicellulose. However, not all constituents contribute to the absorption of moisture. Cellulose, which forms the major constituent of natural fibres, is hydrophilic in nature and it absorbs water molecules. Even though cellulose has a high –OH to C ratio, not all –OH groups are exposed or accessible as cellulose is semicrystalline (Pott 2004). The highly crystalline region of cellulose is virtually inaccessible to water molecules but the water molecules are able to penetrate and gain access into the amorphous region of cellulose. Hemicellulose, on the other hand, is predominantly amorphous (Hansen and Plackett 2008). It is highly accessible to water molecules as it has a high –OH to carbon (C) ratio. Lignin, however, has a low –OH to C ratio and it is hydrophobic in nature (Bismarck *et al.* 2005). As natural fibres absorb water molecules, they swell up owing to water molecules occupying the space in between the microfibrils. This space that water molecules occupy is termed the transient microcapillary network (Hill *et al.* 2009b). The water molecules within natural fibres can either form a monolayer, which associate closely with the available –OH groups, or form a multilayer at which not all water molecules are in intimate contact with available –OH groups. It has been shown that the water molecules in a monolayer are more mobile than the water molecules located in a multilayer (Walker 2006).

It was found that the moisture sorption isotherm of natural fibres follows an International Union of Pure and Applied Chemistry (IUPAC) type II isotherm. Water sorption hysteresis (the difference between the sorption and desorption loop) has also been observed (Pott 2004; Hill *et al.* 2009a; 2009b). The sigmoidal nature of the type II isotherm can be further categorised into three regions. Region 1 is typically assigned to relative humidity (RH) of between 0 and 15%, where the dominant moisture sorption mechanism is monolayer adsorption of water molecules onto the cell wall surface of natural fibres. In region 2, the water molecules form a multilayer in the transient microcapillary network between 15 and 70% RH. Beyond 70% RH (region 3), capillary condensation is the dominant mechanism for water molecules absorption. It must be noted that at low RH, adsorption is the dominant mechanism but at higher RH, absorption is the dominant mechanism. In this context, absorption refers to the uptake of water by natural fibres owing to surface tension forces. The heat required to evaporate absorbed liquid is only slightly higher than the energy required to evaporate a drop of liquid from a flat surface (Walker 2006). Adsorption, on the other hand, is based on the attractive forces between the water molecules and natural fibres as a result of hydrogen bonding or van der Waals forces (Walker 2006).

When discussing the moisture uptake of natural fibres, the concept of the fibre saturation point must be introduced. The fibre saturation point is the moisture content at which absorbed water molecules are removed but the cell walls of the fibres are still saturated (Tiemann 1906) by adsorbed water molecules. This point usually occurs around a moisture content of between 25 and 35% RH (Walker 2006). It can also be calculated by fitting the water sorption isotherm to 100% RH (Hill *et al.* 2009b). However, the fibre saturation point has been shown to be difficult to determine and does not exist experimentally in practice (Hernandez and Pontin 2006; Almeida and Hernandez 2006a; 2006b). The main reason for this lies in the concept of the fibre saturation point. It is assumed that there exist a distinct cut-off point between a fully saturated cell wall and the total removal of absorbed water molecules (Hill *et al.* 2009b). However, at high RH, capillary condensation begins to fill the lumen of natural fibres, causing a sharp increase in the water sorption curve (Walker 2006).

10.2   Methods of quantifying moisture uptake of natural fibres

In general, the moisture uptake behaviour of natural fibres can be measured by two different methods, involving exposing natural fibres to the required

RH, either at constant RH (Baltazar-y-Jimenez and Bismarck 2007) or varying RH as a function of time (Gouanve *et al.* 2006; Hill *et al.* 2009b). In addition to these methods, streaming potential (ζ-potential) measurements, which provide information about the formation of the electrochemical double layer between a solid substrate and an electrolyte solution, could also be used to quantify the moisture uptake of natural fibres. Natural fibres swell in water, causing a decay of the ζ-potential as a function of time (Kanamaru 1960). The difference between the starting ζ-potential (at $t = 0$) and the ζ-potential approaching its asymptotic value (at $t = \infty$) is proportional to the swelling of natural fibres (Kanamaru 1960; Bismarck *et al.* 2000; Bismarck *et al.* 2002; Baltazar-y-Jimenez and Bismarck 2007). The utilisation of streaming ζ-potential to determine the moisture uptake of natural fibres is covered in chapter 7.

10.2.1 Simple weight gain determination of the moisture content of natural fibres

Weight gain measurements utilise humidity chambers or desiccators set up at specified RH using distilled water at room temperature (Bismarck *et al.* 2002; Baltazar-y-Jimenez and Bismarck 2007). The RH of the humidity chamber can be adjusted using salt solutions. In order to obtain more accurate results, the fibre bundles are dried overnight in an oven before placing them in the chambers. The weight difference before and after exposure to the specified RH were then measured at different time intervals and the moisture content calculated using the equation shown:

$$MC = \frac{(m - m_0)}{m_0} \times 100\% \qquad [10.1]$$

where MC = moisture content, m = mass of the sample after exposing to specified RH and m_0 = dry mass of natural fibres.

A variation of this method can be performed by immersing the fibres in distilled water instead of a humidity chamber at specified RH (Sreekala and Thomas 2003). The increase in the weight of the fibres was recorded at specific time intervals. The authors found that the values obtained from this process are highly reproducible. The water uptake by the fibres (in mol percent of water uptake per mass of fibres) can be evaluated using:

$$X(t) = \frac{m_{\text{water}}(t)}{18 m_{\text{initial}}} \times 100\% \qquad [10.2]$$

where $X(t)$ = moles of water absorbed per unit mass of fibres, $m_{\text{water}}(t)$ = equilibrium mass of water at time t and m_{initial} = initial mass of the fibres. This analysis can be extended further by fitting it into the empirical equation [10.3] (Khinnavar and Aminabhavi 1991; Sreekala and Thomas 2003):

$$\log\left(\frac{X(t)}{X(t=\infty)}\right) = \log(k) + n\log(t) \qquad [10.3]$$

where k and n are parameters fitted through linear regression, k is a parameter that depends on the interaction between the solid substrate and the water molecules, and n gives an indication about the types of mass transport phenomena (Fickian or Knudsen).

10.2.2 Dynamic vapour sorption in measurement of the water sorption isotherm of natural fibres

Dynamic vapour sorption (DVS) is very useful in the accurate measurement of sorption isotherms at various temperatures and relative humidity. The vapour phase can also be replaced; instead of water, other organic solvents can be used as long as Antoine's equation (see for instance: Perry and Green 1997) for the particular solvent is available. Antoine's equation describes the pure saturated vapour pressure of a solvent as a function of temperature. With this equation, the partial pressure of a solvent in the vapour phase can be predicted accurately. Using different organic solvents proves to be useful when studying the sorption isotherm of (modified) natural fibres. The use of DVS to study the moisture uptake of natural fibres has been studied extensively (Gouanve *et al.* 2006; Bessadok *et al.* 2007; Bessadok *et al.* 2008; Collet *et al.* 2008; Alix *et al.* 2009; Hill *et al.* 2009b). Such extensive use of DVS is not surprising as this technique is well established. In addition to this, natural fibres can be exposed to varying RH (as a function of time and temperature) to study the effect of different RH *in situ* on the moisture uptake of natural fibres.

The apparatus setup for a DVS instrument is shown in Fig. 10.1. The setup contains two measuring pans (one reference pan and one sample pan) suspended from the arm of an ultra-sensitive microbalance (up to 0.05 μg resolution). The measuring pans are connected to the microbalance by hanging wires and both the pans are situated in their respective chambers. These chambers are located in a temperature-controlled environment. A flow of dry gas along with the correct amount of water vapour is then passed through the chambers to maintain the required RH.

10.3   Moisture uptake behaviour of various natural fibres

10.3.1 Moisture content of natural fibres based on weight gain measurements

Table 10.1 shows the moisture content of various natural fibres evaluated using simple weight gain measurements. The equilibrium moisture content

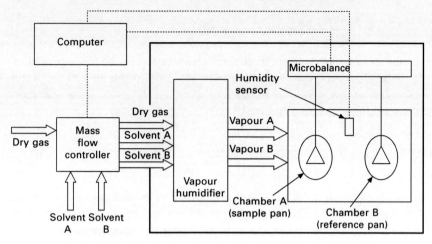

10.1 A schematic diagram showing the apparatus setup for a dynamic vapour sorption surface measurement system.

Table 10.1 Equilibrium moisture content of various natural fibres at 100% RH (unless indicated). Results are obtained from simple weight gain measurement

Fibre	Moisture content (wt%)	References
Flax	7	
Hemp	7	
Abaca fine	7	
Abaca bold	7	
Luffa	6 *	Baltazar-y-Jimenez
Henequen	8	and Bismarck 2007
Sisal	7	
Lechuguilla	7	
Cornhusk	8	
Sisal*	5.8 (neat), 6.5 (dewaxed), 6.5 (2% alkaline treated), 7.5 (5% alkaline treated)	Bismarck *et al.* 2001
Coir*	8.6 (neat), 6.5 (dewaxed), 9.1 (2% alkaline treated), 9.2 (5% alkaline treated)	Bismarck *et al.* 2001
Green flax	42.58	Bismarck *et al.* 2002
Dew-retted flax	26.57	Bismarck *et al.* 2002
Duralin flax	19.22	Bismarck *et al.* 2002
Green flax	3.6 (20% RH), 15.0 (66% RH), 24.0 (93% RH), 42.6 (100% RH)	Stamboulis *et al.* 2000
Duralin flax	2.7 (20% RH), 10.8 (66% RH), 9.0 (93% RH), 14.44 (100% RH)	Stamboulis *et al.* 2000

*Moisture content of fibres at 65% RH.

of different natural fibres is around 7 wt% at 100% RH. It is possible to increase the moisture content of sisal fibres from 5.8 wt% to 6.5 wt% by dewaxing the fibres. This can be done by either heating the fibres in a mixture of ethanol and benzene (a ratio of 1:2) for 72 h at 50 °C (Bismarck *et al.* 2001) or Soxhlet extraction in acetone for 1 h (Mukherjee and Satyanarayana 1984; Juntaro *et al.* 2008) to remove waxy substances. It seems, however, that dewaxing of coir fibres reduces its moisture content. Unfortunately, no explanation was given by the authors. Alkaline treatment is known to remove fatty acids from the surfaces of natural fibres (Geethamma *et al.* 1995). As a result, the moisture uptake of alkaline treated sisal and coir fibres increased when compared with neat fibres. Bismarck *et al.* (2001) also observed an increase in the thermal degradation temperature of sisal and coir fibres. This is thought to be due to the removal of compounds, such as wax or fatty acids, that decompose earlier than the major constituent in natural fibres.

Bismarck *et al.* (2002) investigated the moisture uptake behaviour of flax fibres extracted from flax stems by various treatments; namely green flax, dew-retted flax and Duralin flax. Dew-retting is a fibre extraction process that relies on the natural colonisation of aerobic fungi in the stalks and stems of fibres just after harvesting. The process to obtain Duralin flax is developed by CERES B.V., which uses deseeded straw of flax fibres as starting material. It turns out that the use of the straw of flax fibres is beneficial for both strength and reproducibility (Stamboulis *et al.* 2000). More importantly, no retting process is required to extract natural fibres from the straw. A description of the Duralin process can be found in the literature (Stamboulis *et al.* 2000; Stamboulis *et al.* 2001; Bismarck *et al.* 2002). Green flax takes up 40% more moisture than dew-retted flax. Duralin flax, on the other hand, shows the lowest moisture uptake. This is a result of the Duralin process, which depolymerises hemicellulose and lignin into lower molecular weight compounds. These compounds subsequently cure into a water resistant resin (Stamboulis *et al.* 2000), which leads to the lower moisture uptake of Duralin flax than green flax.

10.3.2 Moisture content of various natural fibres based on DVS

A typical water sorption curve obtained by DVS is shown in Fig. 10.2. The difference between the sorption and desorption loop is termed water sorption hysteresis. Hill *et al.* (2009b) studied the water sorption behaviour of various natural fibres. From Table 10.2, it can be seen that jute, coir and Sitka spruce (a type of wood fibre) show higher moisture content at 95% RH than hemp, flax and cotton. One of the major differences between these two groups of fibres is their lignin content. Jute, coir and Sitka spruce have a higher lignin content than hemp, flax and cotton (Bismarck *et al.* 2005).

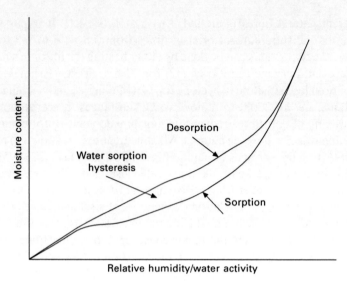

10.2 A typical water sorption curve from dynamic vapour sorption.

Table 10.2 Moisture uptake of various natural fibres measured using DVS at 95% RH. Adapted from Hill *et al.* 2009b

Fibres	Moisture uptake (wt%)	Maximum hysteresis* (wt%)
Jute	25	3.1
Coir	31	3.3
Flax	20	1.6
Sitka spruce	23	3.0
Hemp	25	1.8
Cotton	14	2.0

*Maximum hysteresis occurs at different RH for different types of fibres.

The high moisture uptake of highly lignified natural fibres, such as coir and jute fibres, might be a direct result of the amorphous nature of lignin. Even though amorphous lignin is hydrophobic, it is able to deform to accommodate more water in the cell wall. This argument is also in agreement with the moisture uptake of cotton. Cotton is almost pure cellulose but it has the lowest moisture uptake. The main reason for this observation is that cotton is highly crystalline with low amorphous content. Cotton consists of nearly 90 wt% cellulose molecules with no lignin content (Bismarck *et al.* 2005) and the crystallinity index of cellulose was found to be 80% (Segal *et al.* 1959). Therefore, most of the cellulosic –OH groups are not accessible to water molecules, leading to the observed low moisture uptake.

Table 10.2 also shows that the highly lignified fibres have larger water sorption hysteresis than flax, hemp and cotton fibres. Owing to the presence of amorphous lignin, the water molecules follow different pathways during absorption and desorption; that is, water sorption and desorption occur in two different physical states of the natural fibres (Lu and Pignatello 2004). Tvardovski *et al.* (1997) proposed that the deformation of sorbents is one of the main causes for sorption hysteresis. This concept could be applied to natural fibres. Swelling of natural fibres occurs as a result of the motion of incoming water molecules during water sorption, creating more voids. This swelling effect is confirmed by ζ-potential measurements (Stana-Kleinschek and Ribitsch 1998; Stana-Kleinschek *et al.* 1999; Bismarck *et al.* 2000; Bismarck *et al.* 2001; Bismarck *et al.* 2002; Baltazar-y-Jimenez and Bismarck 2007). Upon desorption of water molecules, there might be a lag between the water molecules leaving the voids and the relaxation of the sorbent to its original state (Lu and Pignatello 2002), which causes the observed hysteresis because sorption and desorption occur in two different physical environments.

10.3.3 Determination of exposed –OH groups of cellulose by using heavy water

Cellulose is a semicrystalline polymer with ordered regions (lower chemical reactivity) and disordered regions (higher reactivity). Therefore, it is of particular interest to characterise the 'accessible' regions of cellulose. This can be done by hydrogen–deuterium exchange. It is well known that deuterium can replace hydrogen in cellulose (Frilette *et al.* 1948; Sepall and Mason 1961). When cellulose is annealed at 260 °C in 0.1M NaOD, all the hydrogen atoms (including the crystalline core of cellulose) will be converted to deuterium atoms (Wada *et al.* 1997). By exposing the deuterium-exchanged cellulose to water under normal condition, the accessible –OD groups will be converted back to –OH, whereas the core of cellulose will stay as –OD groups. By obtaining the IR spectra or measuring the mass increase owing to hydrogen–deuterium exchange of cellulose, the accessibility and crystallite size of cellulose can be obtained (Marrinan and Mann 1954; Horikawa and Sugiyama 2008; Lee *et al.* 2010). Table 10.3 tabulates the exposed –OH to core –OH ratio of cellulose from various sources.

Table 10.3 Accessible OH groups from various cellulose sources

Types of cellulose	Exposed OH to core OH ratio	References
Cotton	0.695	Marrinan and Mann 1954
Viscose rayon	2.125	Marrinan and Mann 1954
Cellulose microfibrils (*Valonia ventricosa*)	0.934	Horikawa and Sugiyama 2008
Bacterial cellulose (*Nata-de-coco*)	0.705	Lee *et al.* 2010

The same analysis can be extended to determine the exposed –OH groups of cellulose in a DVS setup (Lee *et al.* 2010). The authors pre-conditioned highly crystalline bacterial cellulose (BC) for 10 h at 0% RH to remove any adsorbed water molecule. The RH of D_2O was then increased to 90% for 2 h and reduced to 0% for 2 h. This cycle was repeated 10 times to enable the adsorption of D_2O molecule on the surface of BC but avoiding bulk sorption of BC by D_2O. BC was then post-conditioned for 10 h at 0% RH to remove any adsorbed D_2O molecules. The amount of exposed –OH groups can be determined from the mass increase of BC (deuterium is 1.67 $\times 10^{-27}$ kg heavier than hydrogen, see equation [10.4]). This method could potentially be applied to natural fibres to determine the amount of exposed –OH groups or to study the moisture uptake behaviour of natural fibres (Hill *et al.* 2009b).

$$\Delta m = \frac{DSS \times m_i \times A \times m_n}{162\,000} \qquad [10.4]$$

where Δm = mass change after hydrogen–deuterium exchange (mg), *DSS* = degree of surface substitution, m_i = initial mass of sample (mg), A = Avogadro's number and m_n = mass of a neutron (mg).

10.4   Summary

Traditionally, natural fibres such as cotton, hemp or flax are widely used in the textile industry for the production of fabrics and yarns. Currently, there is a growing interest in utilising natural fibres as reinforcement for renewable polymers. One of the major problems of utilising natural fibres in composite materials is the high moisture uptake of natural fibres, which leads to poorer mechanical properties of the natural fibres reinforced composites. Therefore, it is important to study the moisture uptake behaviour of natural fibres.

The moisture uptake of natural fibres is complex owing to the composite nature of natural fibres themselves; the fibres contain amorphous and crystalline cellulose, hemicellulose and lignin. Water molecules do not form a single layer associated with the –OH groups. Instead, multiple layers of water molecules are associated with the –OH groups, depending on the humidity. Water sorption hysteresis is also widely encountered in the moisture uptake of natural fibres. So far, no plausible mechanism has been proposed to describe this phenomenon. In terms of the measurement of water sorption behaviour of natural fibres, there are two methods that are used widely; weight gain measurement and DVS. Both measurements give valuable information regarding the moisture uptake and water sorption of natural fibres.

Simple weight gain measurements describe the moisture content at a fixed relative humidity. This measurement can mimic the real condition where the natural fibres will be utilised. By using DVS, the moisture uptake of natural fibres at different RH (and temperature) is obtained and studied. The results are fitted into various models to further study the water sorption of natural fibres. A novel method based on hydrogen–deuterium exchange using DVS provides information regarding the accessibility of cellulosic –OH groups, and thus further information regarding the water sorption behaviour of natural fibres.

The challenge now is to understand the moisture uptake mechanism of natural fibres by combining different measuring approaches. This could include separating the effects of different chemical constituents on the water sorption of natural fibres at different RH. Because the water sorption of natural fibres is always in equilibrium with the exposed humidity, a thermodynamic approach should be included, along with sorption/desorption isotherm to describe the kinetics and mechanism of the moisture uptake of natural fibres.

10.5   Acknowledgements

The authors would like to thank the UK Engineering and Physical Science Research Council (EPSRC) for funding (EP/F032005/1) and the Deputy Rector's award (Imperial College London) for funding KYL.

10.6   References

Akil, H. M., L. W. Cheng, Z. A. M. Ishak, A. A. Bakar and M. A. A. Rahman (2009). 'Water absorption study on pultruded jute fibre reinforced unsaturated polyester composites'. *Composites Science and Technology* **69**(11–12): 1942–1948.

Alix, S., E. Philippe, A. Bessadok, L. Lebrun, C. Morvan and S. Marais (2009). 'Effect of chemical treatments on water sorption and mechanical properties of flax fibres'. *Bioresource Technology* **100**(20): 4742–4749.

Almeida, G. and R. E. Hernandez (2006a). 'Changes in physical properties of tropical and temperate hardwoods below and above the fiber saturation point'. *Wood Science and Technology* **40**(7): 599–613.

Almeida, G. and R. E. Hernandez (2006b). 'Changes in physical properties of yellow birch below and above the fiber saturation point'. *Wood and Fiber Science* **38**(1): 74–83.

Baltazar-y-Jimenez, A. and A. Bismarck (2007). 'Wetting behaviour, moisture up-take and electrokinetic properties of lignocellulosic fibres'. *Cellulose* **14**(2): 115–127.

Baltazar-y-Jimenez, A., M. Bistritz, E. Schulz and A. Bismarck (2008a). 'Atmospheric air pressure plasma treatment of lignocellulosic fibres: impact on mechanical properties and adhesion to cellulose acetate butyrate'. *Composites Science and Technology* **68**(1): 215–227.

Baltazar-y-Jimenez, A., J. Juntaro and A. Bismarck (2008b). 'Effect of atmospheric air pressure plasma treatment on the thermal behaviour of natural fibres and dynamical mechanical properties of randomly-oriented short fibre composites'. *Journal of Biobased Materials and Bioenergy* **2**(3): 264–272.

Bessadok, A., S. Marais, F. Gouanvé, L. Colasse, I. Zimmerlin, S. Roudesli and M. Métayer (2007). 'Effect of chemical treatments of Alfa (*Stipa tenacissima*) fibres on water-sorption properties'. *Composites Science and Technology* **67**(3–4): 685–697.

Bessadok, A., S. Marais, S. Roudesli, C. Lixon and M. Métayer (2008). 'Influence of chemical modifications on water-sorption and mechanical properties of Agave fibres'. *Composites Part A-Applied Science and Manufacturing* **39**(1): 29–45.

Bismarck, A., I. Aranberri-Askargorta, J. Springer, T. Lampke, B. Wielage, A. Stamboulis, I. Shenderovich and H.-H. Limbach (2002). 'Surface characterization of flax, hemp and cellulose fibers: surface properties and the water uptake behavior'. *Polymer Composites* **23**(5): 872–894.

Bismarck, A., I. Aranberri-Askargorta, J. Springer, A. K. Mohanty, M. Misra, G. Hinrichsen and S. Czapla (2001). 'Surface characterization of natural fibers: surface properties and the water up-take behavior of modified sisal and coir fibers'. *Green Chemistry* **3**(2): 100–107.

Bismarck, A., S. Mishra and T. Lampke (2005). Plant fibers as reinforcement for green composites. *Natural fibers, biopolymers and biocomposites*. A. K. Mohanty, M. Misra and L. Drzal. Boca Raton, CRC Press.

Bismarck, A., J. Springer, A. K. Mohanty, G. Hinrichsen and M. A. Khan (2000). 'Characterization of several modified jute fibers using zeta-potential measurements'. *Colloid and Polymer Science* **278**: 229–235.

Bledzki, A. K. and J. Gassan (1999). 'Composites reinforced with cellulose based fibres'. *Progress in Polymer Science* **24**(2): 221–274.

Collet, F., M. Bart, L. Serres and J. Miriel (2008). 'Porous structure and water vapour sorption of hemp-based materials'. *Construction and Building Materials* **22**(6): 1271–1280.

Dhakal, H. N., Z. Y. Zhang, and M. O. W. Richardson (2007). 'Effect of water absorption on the mechanical properties of hemp fibre reinforced unsaturated polyester composites'. *Composites Science and Technology* **67**(7–8): 1674–1683.

Eichhorn, S. J., C. A. Baillie, N. Zafeiropoulos, L. Y. Mwaikambo, M. P. Ansell, A. Dufresne, K. M. Entwistle, P. J. Herrera-Franco, G. C. Escamilla, *et al.* (2001). 'Review: Current international research into cellulosic fibres and composites'. *Journal of Materials Science* **36**(9): 2107–2131.

EU (1999). Directive 1999/31/EC of the European Parliament and of the Council 26 April 1999 on the Landfill of Waste. *Official Journal of the European Communities* (16/7/1999): L 182/1.

EU (2000). Directive 2000/53/EC of the European Parliament and of the Council of 19 September 2000 on end-of-life vehicles. *Official Journal of the European Communities* (21/10/2000): L 269/34.

EU (2003). Directive 2002/96/EC of the European Parliament and of the Council of 27 January 2003 on waste electrical and electronic equipment (WEEE). *Official Journal of the European Union* (13/2/2003): L 37/24.

Frilette, V. J., J. Hanle, and H. Mark (1948). 'Rate of exchange of cellulose with heavy water'. *Journal of the American Chemical Society* **70**(3): 1107–1113.

Geethamma, V. G., R. Joseph and S. Thomas (1995). 'Short coir fiber-reinforced natural-rubber composites – effects of fiber length, orientation, and alkali treatment'. *Journal of Applied Polymer Science* **55**(4): 583–594.

Gouanve, F., S. Marais, A Bessadok, D. Langevin, C Monvor and M. Métayer (2006). 'Study of water sorption in modified flax fibers'. *Journal of Applied Polymer Science* **101**(6): 4281–4289.

Hansen, N. M. L. and D. Plackett (2008). 'Sustainable films and coatings from hemicelluloses: a review'. *Biomacromolecules* **9**(6): 1493–1505.

Hernandez, R. E. and M. Pontin (2006). 'Shrinkage of three tropical hardwoods below and above the fiber saturation point'. *Wood and Fiber Science* **38**(3): 474–483.

Hill, C. A. S., A. J. Norton, and G. Newman (2009a). Natural fibre insulation materials – the importance of hygroscopicity in providing indoor climate control. *Proceedings of the 11th international conference on non-conventional materials and technologies*, 6–9 September 2009 Bath, UK.

Hill, C. A. S., A. J. Norton and G. Newman (2009b). 'The water vapor sorption behavior of natural fibers'. *Journal of Applied Polymer Science* **112**(3): 1524–1537.

Horikawa, Y. and J. Sugiyama (2008). 'Accessibility and size of Valonia cellulose microfibril studied by combined deuteration/rehydrogenation and FTIR technique'. *Cellulose* **15**(3): 419–424.

Huda, M. S., L. T. Drzal, A. K. Mohanty and M. Misra (2008). 'Effect of fiber surface treatments on the properties of laminated biocomposites from poly(lactic acid) (PLA) and kenaf fibers'. *Composites Science and Technology* **68**(2): 424–432.

Huda, S., N. Reddy, D. Karst, W. Xu, W. Yang and Y. Yang (2007). 'Nontraditional biofibers for a new textile industry'. *Journal of Biobased Materials and Bioenergy* **1**(2): 177–190.

Juntaro, J., M. Pommet, G. Kalinka, A. Mantalaris, M. S, P. Shaffer and A. Bismarck (2008). 'Creating hierarchical structures in renewable composites by attaching bacterial cellulose onto sisal fibers'. *Advanced Materials* **20**(16): 3122–3126.

Juntaro, J., M. Pommet, A. Mantalaris, M. Shaffer and A. Bismarck (2007). 'Nanocellulose enhanced interfaces in truly green unidirectional fibre reinforced composites'. *Composite Interfaces* **14**(7–9): 753–762.

Kanamaru, K. (1960). 'Wasseraufnahme in ihrer Beziehung zur zeitlichen Erniedrigung des Z-Potentials von Fasern in Wasser'. *Kolloid-Z* **168**(2): 115–121.

Khinnavar, R. S. and T. M. Aminabhavi (1991). 'Diffusion and sorption of organic liquids through polymer membranes. 1. Polyurethane versus normal-alkanes'. *Journal of Applied Polymer Science* **42**(8): 2321–2328.

Lee, K.-Y., F. Quero, *et al.* (2010). 'Surface modification of bacterial cellulose nanofibrils with organic acids'. Submitted for publication.

Lu, Y. F. and J. J. Pignatello (2002). 'Demonstration of the "conditioning effect" in soil organic matter in support of a pore deformation mechanism for sorption hysteresis'. *Environmental Science & Technology* **36**(21): 4553–4561.

Lu, Y. F. and J. J. Pignatello (2004). 'History-dependent sorption in humic acids and a lignite in the context of a polymer model for natural organic matter'. *Environmental Science & Technology* **38**(22): 5853–5862.

Marrinan, H. J. and J. Mann (1954). 'A study by infrared spectroscopy of hydrogen bonding in cellulose'. *Journal of Applied Chemistry* **4**: 204–211.

Mukherjee, P. S. and K. G. Satyanarayana (1984). 'Structure and properties of some vegetable fibers. 1. Sisal fiber'. *Journal of Materials Science* **19**(12): 3925–3934.

Perry, R. H. and D. W. Green (1997). *Perry's Chemical Engineering Handbook*. New York, McGraw-Hill.

Pommet, M., J. Juntaro, *et al.* (2008). 'Surface modification of natural fibers using bacteria: depositing bacterial cellulose onto natural fibers to create hierarchical fiber reinforced nanocomposites'. *Biomacromolecules* **9**(6): 1643–1651.

Pott, G. T. (2004). Natural fibers with low moisture sensitivity. *Natural fibers, plastics and composites*. F. T. Wallenberger and N. Weston. Norwell, Kluwer Academic: 105–122.

Rashdi, A. A. A., S. M. Sapuan, *et al.* (2009). 'Water absorption and tensile properties of soil buried kenaf fibre reinforced unsaturated polyester composites (KFRUPC)'. *Journal of Food Agriculture & Environment* **7**(3–4): 908–911.

Sahoo, S., A. Nakai, *et al.* (2007). 'Mechanical properties and durability of jute reinforced thermosetting composites'. *Journal of Biobased Materials and Bioenergy* **1**(3): 427–436.

Satyanarayana, K. G., G. G. C. Arizaga, *et al.* (2009). 'Biodegradable composites based on lignocellulosic fibers–an overview'. *Progress in Polymer Science* **34**(9): 982–1021.

Segal, L., J. J. Creely, *et al.* (1959). 'An empirical method for estimating the degree of crystallinity of native cellulose using the x-ray diffractometer'. *Textile Research Journal* **29**: 786–794.

Sepall, O. and S. G. Mason (1961). 'Hydrogen exchange between cellulose and water. II. Interconversion of accessible and inaccessible regions'. *Canadian Journal of Chemistry* **39**: 1944–1955.

Sreekala, M. S. and S. Thomas (2003). 'Effect of fibre surface modification on water-sorption characteristics of oil palm fibres'. *Composites Science and Technology* **63**(6): 861–869.

Stamboulis, A., C. A. Baillie, *et al.* (2000). 'Environmental durability of flax fibres and their composites based on polypropylene matrix'. *Applied Composite Materials* **7**(5–6): 273–294.

Stamboulis, A., C. A. Baillie, *et al.* (2001). 'Effects of environmental conditions on mechanical and physical properties of flax fibers'. *Composites Part A–Applied Science and Manufacturing* **32**(8): 1105–1115.

Stana-Kleinschek, K. and V. Ribitsch (1998). 'Electrokinetic properties of processed cellulose fibers'. *Colloids and Surfaces A–Physicochemical and Engineering Aspects* **140**(1–3): 127–138.

Stana-Kleinschek, K., S. Strnad, *et al.* (1999). 'Surface characterization and adsorption abilities of cellulose fibers'. *Polymer Engineering and Science* **39**(8): 1412–1424.

Tiemann, H. D. (1906). 'Effect of moisture upon the strength and stiffness of wood'. *US Forest Service Bulletin* **70**.

Tvardovski, A. V., A. A. Fomkin, *et al.* (1997). 'Hysteresis phenomena in the study of sorptive deformation of sorbents'. *Journal of Colloid and Interface Science* **191**(1): 117–119.

Wada, M., T. Okano, *et al.* (1997). 'Synchrotron-radiated x-ray and neutron diffraction study of native cellulose'. *Cellulose* **4**(3): 221–232.

Walker, J. C. F. (2006). *Primary Wood Processing: Principles and Practice*. The Netherlands, Springer.

11
Creep and fatigue of natural fibre composites

M. MISRA, S. S. AHANKARI and A. K. MOHANTY,
University of Guelph, Canada and A. D. NGO, University of
Quebec, Canada

Abstract: The growth in applications of natural fibre polymer composites (NFPCs) has increased the importance of understanding time-dependent viscoelastic properties such as creep resistance, stress relaxation, and fatigue, all of which are covered in this chapter. The fundamentals of measurement of these properties are discussed, and the use of short- and long-term prediction of creep resistance of the NFPCs to achieve adequate long-term performance is outlined. The effects of the interfacial interaction between natural fibres and polymers on fatigue and stress relaxation properties of the NFPCs are explored.

Key words: natural fibre polymer composites, creep, life prediction, stress relaxation, fatigue.

11.1 Introduction

The present generation demands technological advancement as well as an environmental awareness for the sustainable development of engineering components. This concern has created an enormous interest in employing natural fibres as reinforcements in polymer composites. The market for natural fibre reinforced polymer composites (NFPC) has experienced tremendous growth in recent years as it is exposed to many different noncritical applications. Natural fibre/plastic composites, which were once relegated to the world of thermosets (phenolics), have now entered the market for products such as flooring, furniture, automotive interior parts, domestic appliances, decking and marine parts.[1-4]

In 2000, the commercialization of these natural fibre/plastic composites in North America surpassed $150 million. By the end of 2010, however, it will reach nearly $3 billion in sales, with a projected growth rate of 60% per year.[5] With this striking growth rate, a comprehensive analysis of the natural fibre/polymer composites market should be considered with regard to the following information:

(a) assessment of the current (2010) and expected future (2020) of the global market;
(b) identification of the key growth segments for these composites, with an emphasis on the ecological and environmental benefits;

289

(c) identification of the suppliers contributing at all levels to preserve the value chain from fibre suppliers to composite product manufacturers.

A good indication for natural fibre reinforced polymer composites is that the search for the key segments for their growth and exploration is being recognized with full dynamism. They are being increasingly used in structural and other applications, such as construction beams, bolted plates, acoustics materials, hoses, gaskets and mounts. Natural fibres have appealing features such as low density, biodegradability,[6] sound mechanical properties and low cost. Short-term and time-independent mechanical properties (strength, modulus, etc.) can be used for a material for a given particular application, but are not enough to determine the effective life cycle of any product. Polymer composites exhibit time-dependent mechanical properties and the changes in stress/strain that occur over time are still observed, even decades later. The increasing number of applications of natural fibre composites have increased the importance of focusing on time-dependent viscoelastic properties, such as creep resistance, stress relaxation.

Creep is the permanent deformation that occurs when a material is under constant stress and is a function of time, temperature and material properties. The term stress relaxation refers to the way in which viscoelastic polymers relieve the stresses under constant strain in a nonlinear, non-Hookean fashion. Creep deformation can go beyond the creep limit, leading to the failure of the component, especially in applications with long-term loading. A proper understanding, measurement, and short and long-term prediction of the creep resistance of the natural fibre reinforced polymer composites (NFPC) component is required in order to design a component for an adequate long-term performance.

11.2   Fundamentals of the creep test

In the creep test, the stress is applied suddenly and is kept constant throughout the test, while the strain is measured as a function of time. The result is generally expressed as a time-dependent compliance. The recovery test generally follows the creep test and centres on the study of how the material relaxes when the material is suddenly unstressed. The stress relaxation test is the opposite of the creep test. In the stress relaxation test, the specimen is strained to a constant elongation and the stress required to maintain the same elongation is measured as a function of time. This result is generally expressed as a time-dependent modulus. A schematic representation is displayed in Fig. 11.1.

Polymers are viscoelastic, i.e., with the application of a load, stress and strain become functions of time. When a polymer is subjected to a constant load, it deforms elastically (time-independent deformation) as shown in Fig. 11.2. Afterwards, it will continue to deform slowly with time until

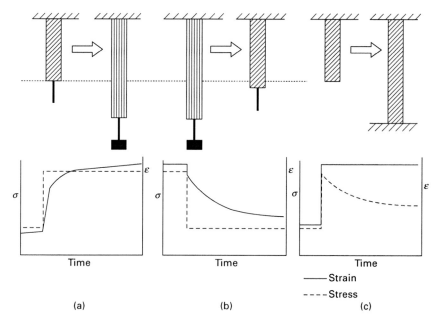

11.1 Typical curves displaying the variation of stress and strain with time in (a) creep test, (b) recovery test and (c) stress relaxation test.

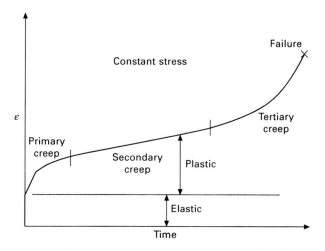

11.2 Variation of strain with time for a given viscoelastic material at constant stress.

yielding or rupture causes failure. The creep region is divided into three main regions: primary, secondary, and tertiary. The primary stage is an early stage of loading wherein the creep rate decreases with time. Then it enters the secondary region where the strain rate is constant with time (steady strain creep rate) followed by a rapid increase in the strain rate in the tertiary zone and fracture. The standard test for creep characterization is ASTM D2290.

A general creep recovery graph for a given polymer displays the following regions:

(a) instantaneous time-independent elastic deformation;
(b) linear viscoelastic region (compliance is independent of stress);
(c) equilibrium region/nonlinear viscoelastic region (compliance is dependent on stress);
(d) recovery region.

A creep study in the linear viscoelastic region is generally carried out in order to identify the polymer properties at equilibrium such as modulus (E), viscosity (η), and compliance (D). It is important to distinguish the region in which the viscoelastic properties are linear.

11.2.1 Linear and nonlinear viscoelasticity

For a given constant load σ_1, the measurement of time dependent strain ε_1 is shown in Fig. 11.3. The specimen is allowed to recover and a greater stress σ_2 is applied. The time dependence of the strain ε_2 is also shown in Fig. 11.3.

At a given time t_1 and t_2 after stretching, if ε_1 and ε_2 are linear with the representing stresses, σ_1 and σ_2, the stress strain relationship can be given by:

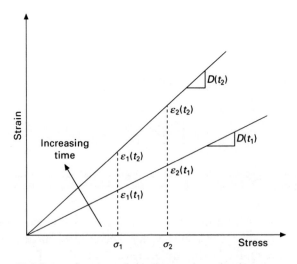

11.3 Determination of the linear viscoelastic creep.

$$\frac{\varepsilon_1(t_1)}{\sigma_1(t_1)} = \frac{\varepsilon_2(t_1)}{\sigma_2(t_1)}; \frac{\varepsilon_1(t_2)}{\sigma_1(t_2)} = \frac{\varepsilon_2(t_2)}{\sigma_2(t_2)}$$

Thus, for any arbitrary time t, the strains at the two stresses can be expressed as:

$$\frac{\varepsilon_1(t)}{\sigma_1(t)} = \frac{\varepsilon_2(t)}{\sigma_2(t)}$$

The strains in the above experiments are proportional to the applied stresses. In general, for a given applied stress σ, creep compliance $D(t)$ is a ratio of strain to stress at a given time t.

$$D(t) = \frac{\varepsilon(t)}{\sigma}$$

This property is described as linear viscoelasticity. In the linear range, the compliance is independent of the stresses employed in the creep test (σ_1, σ_2 or any other stress values). The strain limits in which a specimen is linear viscoelastic can be ascertained by a simple creep test. A transition from the linear to the nonlinear viscoelastic region is shown in Fig. 11.4.

Most polymers usually demonstrate a linear viscoelastic property at low stresses (maximum strain corresponding to ~0.005 or less). At higher stress levels, the material establishes nonlinear viscoelastic behaviour and does not keep the same linear relation between stress and strain. Nunez *et al.* carried out short term and long-term creep tests on polypropylene/wood flour composites. The creep deformation for strain–time plots was linear up to 10 MPa.[7] This signified that the maximum applied stress that works in the linear viscoelastic range is 10 MPa.

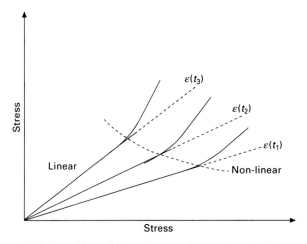

11.4 A transition of linear to nonlinear viscoelastic creep.

11.2.2 Superposition principles

The two superposition principles are significant in anticipating the creep behaviour of plastic materials when they are in the linear viscoelastic range. First, the Boltzmann superposition principle (BSP), depicts the response of a material to various loading histories. The second is the time–temperature-superposition principle, also known as the Williams–Landel–Ferry (WLF) equation, that identifies the equivalence of time and temperature. This will be discussed in the next section.

The Boltzmann superposition principle is applied and limited to the linear viscoelastic region of the polymers.[8-10] It states that the response of a polymer to a given loading is independent of the responses to any other loadings already present in it. Thus, each loading step forms an independent contribution to the final strain, so that the cumulative strain is obtained through the addition of all the contributions.

At time τ_1, stress σ_1 is applied and the strain induced is given by:

$$\varepsilon_1(t) = \sigma_1 D(t)$$

According to linear viscoelasticity, the compliance $D(t)$ is independent of stress, i.e. $D(t)$ is the same for all stresses at a particular time. If the stress increment $\sigma_2 - \sigma_1$ is applied at time τ_2 and the strain increases due to the stress increment, $\sigma_2 - \sigma_1$ is:

$$\varepsilon_2(t) = D(t - \tau_1)(\sigma_2 - \sigma_1)$$

Likewise, the strain increase attributed to $\sigma_3 - \sigma_2$ can be given by:

$$\varepsilon_3(t) = D(t - \tau_3)(\sigma_3 - \sigma_2)$$

For creep, the total strain may be expressed by:

$$\varepsilon(t) = \varepsilon_1(t) + \varepsilon_2(t) + \varepsilon_3(t)$$

or

$$\varepsilon(t) = D(t - \tau_1)\sigma_1 + D(t - \tau_2)(\sigma_2 - \sigma_1) + \ldots + D(t - \tau_i)(\sigma_i - \sigma_{i-1})$$

where $D(t) = 1/E(t)$ is the compliance, a characteristic of the polymer at a given temperature and initial stress. In other words, the BSP shows that the effect of a compound cause is the sum of the effects of the individual causes.[11] Another outcome of the BSP is that the recovery is a mirror image of the preceding creep. i.e. creep and recovery are similar in magnitude.[12]

11.3 Life prediction of natural fibre composites using long-term creep analysis

As polymers are viscoelastic, polymer composites are hardly used for structural applications, because of concerns over their long-term durability and dimensional stability. Long-term strength, stiffness, and deformation can be obtained by accelerating the testing conditions, such as by creating higher temperature, stress and humidity than the service conditions. An extrapolation of the data can be achieved by employing a prediction model. The precision of the creep anticipation depends on the accuracy of the model–reality correlation and the validity of the model in the extrapolation range of temperature, stress and time. To date, various creep models have been used to predict the service life of polymers and composites, such as mechanical analogs, Findley's model, Schapery's model, hereditary integrals, thermal activation theory, time–temperature–stress superposition principle and power law.

The approach of the creep study of natural fibre reinforced plastics is generally carried out in the following steps:

(a) Experimental characterization of creep properties of the composites.
(b) Modelling the measured creep data.
(c) Prediction of the long-term creep behaviour of the composites through short term test measurements.

Experimental analysis emphasizes the parameters and factors affecting the creep behaviour of natural fibre composites.

11.3.1 Factors affecting creep of natural fibre composites

A combination of elastic deformation and viscous flow (i.e. viscoelastic deformation) generally causes creep in polymers. Creep in thermoplastic materials can be ascribed to various material properties, such as crystallinity, molecular orientation[13], and environmental and external factors such as applied stress, temperature and humidity.

The creep behaviour of the composite is governed by either the polymer matrix or the reinforcing fibres, depending upon the amount of fibres and fillers and the orientation of the reinforcing fibres within the polymer matrix. For example, the creep deformation of composites which rely upon the reinforcement of long unidirectional fibres is influenced by the creep of the fibres when they are aligned in the direction of loading. On the other hand, it is determined by the matrix creep if the fibres are discontinuous or randomly oriented. The creep mechanism is also greatly affected by the kind of interaction that occurs between the matrix and the fibre. The better the adhesion between the matrix and the fibre, the more effective will be the stress transfer and hence the creep deformation is less.

11.3.2 Effect of filler loading, stress and temperature

Wood fibres are generally preferred to inorganic fibres for noncritical applications because they have certain advantages such as lightweight fibres, ease of processing and high-volume low-cost composite production. A creep study becomes a necessity if the wood fibre reinforced thermoplastics need to be considered for long-term loading applications. Nunez *et al.* observed the effect of the following variables on the creep behaviour of polypropylene/wood flour composites: filler content (which varied from 0 to 60%), compatibilizing agent polypropylene maleic anhydride copolymer (PPMAN), and temperature.[7] Creep deformation reduced with: increasing wood flour content (except at much higher filler concentrations where wetting and filler dispersion problems start), the addition of PPMAN, and low temperatures. They modelled the creep compliance using Burger's model and the power law equation and ascertained that Burger's model provides a good description of the linear viscoelastic behaviour of the composite. The effect of the concentration of untreated wood flour on PP composites is displayed in Fig. 11.5. With an increased woodflour content, as expected, deformation decreases owing to the enhanced rigidity of the composites. However 60 wt% woodflour composite is an exception to this as the matrix cannot wet the filler completely rendering the filler–filler interaction weaker and therefore allowing it to be broken easily during deformation.

Cellulosic fibres, being predominantly polar, easily absorb moisture. Thus, studies have focused more on enhancing the interfacial adhesion between chemically and physically incompatible phases.[14] The interfacial adhesion can be modified in several ways:

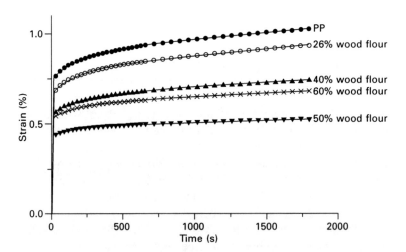

11.5 Creep of neat PP and untreated woodflour composites prepared with various filler concentrations at 20 °C. Symbols represent experimental data and continuous lines correspond to the best fit of the Burger's model with four parameters. Reprinted with permission from Nunez *et al.*[7]

(a) modifying the thermoplastic matrix by the addition of coupling/adhesion promoting agents;

(b) chemical treatment of natural fibres prior to composite preparation.

A sizing agent, used mainly in the papermaking industry, was indicated to be an appealing method of surface modification for natural fibres.[15,16] Researchers have also pointed out that, lignin, a constituent of natural fibres, can be employed as a coupling agent.[17,18] Lignin carries polar (hydroxyl) groups, nonpolar hydrocarbon, and benzene rings and can be used to achieve compatibility between natural fibres and the thermoplastic matrix.

The effect of variation in the interfacial interaction between the matrix and the fibre on the creep behaviour of the composite also has been studied (see Fig. 11.6).[7] When 5% PPMAN is introduced into the composite system, the deformation is reduced in comparison with untreated composites. With increasing PPMAN content, the extent of creep deformation reduced further owing to the better dispersion of the woodflour and the improved interfacial adhesion. With 10% PPMAN, a reduction of 45% creep deformation compared with neat PP was observed.

As PP is partially crystalline, the crystalline phase hinders the mobility of the amorphous phase. With increasing temperature, the movement of the amorphous part increases, sidelining the reinforcing effect of the crystalline phase, and increasing the extent of creep deformation, as shown in Fig. 11.7. The strains measured at the end of the test, at 50 and 80 °C, are 2.15 and 3.93 times the strain recorded at 20 °C. At 80 °C, the creep curve enters the third zone (Fig. 11.2), where the strain rate increases with time.

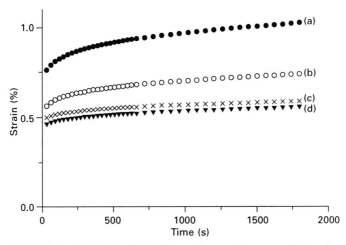

11.6 Effect of the interfacial interactions of the creep behaviour of the composites prepared with 40% woodflour: (a) ● neat PP, (b) ○ untreated woodflour, (c) × untreated woodflour and 5% PPMAN, (d) ▼ untreated woodflour and 10% PPMAN. Reprinted with permission from Nunez *et al.*[7]

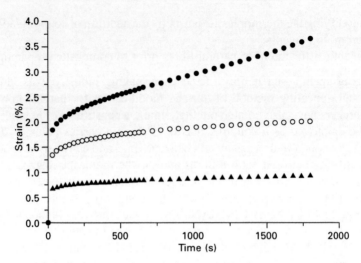

11.7 Creep of composites containing 26% of untreated woodflour at various temperatures: ● 80 °C, ○ 50 °C, ▲ 20 °C. Reprinted with permission from Nunez *et al.*[7]

Wood and thermoplastics are viscoelastic materials and both respond differently to creep behaviour.[19–21] The water/moisture uptake of wood–plastics composites is probably the only limiting factor to exploring their applicability in the construction of outdoor facilities. Wood fibre reinforced composites can take up a large amount of water because wood, which consists mainly of cellulose fibres, contains numerous hydroxyl groups that are strongly hydrophilic. The rate of moisture absorption in natural fibre–thermoplastic composites depends on the type of fibre and matrix, temperature, and the variation in water distribution within the composite.[22–24]

The effect of water absorption on the creep behaviour of wood plastic composites has been studied by Kazemi Najafi *et al.*[25] Employing up to 80% of medium density fibreboard flour as a natural fibre into recycled HDPE, it was observed that the creep strain decreased with increasing lignocellulosic flour, whereas water absorption enhanced the creep deformation of the composite. It was shown that the creep strain increased with immersion time owing to the additive effect of the absorbed water on fibre–matrix debonding, making the relaxation of the molecules easier at a higher moisture content. Bledzki and Faruk conducted a short term flexural creep test to study the creep behaviour of wood fibre PP composites.[26] Creep resistance was found to be increased by the addition of an MAH-PP compatibilizer, with decreasing temperature and also with increasing wood fibre content. Pooler and Smith[27] have proposed wood plastic composites as an alternative to treated timber for waterfront structures because of their ability to resist water absorption and degradation without chemical treatment. With increasing wood fibre concentration, the moisture content in the composites increases.[27] Fibre length and geometry play a significant role in deciding the equilibrium moisture content (volume expansion). Bledzki and Faruk[26] have also noted that long wood fibre-reinforced composites are more hygroscopic (nearly twice) than the hardwood composites at higher concentrations of fibre (Fig. 11.8). The use of compatibilizer MAH-PP reduced the hygroscopicity and significantly affected the impact properties. With increasing wood fibre content, the damping index was observed to increase, and this correspondingly lowered the impact strength of the composite. Long wood fibre reinforced PP composites displayed higher impact resistance than their hardwood fibre reinforced counterparts. The compatibilizer MAH-PP improves fibre dispersion in the matrix and facilitates the interfacial adhesion between the phases. It was observed that the MAH-PP treated composites exhibited an enhanced modulus and decreased creep in the composite.

Wood can be used as a substitute filler to reinforce traditional thermoplastics and biopolymers.[28–30] However, environmental problems such as deforestation, forest degradation and the impact on global warming and biodiversity has prompted studies on how to utilize other lingo-cellulosic materials in the form of agro residues from annual crops.[31] These field crop residues and/ or agricultural byproducts, such as cereal straw, rice husk, corn stalk, corn cob, soy stalk and bagasse, are very cheap, voluminous and easily available as raw materials.[32] Xu and coworkers investigated the creep behaviour of bagasse/PVC (recycled and virgin) and bagasse/HDPE (recycled and virgin) composites.[33, 34] Bagasse is a byproduct obtained from sugarcane processing. Acha *et al.* studied the creep and dynamic mechanical behaviour of bi-directional jute fabric/PP composites.[35] The interfacial adhesion between the polar fibres and the nonpolar matrix was improved by two different methods:

(a) Addition of coupling agents, such as MAPP and lignin; and
(b) chemical modification of the fibre surface such as esterification of commercial alkenyl succinic anhydride.

(a)   (b)

11.8 Micrographs of hard wood fibre (a) and long wood fibre (b). Reprinted with permission from Bledzki *et al.*[26]

It was observed that the creep deformation decreased with increasing jute content in PP (see Fig. 11.9). The Burger model fitted well to compositions of varying jute concentration in the creep part of the curve. The equation used for theoretical calculations of creep deformation is:

$$\varepsilon(t) = \frac{\sigma_0}{E_0} + \frac{\sigma_0}{E_1}(1 - e^{-tE_1/\eta_1}) + \frac{\sigma_0}{\eta_0}t$$

where, σ_0 is applied stress, t is time, E_0, and η_0 are the modulus of the spring and the viscosity of the dashpot for the Maxwell element, respectively, whereas E_1 and η_1 are the same for Voigt element. A drop in the immediate deformation and the slope of the steady-state creep zone was also observed.

Figure 11.10 displays the creep–recovery graphs for various fibre/matrix treated composites. Surprisingly, the esterified fabrics reinforced and lignin modified PP matrix composites showed the highest creep deformation, whereas both the samples prepared with PPMAN as a compatibilizer displayed the lowest values. Figure 11.11 is the attestation of these results: Fig. 11.11a shows the smoother surface of the esterified fabrics compared with the untreated one (Fig. 11.11c), but even after that the creep deformation is still higher. Figure 11.11c displays no adherence of the PP matrix to the untreated fibres, confirming poor interfacial adhesion between the phases.

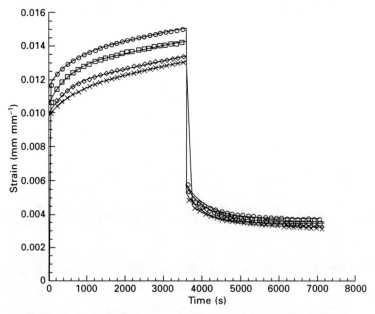

11.9 Experimental creep-recovery curves (symbols) and fitting with the four-element model (lines) for PP/jute composites, jute content (wt%): ○ 9; □ 17; ◇ 25; × 30. Reprinted with permission from Acha *et al.*[35]

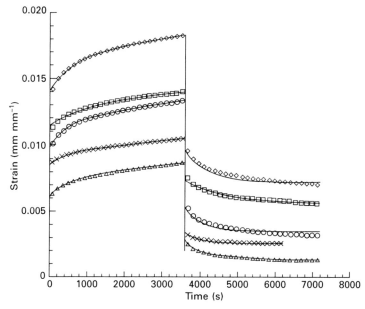

11.10 Experimental creep-recovery curves (symbols) and fitting with the four-element model (lines) for composites made from 25 wt% jute: ○ neat PP with jute; □ PP modified with 5 wt% lignin; ◇ esterified jute fabrics and neat PP; × PP modified with 5% Epolene E-43 wax; △ PP modified with Epolene G3003. Reprinted with permission from Acha *et al.*[35]

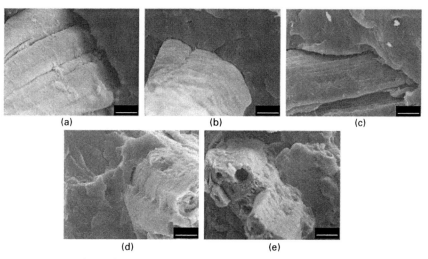

11.11 Scanning electron micrographs for composites made from 25 wt% jute: (a) esterified jute fabrics and neat PP; (b) PP modified with 5 wt% lignin; (c) neat PP with jute; (d) PP modified with 5% Epolene E-43 wax; (e) made from PP modified with Epolene G3003. Reprinted with permission from Acha *et al.*[35]

On the contrary, it can be seen that the PP has distributed evenly onto the fibre surface (as there is no gap present at the interface) as shown in Fig. 11.11d and 11.11e, which shows the better wettability of the fibres to the matrix. Better interfacial interaction is reflected in the better creep resistance of the composite material.[36] It was observed that both the coupling agents G3003 and E-43 wax worked well: G3003 induced better creep performance, possibly because it has a higher molecular mass than E-43 wax, molecular weight (MW) ~ 52,000 and 9,100, respectively. A mass coupling agent with higher MW can form better and tighter entanglements with the PP matrix. Where the coupling agent has a lower MW, the creep deformation is expected to be higher as it acts as a lubricant. Another point to note is that the maleic anhydride content of G3003 is lower than that of E-43 wax. This leads to the conclusion that a higher concentration of maleation is not really required to saturate the fibre surface –OH groups. Acha et al.[35] have further analyzed the effect of the amount of the coupling agent applied on the interfacial adhesion and hence on the creep performance of the composite (Fig. 11.10). The creep deformation for 3 and 5 wt% is nearly the same, making it clear that 5 wt% of G3003 coupling agent is sufficient to saturate the fibre surface. However, further increase in the compatibilizing agent (7 and 9 wt% respectively) demonstrated an increase in the creep deformation as shown in Fig. 11.12. The maleated PP is less regular than the virgin PP. With an increasing concentration of maleated PP, the degree of grafting

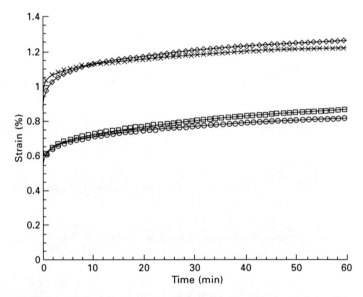

11.12 Experimental creep graphs composites for containing 25 wt% of jute and various concentrations of Epolene G3003: ○ 3%; □ 5%; ◇ 7%; × 9%. Reprinted with permission from Acha et al.[35]

increases, which correspondingly reduces its ability to crystallize.[37] In semicrystalline polymers such as PP, creep resistance increases with the increasing percentage of crystallinity,[38–40] which effectively reduces with excessive compatibilizing agent. It was expected that the anhydride chains attached to the fibres were either too short or else the polarity of fibres was still much higher than that of the PP matrix. At the same time, the increased smoothness of the fibres reduced mechanical interlocking, causing a reduction in the interfacial strength of the composite. Lignin modified PP composites also showed a higher creep deformation, suggesting that lignin is a poor compatibilizer for the PP–jute system.

11.4   Creep modelling

11.4.1 Four-element Burgers model

The easiest way to model the combined viscous and elastic behaviour of a polymer is to employ the mechanical analogs: dashpots (viscous behaviour) and springs (elastic behaviour). The four-element Burgers model is more often used for modelling the creep behaviour of natural fibre reinforced polymer composites.[7, 35]

The various combinations of the spring (elastic deformation) and dashpot (viscous flow) can be correlated with the creep–recovery graph of polymers. The Maxwell model with the spring and dashpot attached in a series, does not work well for creep. It deforms continuously with time as the dashpot responds for loading. The Kelvin–Voigt model with the spring and dashpot in parallel is another simple representation of creep. It displays the time-dependent response of the strain as the viscous flow in the dashpot is opposed by the spring flow. It does not represent instantaneous elastic deformation as well as continued flow under equilibrium. To resolve the problems of the Maxwell and Kelvin–Voigt models, the four-element model developed combines the spring and dashpot in a series and parallel combination as shown in Fig. 11.13. This model exhibited reasonable success in simulating the linear viscoelastic behaviour of polymers. It displays all the regions that are generally observed in real polymers including time-independent elastic deformation, a levelled-off equilibrium region and a recovery region too. For the analysis of the creep–recovery graph, stress and strain curves are generally plotted against time and the data is fitted into the four-element model as shown in Fig. 11.14. It can alternatively be quantitatively analyzed for properties such as modulus, viscosity, irrecoverable creep and retardation time.

One can see that the elastic, viscous, and viscoelastic behaviour of the polymer sample can be easily differentiated. The creep experiment starts when the stress σ_0 is applied to the polymer at time $t = 0$. With the application

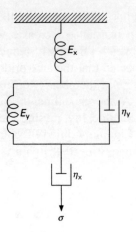

11.13 A schematic representation of the four-element Burgers model.

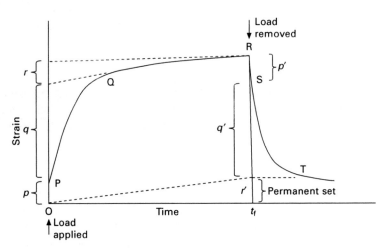

11.14 Instantaneous deformation (Maxwell element E_x); PQ linear viscoelastic region (creep Kelvin element); QR equilibrium region (creep Maxwell element), RS instantaneous recovery (Maxwell element E_x), ST delayed recovery (Kelvin element).

of stress (σ_0), the polymer shows some instantaneous deformation that corresponds to σ_0/E_x (OP), where E_x is the independent spring constant of the model as shown in Fig. 11.13. The spring elongates immediately and locks itself after a certain degree of extension. After the time-independent deformation, the separate dashpot (η_x) and the complete Voigt element (E_y and η_y) start responding. At the end, when the stress is relieved completely, the independent spring recoils immediately to the same extent ($p = p'$). The share of the independent dashpot of the total deformation can be correlated with the slope of the strain–time graph when it reaches the equilibrium flow region. The gradient of the line (QR) will be equal to σ_0/η_x. Using that gradient, one can determine the permanent set of the sample by extending the time to t_f, where the creep test ends. The permanent set in the sample $= \sigma_0 t_f/\eta_x$ ($r = r'$). This independent dashpot does not recover as there is no restraining force acting on it. The permanent set shows the permanent deformation left in the material and that which can be correlated with this independent dashpot. This dashpot represents the molecules slipping past each other. The middle graph corresponding to the linear viscoelastic region (PQ) can be represented by the voigt element of the four-element model. The equations for compliance and strain measurement, respectively, take the form:

$$D(t) = \frac{1}{E_0} + \sum_{i=1}^{n-1} \frac{1}{E_i}[1 - e^{-t/\tau_i}] + \frac{1}{\eta_n}t$$

where $D(t)$ is the creep compliance.

$$\varepsilon = \varepsilon_1 + \varepsilon_2 + \varepsilon_3 = \frac{\sigma}{E_1} + \frac{\sigma}{E_2}[1 - e^{(t/\tau_i)}] + \frac{1}{\eta_3}t$$

where ε_1, ε_2 and ε_3 are the elastic, viscoelastic and viscous behaviour contributions, respectively.

Wong and Shanks (2008) have quantitatively resolved the creep–recovery behaviour of a natural fibre–biopolymer composite system into elastic deformation, viscoelastic deformation and viscous flow (see Fig. 11.15).[41] It was observed that the total creep strain of the unmodified PLA–flax and

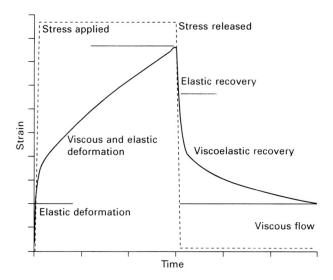

11.15 Typical polymer deformation profile showing the elastic, viscoelastic, and viscous flow contributions.

PHB–flax composite displayed a significant amount of creep. With the use of plasticizers such as tributyl citrate, triethyl citrate and glyceryl triacetate and the interfacial modifier, thiodiphenol, the total creep is effectively reduced in comparison with the unmodified composite system. After quantification of the various deformations, they found that the decrease in the total creep of the modified composites employing either plasticizers or modifiers is accompanied by a decrease in the permanent strain and an increase in the recoverable strain. On the contrary, the addition of a toughening agent, hyperbranched polyester, increased the permanent strain and reduced the recoverable strain, enhancing the overall creep strain of the composites.

With increasing stress levels from 5 to 15 MPa, it was observed that the total creep strain of the flax–PLLA and flax–PHB unmodified composites increased substantially (Fig. 11.16). Compared with the unmodified system, the use of plasticizers and thiodiphenol reduced the total creep much more. Plasticized composites have the ability to sustain higher loads without suffering a significant increase in the creep strain of the composites.[41] The incorporation of toughness modifier, hyperbranched polyester (HBP), increased the creep behaviour. Thiodiphenol was the best additive in terms of reducing the total creep strain, decreasing 83% of the creep with just 5% of it in the composite system. The suppression of the creep was explained by the reduction in the viscous flow contribution of the total creep and the significant increase in the elastic and viscoelastic deformation.

11.4.2 Time–temperature superposition

The determination of the long-term performance of reinforced composites has often been hindered by the expensive and time-consuming experimentation necessary to obtain reliable results. The time–temperature superposition (TTS) principle, the most useful extrapolation technique applied to almost all types of plastics, was originally developed in the mid-1950s.[42,43] In the late 1970s this method was expanded for use with fibre-reinforced composites.[42,44–45] Time–temperature superposition is based on the assumption that the effect of temperature on the time-dependent behaviour of a material is equivalent to a stretching or shrinking of the real time for temperatures above or below the reference temperature. Thus when plotted on graphs where the abscissa is defined as log time, the individual creep/stress relaxation graphs obtained at different temperatures can be shifted to the left or right to obtain a continuous master graph which spans a much longer time period than was actually employed in testing. The master graph is then used to predict the long-term behaviour of the composites.

Long-term experiments are replaced by short-term tests at higher temperatures. The shifting distance is generally termed the shift factor. The polymers that follow TTS extrapolation are thermorheologically simple

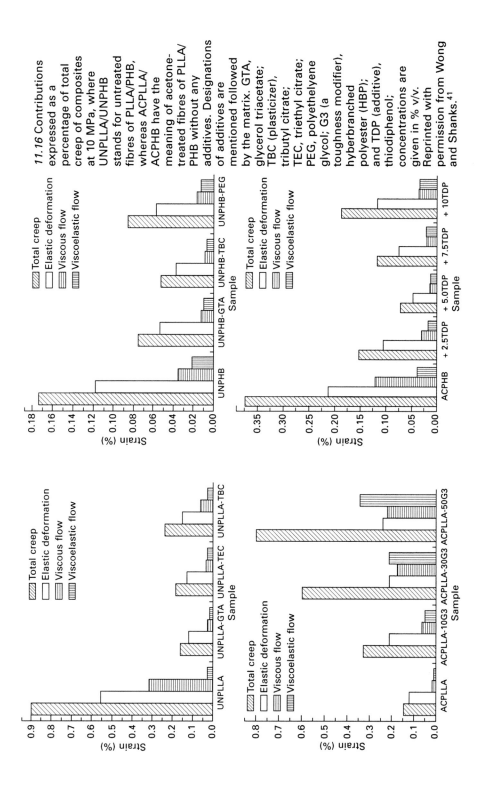

11.16 Contributions expressed as a percentage of total creep of composites at 10 MPa, where UNPLLA/UNPHB stands for untreated fibres of PLLA/PHB, whereas ACPLLA/ACPHB have the meaning of acetone-treated fibres of PLLA/PHB without any additives. Designations of additives are mentioned followed by the matrix. GTA, glycerol triacetate; TBC (plasticizer), tributyl citrate; TEC, triethyl citrate; PEG, polyethelyene glycol; G3 (a toughness modifier), hyberbranched polyester (HBP); and TDP (additive), thiodiphenol; concentrations are given in % v/v. Reprinted with permission from Wong and Shanks.[41]

and the rest of them are thermorheologically complex. It is evident that the superposition theory is based on the molecular behaviour of polymers and hence uses the term activation energy (E_a) to formulate the equations, such as the Arrhenius equation:[46]

$$\ln \alpha_T = \frac{E_\alpha}{R}\left(\frac{1}{T} - \frac{1}{T^0}\right)$$

where, α_T is the horizontal shift factor, R is the universal gas constant, T^0 is the reference temperature (K), and T is the temperature at which the test is performed (K). Regression analysis is generally carried out employing this equation to evaluate the activation energy for different materials from the experimental data.

One more empirical equation is generally used for TTS: the William–Landel–Ferry (WLF) equation. This equation gives a correlation of temperature with time. For example, if the information at higher temperatures is available, then we can interpolate/extrapolate this data to longer times to correspond with the lower temperature data. The WLF equation takes the following form

$$\log \alpha T = \frac{17.4(T - T_g)}{51.6 + T - T_g}$$

where T is the temperature at which the test is carried out. Even though the TTS theory was developed earlier and mainly for amorphous polymers, Neilsen and Landel mentioned that it could also be applied to semicrystalline polymers.[47] The changes in the degree of crystallinity with temperature and hence the changes in the modulus of elasticity with temperature in semicrystalline polymers correspond to a part of the vertical shift factor. Neilsen and Landel concluded that the vertical shift factors are mostly empirical with hardly any theoretical validation. Ward argued that the TTS theory cannot be applied to crystalline polymers as their thermal behaviour is rather complex.[8] Faucher modified the statements and stated that the theory could be applicable to crystalline polymers provided the polymer structure was maintained in the testing condition, and that it was possible with sufficiently low strain.[48]

There have been many studies on the applicability of the Arrhenius equation or WLF equation to apply TTS to the creep data.[49-53] Pooler et al.[27] applied TTS to wood fibre/HDPE composites and emphasized that only a single horizontal shifting is enough to superimpose the creep data. With the use of DMA experiments, storage modulus graphs were only used to ascertain the shift factors, marginalizing other viscoelastic parameters. The TTS theory was also applied by Tajvidi et al.[54] to the kenaf fibre/HDPE composite. They observed that the frequency sweep data could be superimposed well with the two-dimensional minimization method to satisfy the TTS requirements. Shift factors conformed to an Arrhenius equation for temperature dependence.

11.4.3 Findley's power law model

Various researchers have suggested empirical relations to describe the viscoelastic behaviour of fibre-reinforced composites. The Findley power law model is one of the most widely employed analytical models for elaborating the viscoelastic behaviour of fibre-reinforced polymer composites.[55] It has the form:

$$\varepsilon(t) = at^b + \varepsilon_0$$

where ε_o is the instantaneous (elastic) strain, $\varepsilon(t)$ is time dependent creep strain at time t, a is the amplitude of transient creep strain and b is the time exponent.

The Findley power law model, one of the simple empirical mathematical models, was used to simulate the creep behaviour of reinforced wood–plastic deck boards.[56] The power law model formulated by Findley has become a more popular analytical model for describing the viscoelastic behaviour of fibre-reinforced polymer composites under a constant stress. This model has been used in many studies[33,57–59] and documented for predicting the time-dependent behaviour of fibre-reinforced polymer composites. The model has also been recommended by the American Society of Civil Engineers (ASCE) *Structural plastic design manual* for the analysis of fibre reinforced plastic composites (FRPC).[60]

11.4.4 Two-parameter power law model

In general, creep strain is presented as relative creep (ε_r); a percentage of elastic strain in the material. The equation becomes

$$\varepsilon_r = A_0 t^n$$

where A_0 is the slope of the power law model. Taking logarithms to both sides to obtain a linear relationship between A_0 and n,

$$\log \varepsilon_r = \log A_0 + n \log t$$

The two-parameter power law model is also used for the creep modelling of kenaf fibre/HDPE composites.[54]

11.4.5 Prediction of long-term creep behaviour of the composites

During the short-term testing of the polymer, we generally do not take physical ageing into consideration owing to the short length of time involved.

Compliance makes the value lower than expected, when ageing of the material occurs.[61-64] So, it is really important to validate the data obtained from the short term test. Thus, the master graph generated from the short-term creep test may not depict the actual creep behaviour of the composite in the long term. Sometimes, to cross-check the data obtained from the master graph, long-term experiments are also conducted.

Xu *et al.* employed several creep models (like Findley's Power Law Model, Burger's Model, two-parameter power law model) for bagasse/PVC composites.[34] All of these models were capable of fitting the creep data but the Burger's model and two-parameter power law model yielded the most consistent parameters. The time–temperature superposition model predicted the creep behaviour better from the short term tests for PVC composites than from those for the HDPE composites. The three-day creep test (as displayed in Fig. 11.17) for the composites showed a similar trend of the creep curves to those in the 30 min tests (not shown here). It was found that the Burger's model fitted well for all composites over the entire long-run test. But the parameters for the 30 min test were very different from those for the long-duration creep test. Based on the sum of the squares, Findley's power law model was also found to fit well over the entire range but the parameters were different and sensitive to different graph shapes and did not maintain consistency. A comparative study was made of creep prediction with the short-term (30 min.) test employing TTS and the long-run experimental results. It was observed that the TTS theory predicted well the creep data for the bagasse/PVC composites and overestimated it in the case of the

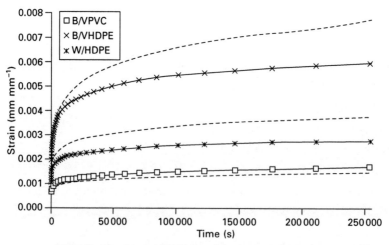

11.17 Comparison of TTS predicted creep with experimental data for various composites, where B/V is bagasse/virgin and W is wood. Experimental data are shown as symbols; solid lines are the Burger's model prediction and dashed lines are predictions from TTS. Reprinted with permission from Xu *et al.*[34]

HDPE composites. This inconsistency in the results for various composites indicated a need for further study and modification of the TTS theory before applying it to natural fibre reinforced polymer composites.

11.5   Nonlinear viscoelastic response

Many polymers display linear viscoelastic behaviour for small deformations or stresses. The theory of linear viscoelasticity generates a simple mathematical correlation of stress–strain–time parameters. The stress or strain where the deformation behaviour is estimated as linear, however, is small enough compared with the total stress or strain before yielding or fracture. Furthermore, at high temperatures, one can observe the nonlinear deformation even at low strains (less than 0.2%). Therefore, theories adopting linear viscoelastic behaviour do not provide a generalized solution for the wide range of time, loading conditions and temperatures. For example, solid lumber was originally mostly used as a material for naval waterfront facilities.[65] However, lumber degrades over time and, therefore, requires chemical treatment to effectively delay the course of degradation. This apparently increased expenditure on the repair and replacement of these structures. In addition to that, the lowering of the water quality owing to the leaching of the chemically treated lumber is a worrying environmental factor. An alternative for this is to use a material that can withstand the static as well as time-dependent loading during its service period (i.e. the material must have high fatigue and creep resistance).

The following approaches are generally used to address nonlinear viscoelasticity. The first approach generally uses continuum mechanics to derive the constitutive equations for nonlinear viscoelasticity. The second approach uses thermodynamics theory for irreversible processes. Rangaraj et al.[65] extruded wood flour with high density polyethylene (HDPE) and carried out a series of creep and recovery tests. After crossing a threshold value of creep stress amplitude and duration, it was observed that there is a permanent deformation along with a reduction in stiffness (i.e. damage) in the material. The damage and permanent strains were found to depend on creep stress amplitude and duration. To correlate the damage and permanent deformation effects, they developed a nonlinearly viscoelastic model and equated the results favourably with the experimental creep/recovery data.

Pramanick et al.[66] worked on developing a model to study the nonlinear behaviour of creep in polymer composites of the two-phase system where they used an extended 'theory of mixtures' to describe the creep phenomenon. It is interesting to mention that the effect of temperature and stress on the creep deformation of the individual constituents can be fitted in combination in a single analytical function where the interaction is additive. They approximated stress and temperature related shift factors in terms of the activation energy of the constituent phase. As the temperature shift interferes with the stress shift, they incorporated a relationship between these shifts in the model to make the stress equivalency of temperature possible. Rice husk–HDPE composites were experimentally analyzed with this model, which was validated in the linear as well as the nonlinear viscoelastic regions of creep for rice husk–HDPE composites with power-law-Boltzmann's superposition principles.[67] It was suggested that the model could also incorporate environmental variations such as relative humidity, time, stress and temperature. Lin *et al.*[68] carried out the creep and recovery testing in the flexural bending mode for 60% wood flour reinforced extruded HDPE composites to determine the creep related material constants for the nonlinear viscoelastic model.

11.6   Stress relaxation

George *et al.*[69] investigated the stress relaxation behaviour of short pineapple-fibre reinforced polyethylene composites. The stress relaxation of the composite decreased with the incorporation of stiff natural fibres in the polymer matrix. The fibre–matrix bonding at the interface also plays a significant role in the overall performance. Various surface modification techniques were developed to enhance the interfacial adhesion and hence decrease the relaxation of stress.[70] George *et al.* observed that the stress relaxation surprisingly increased with the increasing length of the fibre.[70] They reasoned that with increasing fibre length, the possibility of the bending/curling of fibres as well as of greater fibre–fibre interaction would be higher, which would cause stress to be transferred insufficiently. An optimum balance of fibre length is a crucial parameter as it may lower the relaxation modulus owing to insufficient stress transfer or else because of the pullout of fibre from the matrix when deformed.[71] Mechanical and viscoelastic properties decrease after exposure to water depending on the time of water immersion, fibre loading and fibre surface modification.[72] Fibre orientation and strain level create an impact on the stress relaxation behaviour of the composite.[69,73]

One can obtain a relaxation master graph the same way as for a creep graph, spanning a much longer period of time. With log (time) as the abscissa, the individual relaxation graphs obtained at various elevated temperatures and/ or strain can be shifted to generate a continuous master graph. To achieve a master graph for both time and stress superposition, vertical shifts can also be applied in concurrence with horizontal shift. Therefore, a process can be formulated where a series of short-term relaxation tests are performed at elevated temperatures/strains leading to the generation of a family of graphs for an established composite system. This technique was used successfully to anticipate the stress relaxation behaviour of polymer composites in the long-term utilizing horizontal and vertical shift in the graph concurrently.[74,75] TTS has taken advantage of the relationship between temperature and viscoelastic behaviour in fibre-reinforced composites.[42,44] Findley's power law model has also been used successfully by many researchers to model short as well as long-term relaxation behaviour.[76]

The increasing applications of short-fibre composites for static as well as dynamic loadings, for instance, hoses, V belts, tubes and cables, has increased the importance of stress relaxation measurements, particularly for rubber based composites. The interface interaction between the rubber and the filler can be observed through the stress relaxation measurements. Cured elastomers undergo a marked relaxation when stretched at a given deformation.[77] Stress in the deformed component relaxes in proportion to a logarithm of the period. Bhagawan et al.[78] detected a two-stage relaxation pattern for the short jute fibre–nitrile rubber composites. Flink and Stenberg[79] mentioned that the slope of the stress relaxation graph provides a better way to study the interfacial adhesion between the phases while carrying out a study on short cellulose fibre–natural rubber composites. A plot of $E(t)/E(t = 0)$ against $\log(t)$, where $E(t)$ is the stress at a given time and $E(t = 0)$ is the initial stress, reveals the interfacial interaction.

Kutty and Nando[80] studied the stress relaxation behaviour of short fibre-thermoplastic elastomer composites and the effect of fibre loading, orientation, strain level, strain rate and temperature. A two-stage relaxation process for unfilled thermoplastic polyurethane and a three-stage relaxation process for the short kevlar fibre reinforced polyurethane composites were reported. The orientation of fibres in the polymer matrix can be either longitudinal or transverse as shown in Fig. 11.18. With variation in the fibre loading, it was observed that the initial rate of relaxation in the transverse direction (except

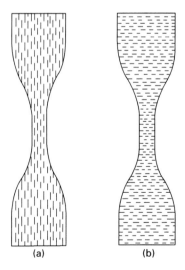

(a)   (b)

11.18 Schematic representation of orientation of fibres in relaxation test specimen: (a) longitudinal orientation, (b) transverse orientation. Reprinted with permission from Kutty and Nando.[80]

at 40 phr) is nearly 100% more in the longitudinal direction (see Fig. 11.19). Even the second and the third slopes are higher in the transverse direction. One can also see the consequences of fibre loading on the stress relaxation phenomenon which reveals that it has similar effects on both orientations of the fibres. It is observed that the initial rate of relaxation is dramatically reduced to 34% with the incorporation of only 10 wt% fibres (see Fig. 11.19). With a further increase in the fibre content, the rate of decrement was not significant.

Figure 11.20 displays the effect of various strain levels on the stress relaxation plot of neat thermoplastic polyurethane (at a constant strain rate of 0.16 s^{-1}). Two different slopes of the plot at any given strain level shows that there are two different relaxation mechanisms operating in a system. The first mechanism is for shorter period (~ 200 s). The change in the slope suggests the point of time at which the changeover in the mechanism occurs. For the first relaxation mechanism, the rate of relaxation is independent of the strain levels up to 100% as it is also observed[81] in many vulcanized rubbers such as natural rubber (NR), styrene–butadiene rubber (SBR), and isoprene rubber (IR).[82] But at 500% strain, the relaxation rate is increased by ~ 65% which is accredited to the build-up of the crystalline phase as occurs in natural rubber.[82] The increase in the second slope and the shift in

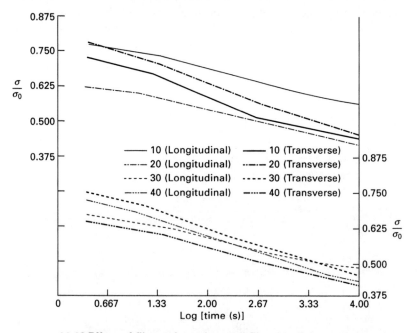

11.19 Effect of fibre orientation and fibre loading on stress relaxation of Kevlar fibre reinforced polyurethane composites. Reprinted with permission from Kutty and Nando.[80]

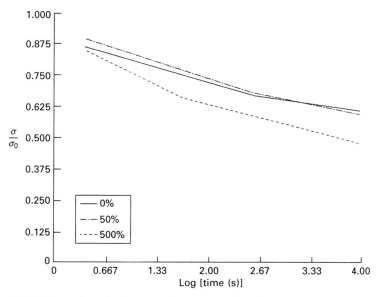

11.20 Effect of strain level on the stress relaxation of neat thermoplastic polyurethane (strain rate, 0.16 s^{-1}). Reprinted with permission from Kutty and Nando.[80]

the changeover point to shorter times at higher strains indicates that the first relaxation mechanism takes over at a faster rate at higher strain levels.

Varghese *et al.*[83] carried out the stress relaxation study on untreated and acetylated short sisal fibre reinforced natural rubber composites and reported that the acetylation improves adhesion between the fibre and the matrix. The rate of stress relaxation might be a better way to rank the adhesion of the fibres/fillers with the matrix. They further investigated the effect of strain level, fibre loading, fibre orientation, and temperature on the stress relaxation behaviour of the same chemically treated composite. Ageing is also one of the parameters that changes the stress relaxation behaviour of the polymer composites (see Fig. 11.21). Varghese *et al.*[84] aged the NR–sisal fibre composite specimens at 0 and 100 °C for 4 days and then tested them. It was noted that the composites employing no bonding agent (Mix D) as well as with bonding agent (Mix C) showed a decreased rate of initial relaxation with ageing. The relaxation graph for the composites with bonding agent demonstrates the maximum crossover time and contribution to the initial relaxation. In the second slope of the curve, the relaxation rate increases sharply owing to chain/crosslink scission.

11.7   Fatigue

The other mechanism that exhibits a form of property deterioration during the service life of fibre reinforced composites is 'fatigue'. Fatigue failure in

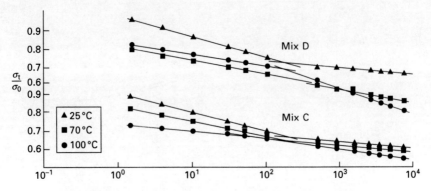

11.21 Stress relaxation graphs of mixes C and D at 30% elongation (after ageing). The composition of mix C is: natural rubber 100 phr, acetylated sisal fibre 15 phr, resorcinol 3.75 phr, hexamethylenetetramine 2.4 phr, zinc oxide 5 phr, stearic acid 1.5 phr, and sulfur 2.5 phr. Mix D does not contain resorcinol or hexaethylenetetramine but the remaining ingredients are the same. Reprinted with permission from Varghese *et al.*[83]

unidirectionally oriented fibre reinforced composites takes place throughout the volume of the material and not in a localized fashion. Phenomena such as debonding at the interface, fibre breakage, matrix cracking, delamination and transverse ply cracking occur either interactively or independently depending on the manufacturing method adopted. The predominance of a particular phenomenon over others depends on the material variables and testing conditions involved. In general, matrix cracking and delamination occurs early, whereas fibre–matrix debonding and fibre fracture starts during the beginning of component life but accumulates rapidly towards the end, leading to the final failure.[85] In comparison with metals, composite materials display greater variation in terms of their fatigue properties because of the greater number of variables involved. Unfortunately, most of the research involved has focused on synthetic fibre/resin composites.[86–93] The study of natural fibre reinforced composites, and their ability to withstand complex forces, and material, as well as external variables, with a view to employing them in structural applications is timely but there are few reported studies.

The main constituent of natural fibres is cellulose.[94] The basic unit of cellulose, anhydro-D-glucose, contains three hydroxyl (–OH) groups. These hydroxyls form inter/intramolecular hydrogen bonds and also bond with the hydroxyl groups from damp air and become hydrophilic. The moisture content of the natural fibres varies between 3 and 13%.[95] A problem with natural fibres is that there are more variables that determine their final properties and performance. The physical and mechanical properties of natural fibres are governed by factors such as the fibre diameter, the structure of the fibre, whether the fibres are collected from the stem, leaf or seed and the growing conditions of the plant.[96] The structure of the fibre itself has many variables, including the amount of crystalline fibrils, the spiral angle, supramolecular structure (degree of crystallinity), degree of polymerization, type of cellulose, defects, orientation of the chains of noncrystalline cellulose and crystalline fibrils, pore size and pore volume, decide the final structure and hence the properties of the fibres.[94,97,98] The conclusion is that the static properties of natural fibre reinforced composites display a great variation in their results because of the dimensional variations and hydrophilicity of the fibres and hence the fluctuation in fatigue properties is expected to be much higher than that in synthetic fibre–resin composites.

The incorporation of hydrophilic natural fibres in traditional thermoplastics weakens the interfacial bonding between them with time leading to the deterioration of the mechanical properties (ageing phenomenon). Waxy residues on the fibre surfaces also introduce poor bonding characteristics to these fibres.[99] The effects of this time-dependent declination of the mechanical properties owing to the weakening of the interfacial bonding is an important area of study for applications that require the composites to withstand long-term exposure of the composites. Alternative options for enhancing the mechanical properties of natural fibre reinforced polymer composites were focused on: the use of effective coupling agents to increase the hydrophobicity of the natural fibres, hybridization of natural fibres with more environmental ageing resistant synthetic fibres such as carbon and glass fibres.

To enhance the interface interaction between the fibre and the matrix and hence to improve the properties of the composites, natural fibres are either physically or chemically treated. The physical methods such as thermal treatment,[100] calendaring,[101,102] stretching,[103] and hybrid yarn production[104,105] involve only the surface modification and do not alter the chemical composition of the composite. Chemical treatment of the fibres mostly uses a coupling agent that generally reacts with the fibre as well as with the polymer. Coupling agents form chemical bonds (either covalent or hydrogen bonding) and improve the adhesion at the interface. In general, the natural fibres are treated with compounds containing isocyanates,[106,107] methyl,[105,108,109] and organosilane groups.[110] Graft copolymerization is the other regular method used for natural fibre reinforced polymer composites. The alkalization of fibres with NaOH solution[111,112] has resulted in a substantial improvement in the strength of the composites.

11.8   Factors affecting the fatigue life of natural fibre composites

11.8.1 Internal/material variables

In general, the fatigue properties of composites depend on the polymer type (brittle or soft matrix)/structure, MW, crosslinking, type of filler, and its content. For polymer composites with a brittle matrix such as epoxy, it is the interfacial strength parameter that plays an important role in the fatigue behaviour of composites. The interfacial strength controls the crack propagation (crack bridging, debonding) phenomenon. With poor interfacial bonding between the phases, debonding and frictional sliding occur readily upon crack extension, allowing the long fibres to remain intact and bridge the crack. Contrarily, a stronger interface suppresses sliding and contributes to fibre fracture instead of bridging the crack.[113,114] Microcracking in toughened epoxy/polyester as well as in thermoplastic matrices is also a dominant energy dissipation mechanism precluding plastic deformation (which readily occurs in bulk matrix).[115]

Effect of filler, whether treated or untreated, and content

Gassan investigated the effect of the fibre and the interface parameters affecting the tension–tension fatigue behaviour of various natural fibre reinforced plastics.[116] He selected the combinations of fibres (jute and flax in yarns and woven form) and matrix (epoxy resin, polyester resin, and polypropylene) so as to study the individual effects of fibre type, textile architecture, interphase properties, fibre properties and their content on the fatigue behaviour of natural fibre composites. A variation in the damping of composites was correlated with the individual effect of these parameters against an increasing applied load. It was concluded that natural fibre reinforced plastics employing higher fibre strength and modulus, firmer interfacial adhesion between the phases or higher amounts of fibre in the matrix demonstrated higher critical loads for damage initiation as well as higher failure loads. He also observed the reduction in damage propagation with the above effects.

Figure 11.22 depicts the effect of the type of the natural fibre incorporated in a given polymer matrix. One can see that the specific damping capacity (energy loss/strain energy) increases significantly suggesting that the start of the damage initiation is at 45 N mm^{-2} for the jute–epoxy composite. With further increase in the mean stress (keeping the stress ratio R and frequency constant), specific damping capacity increases owing to damage propagation and the composite finally fractures at 120 N mm^{-2}. A similar type of behaviour is observed in flax–epoxy composites with much higher specific damping capacity values. The cellulose content in jute and flax fibres was reported to be 61.4 and 78.5 wt.%, respectively.[117] The energy required for fibre damage is far more than expected. This suggests that the greater part of the energy is lost in the internal friction owing to the viscoelastic deformation of the fibre itself. It is also reported that the damage of the fatigued fibres mainly results from micromechanical degradation followed by the structural breakdown of the microfibrils.[118]

Gassan[116] proposed that the cellulose microfibrils transfer the stress

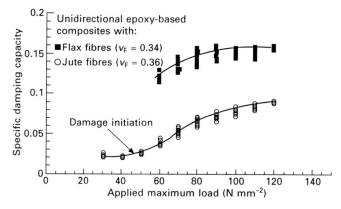

11.22 Specific damping capacity versus applied maximum load for unidirectional flax and jute epoxy composites (10^4 load cycles/load level). Reprinted with permission from Gassan.[116]

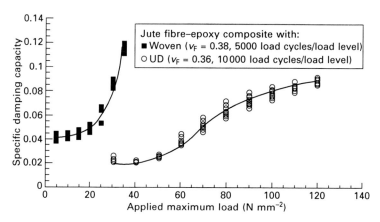

11.23 Specific damping capacity versus applied maximum load for woven and unidirectional (UD) reinforced jute–epoxy composites. Reprinted with permission from Gassan.[116]

from one layer to the other layer and decrease the role of the cementing matrix (hemicellulose and lignin) as a load carrying component (similarly to the laminate composite fibres). It is the cellulose microfibrils that mainly determine the delamination crack growth resistance.

Figure 11.23 reveals the effect of the natural fibre architecture on the fatigue mechanism of the composites. With woven fibres in the brittle epoxy matrix, the rapid damage development during the initial loading cycles is mainly accredited to the matrix cracking along the off-axis plies and perpendicular to the loading direction. These microcracks coalesce with time and grow along the interfaces and delaminate (if present). Finally, the composite fails with the fracture of the fibres.[119] In contrast to this, interfacial debonding is the dominant mechanism of composite failure in unidirectional composites.[120] Crack propagation in [0,90] composites primarily depends upon the interfacial interaction between the matrix and the fibre. With a relatively weaker interface, debonding and frictional sliding occurs. A strong interface between the phases suppresses the interface sliding and contributes to fibre fracture[113,114,121,122] increasing the critical stress for damage propagation.

In natural fibre reinforced polymer composites, the interaction between the fibres and the matrix was improved by the use of coupling agents.[123–125] The use of MAH-PP copolymer as a coupling agent in jute–PP composites demonstrated reduced damage propagation with increased critical loadings.

Effect of physical/chemical modification of fibre/matrix on fatigue mechanism

A fatigue phenomenon is more common with decreasing fibre modulus in unidirectional composites.[126,127] Shrinkage of fibres during alkali treatment has a significant impact on the fibre structure and hence on the mechanical properties of the fibres and consequently on the composites.[128] Alkali (NaOH) treated flax fibres having various states of allowed shrinkage showed tremendous variation in their yarn tensile modulus (changes from 1.5 to 0.2 times compared with their untreated counterpart when the shrinkage varied from 0 to 26%).[116] Alkaline treatment alters the Hermans orientation factor (0.96 for zero-shrunk fibres and 0.653 for 26% shrunk fibres). Gassan *et al.*[128] highlighted the effect of fibre shrinkage on the fatigue properties of flax–epoxy unidirectional composites (Fig. 11.24). They observed a higher specific damping capacity at a given loading for increasing shrinkage of

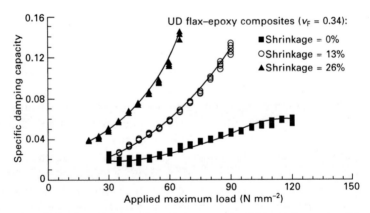

11.24 Specific damping capacity versus applied maximum load for UD epoxy composites which contain flax fibres of various fibre tensile strength and modulus (10^4 load cycles/load level). Reprinted with permission from Gassan.[116]

the fibres. Impact stiffness also increases with decreasing shrinkage. With decreasing shrinkage, i.e. increasing orientation of the fibrils, it was observed that the damage initiation in composites was 40–50% higher than in the composites with shrunk fibre.[128] In other words, the stress required to initiate the damage in a composite system was reduced with increasing shrinkage. The lowered stress for matrix cracking can be ascribed to the decreased fibre modulus and increased fibre diameter owing to shrinkage as mentioned in the following equation.[129]

$$\sigma_c = \left[\frac{6E_f f^2 \tau E_c^2 \Gamma_m}{E_m^2 (1-f)R} \right]^{1/3}$$

where σ_c is the stress required to initiate the crack (in the matrix), f is the fibre volume fraction, τ is the interfacial shear stress, E_c is the modulus of the composite system, Γ_m is the fracture energy for the matrix and E_m is the modulus of the matrix.

It was observed that the noncellulosic components (hemicelluloses and lignin) of the natural fibres can be removed by alkali treatment.[130,131] With the removal of hemicellulose, the empty space between the fibrils helps them to rearrange themselves in the direction of loading. This fibril rearrangement increases the load sharing capability and hence the tensile/flexural strength of the composite. Some studies reported an increase in % crystallinity of the fibres through alkali treatment.[132–135] The fibres treated with NaOH decrease the spiral angle and increase the randomness of molecular orientation, which increases the modulus of the fibre.

Gassan and Bledzki[123] modified jute fibre with alkali, MAH-PP and silane, and found that surface modification had obvious effects on the fatigue behaviour of jute fibre reinforced polypropylene composites.[123,136,137] To investigate the influence of MAH-PP content on the fatigue behaviour (correlated with specific damping, dynamic modulus, hysteresis, and accumulated dissipated energy) of jute–PP composites, Gassan and Bledzki carried out testing in InDyMat (intelligent dynamic mechanical testing). With improved fibre–matrix adhesion, damage resistance was increased to higher maximal stresses and the progress (rate) of damage was also reduced.[136]

Figure 11.25 clearly shows that it is the fibre–matrix adhesion that reduces the progress of the damage with increasing fibre content. Untreated jute–PP composites hardly display any change in accumulated dissipated energy whereas the MAH-PP coupling agent improved the stress transfer and hence the increase in dynamic strength.

For a similar volume fraction of fibre, one can see ~ 40% increment in the dynamic strength with 0.1 wt.% of coupling agent in jute–PP composites (Fig. 11.26). Furthermore, the damage in both unmodified and modified composites is not abrupt but increases continuously with stress.

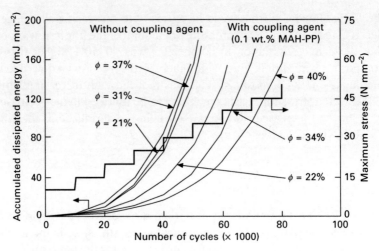

11.25 Influence of fibre content on the specific damping capacity of MAH–PP modified jute–PP composites, test frequency 10 Hz, *R* 0.1, 10^4 cycles/stress level, ϕ fibre content (vol%). Reprinted with permission from Gassan and Bledzki.[136]

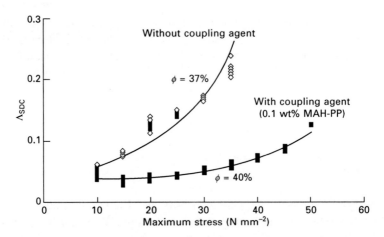

11.26 Influence of surface treatment on the specific damping capacity (SDC) of jute–PP composites, test frequency 10 Hz, *R* 0.1, 10^4 cycles/stress level, ϕ fibre content (vol%). Reprinted with permission from Gassan and Bledzki.[136]

Ray *et al.* carried out an impact fatigue study on 35% jute–vinyl ester composites on untreated and alkali-treated fibres.[138] A longer time for alkali treatment is expected to improve the crystallinity and remove the hemicellulose. It was observed that a 4 h alkali treatment of jute fibres imparted optimized interfacial bonding and fibre strength properties. Figure 11.27 displays clearly that the alkali-treated fibre-reinforced composites show a marked drop

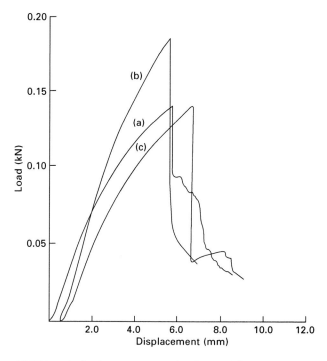

11.27 Load–displacement graph of 35 vol% composites reinforced with jute fibres: (a) untreated, (b) 4-h alkali treated, (c) 8-h alkali treated. Reprinted with permission from Ray *et al.*[138]

in load after the fracture of matrix and fibres. Contrarily, untreated fibres showed just the pullout of the fibres. It was observed that with an 8 h fibre treatment, even though the fibres had superior characteristics, the tensile/flexural strength of composites reinforced with these fibres was lower. With increased treatment time, fibrillation occurs to a great extent and renders the reduced fibre diameter less effective in the property enhancement of composites.

11.8.2 External variables

External variables include the effects of stress amplitude, stress intensity factor, mean stress, temperature, frequency and environment. A systematic study of these variables on the fatigue mechanism of natural fibre reinforced polymer composites is needed to exploit these composites in wide-ranging external surroundings. Some work related to S–N plots and constant life line graphs provide some basic information regarding the effect of these variables on the fatigue behaviour of composites.

S–N graph

Ray *et al.*[138] have plotted (Fig. 11.28) *E–N* type fatigue graphs for alkali-treated and untreated jute fibre reinforced composites, which show a four-region fatigue graph for untreated and a three-zone graph for alkali-treated jute-reinforced vinyl ester composites. It is clear that the *E-N* graph decreases progressively with applied impact energy which indicates that the alkali-treated composites have longer endurances. This is not observed in untreated fibre reinforced composites. For 4-h alkali-treated fibres, the slope in region II is more rapid than in 8-h alkali-treated fibres. This reveals that the damage mechanisms (debonding at the interface, matrix cracking and fibre fracture) are rapid. Nevertheless, it was observed that 8-h alkali-treated fibres displayed improved impact fatigue properties.

Constant life line graphs

In most of the preliminary analyses of the fatigue behaviour of natural fibre composites, results were demonstrated in terms of variation in specific damping capacity[116] and limited *S–N* (stress versus number of cycles to failure) data.[139] Bond and Ansell plotted *constant life diagrams* for carbon fibre/epoxy composites and identified that they followed a parabolic graph shifted towards the tension side of the plot.[140] The shift of the graph was

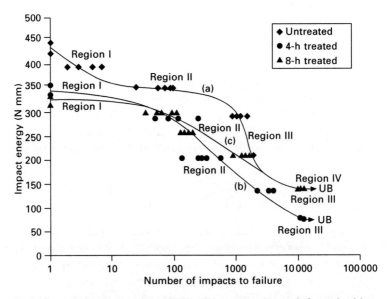

11.28 Impact fatigue graph of 35 vol% composites reinforced with: (a) untreated jute fibres, (b) 4-h alkali-treated jute fibres, (c) 8-h alkali-treated jute fibres. Reprinted with permission from Ray *et al.*[138]

attributed to the relative weakness of the fibre composites under cyclic and compressive static loading. Constant-life lines are capable of predicting the fatigue lives of composites at any R ratio[140,141] and these are a valuable tool for design engineers to apply to components failing owing to fatigue (Fig. 11.29). The area under this constant-life line (a plot of mean stress (x axis) against alternating stress (y axis) makes up safe combinations of mean and alternating stresses.

While analyzing the fatigue behaviour of sisal fibre reinforced thermosetting (polyester or epoxy) composites comprising untreated or alkali treated sisal fibres, Towo and Ansell[142] used S–N curves and dynamic mechanical thermal analysis (DMTA) to investigate variations in the mechanical properties of these composites with the number of fatigue cycles and temperature. Interfacial damage was greater in the composites reinforced with untreated sisal fibres. Better interfacial bonding brought about by the alkali treatment improved the fatigue life of the composites. For the same composite system, Towo and Ansell built constant-life lines from S–N graphs using experimental results.[143] At different stress ratios R (ratio of minimum to maximum cyclic stress), the fatigue failure in tension–tension (stress ratio $R = 0.1$) mode and reversed loading ($R = -1$; i.e. tension–compression) were assessed. The scatter in the number of cycles to failure for the specimen tested under tension–tension

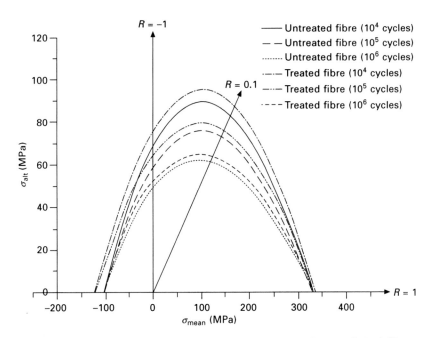

11.29 Constant life diagrams for untreated and treated sisal fibre epoxy matrix composites for 10^4, 10^5, and 10^6 cycles. Reprinted with permission from Towo and Ansell.[143]

loading showed that the natural fibre reinforced composites have highly fluctuating mechanical properties and hence fatigue lives. However synthetic fibre composites also demonstrate variation in their fatigue lives. In spite of this, for alkali-treated sisal fibre reinforced polyester (there was hardly any improvement in epoxy composites possibly owing to similar interfacial bond strengths) composites, an improvement in terms of load carrying capacity and an increased rate of fatigue degradation during cyclic loading was observed. It was reasoned that this was due to brittle micro-failure events occurring in the composites. Alkali-treated composites displayed superiority at lower fatigue lives (up to 10^6 cycles).

Hysteresis loops

Bond and Ansell[140] even studied the hysteresis loops at these R ratios (ratio of minimum to maximum cyclic stress). Hysteresis loops are generally used to measure properties such as loss of stiffness and internal friction during fatigue.[144] These graphs also show fatigue damage development and the nature of the chemical bond between fibre and matrix. Towo and Ansell[143] used hysteresis loop graphs to distinguish the differences in damage development between composites reinforced with untreated sisal fibres and those containing alkali-treated fibres as shown in Fig. 11.30.

Composites endure heavy interfacial damage during tension–compression loading ($R = -1$) with an increasing number of cycles. Hysteresis experiments

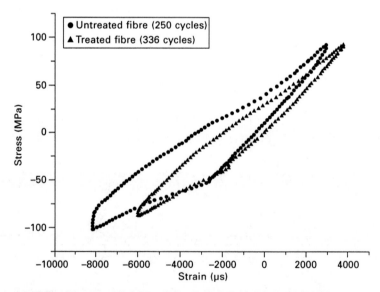

11.30 Final hysteresis loops for untreated and 0.06 M NaOH treated composite at $R = -1$ (fully reversed loading). Reprinted with permission from Towo and Ansell.[143]

display smaller loops during the initial cycles. With an increasing number of cycles, the loop area just before failure was increased by ~447% and 258% for untreated and alkali-treated fibre composites, respectively. Unidirectional fibre composites are weak when under compressive loading and suffer debonding and fibre buckling leading to the formation of kink zones at failure (see Fig. 11.31). In short, the smaller the loop area, the lesser is the energy dissipation and correspondingly the better is the bonding between the fibres and the matrix.

Effect of low-velocity impact

Tong and Isaac carried out various experiments to determine the effects of low-velocity impact on the residual tensile and fatigue properties of nonwoven hemp fibre mat reinforced polyester samples and compared them with ±45° glass fibre reinforced polyester samples (Fig. 11.32).[145] The damage accumulation of the composites was correlated with the degradation of the tensile modulus during fatigue cycling. The following were the noticeable points observed: (a) the necessity of a relatively high pressure on the hemp fibre reinforced composite during curing in order to ascertain a higher volume fraction of the fibre to render the reinforcing effect; (b) with similar fibre weight fractions, the hemp and glass reinforced composites demonstrated similar tensile properties and fatigue lifetimes with hemp composites displaying the ability to withstand marginally higher cyclic stress levels for equivalent numbers of cycles (see Fig. 11.33); (c) hemp fibre reinforced composites fractured in a brittle manner, without any damage such as matrix cracking which was seen in the glass fibre based composite.

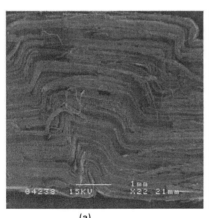

(a)	(b)

11.31 Failure of sectioned sisal fibre epoxy resin composite in tension–compression loading showing (a) wedge-shaped kink failure and (b) single kink failure (R = –1). Reprinted with permission from Towo and Ansell.[143]

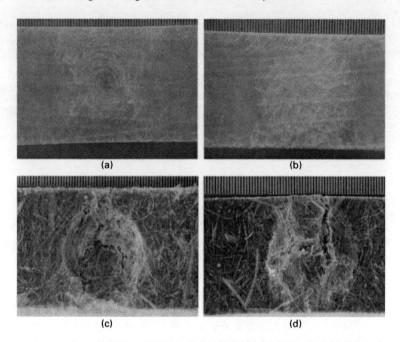

11.32 Low-velocity impact damage of ±45° glass fibre and hemp reinforced composites following impact of energy 5 J: (a) glass fibre impact face; (c) glass fibre, back face; (c) hemp fibre, impact face; (d) hemp fibre, back face (scale bar lines at 0.5 mm intervals). Reprinted with permission from Tong and Isaac.[145]

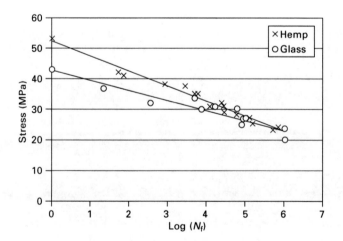

11.33 S–N fatigue lifetime data for hemp fibre mat reinforced polyester and ±45° glass fibre reinforced polyester. Reprinted with permission from Tong and Isaac.[145]

11.9   Wood-based composites

Wood, a ligno-cellulosic material, produced by trees, has a structure that permits the passage of moisture. The fatigue properties of wood were studied earlier in tension, compression and shear mode.[146,147] An investigation into the fatigue performance of wood-reinforced composites was accelerated when these composites began being employed in the construction of bridges (timber bridges), wind turbine blades and panel products. Wood-based composites experience property changes such as a change in the dynamic modulus or a change in the hysteresis loop area in a similar way to traditional fibre composites. Infra-red thermography and environmental scanning electron microscopy are important tools used to investigate the mechanism of damage development in wood/natural fibre reinforced composites.[148]

Designers generally sum up fatigue performance as an additional property along with other general concerns for particle board and plywood applications. For structural applications such as wood roof truss and flooring, framing, and bracing systems in a structure that undergoes cyclic loading, many of the wooden materials fail at the connector junctions by fastener withdrawal, glue line separation or truss plate failure. Research focused on the fatigue properties of wood-based panels showed that in terms of the trends in fatigue life, the nature of property changes bear considerable resemblance to those for fibre reinforced composite materials. An evolution of S–N curves, constant life lines and life prediction models match well for traditional synthetic fibre reinforced composites and natural fibre reinforced composites. A distinguishing point is that the natural fibres have a cellular structure. Even though the natural fibre composites exhibit similar failure mechanisms such as compressive splitting and kinking, the characterization within the cell walls of the natural fibres is rather difficult. However, in spite of the various natural fibre species and the composite products they make available for engineering applications, fatigue performance, in general, depends on moisture content and density.[149,150]

The failure of wood-based components owing to fatigue and under static/quasi-static conditions are difficult to distinguish (Fig. 11.34).[149] With wood-based materials, one must consider the hygroscopic and rheological aspects of the cellulosic structure when suggesting any theory and this must be coupled with the cylindrically anisotropic nature of wood.

11.10   Conclusions

Natural fibres and their composites are readily accepted instead of petroleum-based plastics for noncritical applications because of the combined endeavours of scientists, the government and interested parties in exploring the wide range of applications of NFPCs. These products are not only ecologically

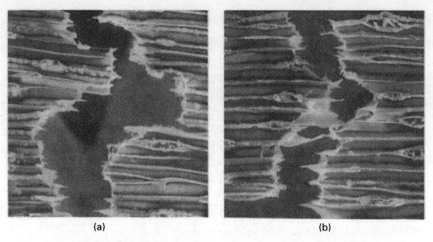

11.34 (a) Static tensile test fracture of southern yellow pine showing the preferred fracture path around crossing cells (rays) or holes (pits). (b) Fatigue tensile failure of the same material at approximately 5000 cycles at 80% of ultimate showing no observable change in fracture morphology (× 100). Reprinted with permission from Kyanka.[149]

beneficial, but also commercially viable, and sustainable. However, there are some concerns over the complete replacement of conventional materials for durability and long-term performance. Furthermore, for outdoor applications, weathering conditions such as humidity, temperature and UV radiation impact adversely on the service life of the product. Several other factors that affect the creep resistance of natural fibre polymer composites are: type of polymer matrix, loading of the natural fibre and other additives. The incorporation of natural fibres into the polymer matrix greatly enhances the creep resistance of polymer composites. The effect of a coupling agent on the creep resistance of polymer composites depends on the modulus of the composite and the coupling effect. Higher temperatures lead to not only larger instantaneous deformation but also to long-term creep rates.

The fatigue properties of composites in general depend on the polymer type (brittle or soft matrix)/structure, MW, crosslinking, type of filler and its content, and, most importantly, on the interfacial strength. It is of paramount importance for design engineers to comprehend the fatigue phenomenon in natural fibre reinforced composites as it actuates design failures. Increasing understanding of property degradation mechanisms during service, and acquiring the ability to anticipate the life of natural fibre composites under defined conditions, will allow designers to develop materials with more reliable and desirable characteristics. Owing to their organic nature and high moisture absorption behaviour, the rate of degradation of the mechanical properties of natural fibre reinforced composites is comparatively high compared with that of synthetic fibre reinforced composites. Research is actively ongoing to employ these eco-friendly, inexpensive natural fibres in polymer matrices to give materials with enhanced mechanical properties that are hardly affected by ageing. With the increased potential of exploring their applications in areas such as motor vehicles, buildings, and in primary structures, a fundamental study of the fatigue properties of natural fibre composites is necessary.

11.11   Acknowledgements

The authors would like to thank Ontario Ministry of Agriculture, Food and Rural Affairs (OMAFRA)-2008 New Directions Research Program Project number SR9211, 'Co-injection molded biodegradable green composites from engineered biomass-based biofibers and bioplastics' for its financial support to carry out this research work.

11.12   Notation

a	amplitude of transient creep strain
A_0	slope of the power law model
b	time exponent
D	creep compliance
E	equilibrium modulus
E_c	modulus of the composite
E_m	modulus of the matrix
$E(t)$	stress at a given time t
$E(t = 0)$	initial stress
f	fibre volume fraction
R	universal gas constant
R	stress ratio
t	time (s)
T	temperature at which the test is performed (K)
T^0	reference temperature (K)
α_T	horizontal shift factor
Γ_m	fracture energy for the matrix
ε	strain
$\varepsilon(t)$	time dependent creep strain at time t
ε_0	instantaneous (elastic) strain,
η	equilibrium viscosity
σ	stress
σ_c	stress required to initiate the crack (in matrix)
τ	interfacial shear stress

11.12.1 Abbreviations

ACPHB	acetone-treated PHB
ACPLLA	acetone-treated PLLA
ASCE	American Society of Civil Engineers
CF	carbon fibre
DMA	dynamic mechanical analysis
DMTA	dynamic mechanical thermal analysis
FRPC	fibre reinforced plastic composites
GTA	glyceryl triacetate
HBP	hyperbranched polyester
HDPE	high-density polyethylene
MAH-PP	maleic anhydride modified polypropylene
NR	natural rubber
NFPC	natural fibres reinforced polymer composites
OPFBF	oil palm fruit bunch fibre
PEG	polyethylene glycol
PHB	poly(3-hydroxybutyrate)
PLLA	poly-(L-lactide acid)
PP	polypropylene
PPMAN	polypropylene maleic anhydride copolymer
TDP	thiodiphenol
PVC	poly(vinyl chloride)
TBC	tributyl citrate
TEC	triethyl citrate
$S\text{–}N$ curve	stress–number of cycles curve
TTS	time–temperature superposition
UD	unidirectional
UNPHB	unmodified PHB
UNPLLA	unmodified PLLA
UV	ultraviolet
WLF	Williams–Landel–Ferry

11.13   References

1   L Thomas, C Steve, F Robert, F Colin. Accelerated weathering of natural fibre-filled polyethylene composites. *Journal of Materials in Civil Engineering*, 2004, **16**(6), 547–555.

2   J L Lopez, M Sain, P Cooper. Performance of natural-fibre-plastic composites under stress for outdoor applications: effect of moisture, temperature, and ultraviolet light exposure. *Journal of Applied Polymer Science*, 2006, **99**, 2570–2578.

3   M Wolcott, P Smith, J Hermanson. 8th international conference on woodfibre-plastic composites, Madison, WI.

4   C Eckert. Opportunities for natural fibres in plastic composites. Proceedings of

the conference on the progress in woodfibre–plastic composite conference, May 25–26, 2000, University of Toronto, Ontario, Canada.

5   D. Brosius. Natural fibre composites slowly take root. *Composites Technology*, Feb. 2006.

6   R M Rowell. Property enhanced natural fibre composite materials based on chemical modification. *Science and technology of polymers and advanced materials*, ed. P N Prasad *et al.*, Plenum Press, New York, 1998, 717.

7   A J Nunez, N E Marcovich, M I Aranguren. Analysis of the creep behavior of polypropylene – woodflour composites. *Polymer Engineering and Science*, 2004, **44**(8), 1594–1603.

8   IM Ward and DW Hadley. *An introduction to the mechanical properties of solid polymers*. Wiley, Chichester, 1993.

9   L E Nielsen and R F Landel. *Mechanical properties of polymers and composites*. M Dekker, New York, 1994.

10 F Rodriguez. *Principles of polymer systems*. Taylor & Francis, Washington, DC, 1996.

11 R S Lakes. *Viscoelastic solids*. CRC Press, Boca Raton, FL, 1999.

12 R J Crawford. *Plastics engineering*, Butterworth–Heinemann, Oxford, 1998.

13 A A Ogale. Creep behaviour of thermoplastic composites, *Thermoplastics Composite Materials*, ed. L. A Charlsson. Elsevier Science Publisher, 205–232.

14 D Bikiaris, P Matzinos, A Larena, V Flaris, C Panayiotou. Use of silane agents and poly(propylene-g-maleic anhydride) copolymer as adhesion promoters in glass fibre/polypropylene composites. *Journal of Applied Polymer Science*, 2001, 81(3), 701–709.

15 N E Zafeiropoulos, D R Williams, C A Baillie, F L Matthews. Engineering and characterization of the interface in flax fibre / polypropylene composite materials. Part I. Development and investigation of surface treatments. *Composites, Part A: Applied Science and Manufacturing*, 2002, **33A**(8), 1083–1093.

16 T Arnson, B Crouse, W Griggs. *The sizing of paper*, ed. W F Reynolds, 2nd ed. Atlanta Tappi Press, 1989.

17 H D Rozman, K W Tan, R N Kumar, A Abubakar, M Ishak, H Ismail. The effect of lignin as a compatibilizer on the physical properties of coconut fibre-polypropylene composites. *European Polymer Journal*, 2000, **36**(7), 1483–1494.

18 W Thielemans, R P Wool. Butyrated kraft lignin as compatibilizing agent for natural fibre reinforced thermoset composites. *Composites, Part A: Applied Science and Manufacturing*, 2004, **35A**(3), 327–338.

19 K J Fridley. Design for creep in wood structures. *Forests Products Journal*, 1992, **42**(3), 23–28.

20 K J Fridley. Creep rupture behaviour of wood. *Agricultural Experiment Station Bulletin* SB-637, Purdue University, 1992.

21 S G Mason. The rheology of paper: a new approach to the study of paper strength. *Pulp Paper Mag. Canada*, 1948, **49**(3), 207–217.

22 I Merdas, F Thominette, A Tcharkhtchi, J Verdu. Factors governing water absorption by composite matrices. *Composites Science and Technology*, 2002, **62**(4), 487–492.

23 H S Yang, H J Kim, H J Park, B J Lee, T S Hwang. Water absorption behavior and mechanical properties of lignocellulosic filler–polyolefin bio-composites. *Composite Structures*, 2006, **72**(4), 429–437.

24 S Kazemi Najafi, E Hamidinia, M Tajvidi. Mechanical properties of composites

from sawdust and recycled plastics. *Journal of Applied Polymer Science*, 2006, **100**(5), 3641–3645.

25   S Kazemi Najafi, H Sharifnia, M Tajvidi. Effects of water absorption on creep behavior of wood–plastic composites. *Journal of Composite Materials*, 2008, **42**(10), 993–1002.

26   A K Bledzki, O Faruk. Creep and impact properties of wood fibre–polypropylene composites: influence of temperature and moisture content. *Composites Science and Technology*, 2004, **64**, 693–700.

27   D J Pooler, L V Smith. Nonlinear viscoelastic response of a wood–plastic composite including temperature effects. *Journal of Thermoplastic Materials*, 2004, **17**, 427–445.

28   S Singh, A K Mohanty. *Composites Science and Technology*, 2007, **67**, 1753–1763.

29   S Singh, A K Mohanty, T Sugie, Y Takai, H Hamada. *Composites: Part A*, 2008, **39**, 875–886.

30   L Jiang, J Huang, J Qian, F Chen, J Zhang, M P Wolcott, Y J Zhu. *Polym. Environ.*, 2008, **16**, 83–93.

31   M Kiguchi. Latest market status of wood and wood plastic composites in North America and Europe. The Second Wood and Wood Plastic Composites Seminar in the 23rd Wood Composite Symposium, Kyoto, Japan, 2007, 61–73.

32   S Panthanpulakkal, M Sain. *Compos. A*, 2007, **38**, 1445–1454.

33   Y Xu. Creep behaviour of natural fibre reinforced polymer composites. PhD dissertation, Lousiana State University, Dec. 2009.

34   Y Xu, Q Wu, Y Lei, F Yao. Creep behavior of bagasse fibre reinforced polymer composites. *Bioresource Technology*, 2010, **101**(9), 3280–3286.

35   A B Acha, M M Reboredo, N E Marcovich. Creep and dynamic mechanical behavior of PP–jute composites: effect of the interfacial adhesion. *Composites, Part A: Applied Science and Manufacturing*, 2007, **38A**(6), 1507–1516.

36   S Houshyar, R A Shanks, A Hodzic. Tensile creep behaviour of polypropylene fibre reinforced polypropylene composites. *Polymer Testing*, 2005, **24**(2), 257–264.

37   A B Acha, M M Reboredo, N E Marcovich. Effect of coupling agents on the thermal and mechanical properties of polypropylene-jute fabric composites. *Polymer International*, 2006, **55**(9), 1104–1113.

38   B E Read, P E Tomlins, G D Dean. Physical ageing and short-term creep in amorphous and semicrystalline polymers, *Polymer*, 1990, **31**(7), 1204–1215.

39   P E Tomlins, B E Read. Creep and physical ageing of polypropylene: a comparison of models. *Polymer*, 1998, **39**(2), 355–367.

40   D S Matsumoto. Time temperature superposition and physical ageing in amorphous polymers. *Polymer Engineering and Science*, 1988, **28**, 1313–1317.

41   S Wong, R Shanks. Creep behaviour of biopolymers and modified flax fibre composites. *Composite Interfaces*, 2008, **15**(2–3), 131–145.

42   M L William, R F Landel, J D Ferry. The temperature dependence of relaxation mechanism in amorphous polymer and other glass-forming liquids. *Journal of the American Chemical Society*, 1955, **77**, 3701–3707.

43   I M Ward. *Mechanical properties of solid polymers*. Wiley Interscience, 1971.

44   J D Ferry. *Viscoelastic properties of polymer*. John Wiley and Sons Inc., 1961.

45   T Yeow, D H Morris, H F Brinson. A new experimental method for the accelerated characterization of composites material. VPI-E-78-3.

46   A D McNaught, A Wilkinson. Arrhenius equation IUPAC Gold Book, 1997.

47   L E Nielsen, R F Landel. *Mechanical properties of polymers and composites*. 2nd ed., Marcel Dekker, New York, 1994.

48   J A Faucher. Viscoelastic behavior of polyethylene and polypropylene. *Transaction of the Society of Rheology*, 1959, **3**, 81–93.

49   I M Ward. *An introduction to the mechanical properties of solid polymers*. 2nd ed., John Wiley & Sons, Chichester, UK, 2004.

50   L E Nielsen, R F Landel. *Mechanical properties of polymers and composites*. 2nd ed., Marcel Dekker, New York, 1994.

51   N K Dutta, E H Graham. Generic relaxation spectra of solid polymers. I. Development of spectral distribution model and its application to stress relaxation of polypropylene. *Journal of Applied Polymer Science*, 1997, **66**(6), 1101–1115.

52   K Nitta, K Suzuki. Prediction of stress-relaxation behavior in high density polyethylene solids. *Macromolecular Theory and Simulations*, 1999, **8**, 254–259.

53   B D Harper, Y Weitsman. Characterization method for a class of thermorheologically complex materials, *Journal of Rheology*, **29**(1), 49–66.

54   M Tajvidi, R H Falk. J C Hermanson. Time–temperature superposition principle applied to a kenaf-fibre/high-density polyethylene composite. *Journal of Applied Polymer Science*, 2005, **97**(5), 1995–2004.

55   N Findley, J S Lai, K Onaran. *Creep and stress relaxation of non-linear viscoelastic materials*. Dover Publications, New York, 1976.

56   L Jiang, M P Wolcott, J W Zhang, K Englund. Flexural properties of surface reinforced wood/plastic deck board. *Polymer Engineering and Science*, 2000, **47**, 281–288.

57   S W Yang, W K Chin. Mechanical properties of aligned long glass fibre reinforced polypropylene. II: tensile creep behaviour. *Polymer Composites*, 1999, **20**(2), 207–215.

58   W N Findley, G Khosla. Application of the superposition principle and theories of mechanical equation of state, strain, and time hardening to creep of plastics under changing loads. *Journal of Applied Physics*, 1955, **26** (7), 821–832.

59   L C Bank, A S Mosallam. Creep and failure of a full size fibre reinforced plastic pultruded frame. *Composites Engineering*, 1990, **2**(3), 217–227.

60   American Society of Civil Engineers. *Structural plastic design manual*. ASCE Publications.

61   C Marais, G Villoutreix. Analysis and modeling of the creep behavior of the thermostable PMR-15 polyimide. *Journal of Applied Polymer Science*, 1998, **69**(10), 1983–1991.

62   D S Matsumoto. Time–temperature superposition and physical aging in amorphous polymers. *Polymer Engineering and Science*, 1988, **28**, 1313–1317.

63   D Dean, M Husband, M Trimmer. Time-temperature-dependent behavior of a substituted poly(paraphenylene): Tensile, creep, and dynamic mechanical properties in the glassy state. *Journal of Polymer Science Part B: Polymer Physics*, 1998, **36**(16), 2971–2979.

64   L C Brinson T S Gates. Effects of physical aging on long term creep of polymers and polymer matrix composites. *International Journal of Solids and Structures*, 1995, **32**(6–7), 827–846.

65   S V Rangaraj, L V Smith. The nonlinearly viscoelastic response of a wood–thermoplastic composite. *Mechanics of Time-Dependent Materials*, 1999, **3**(2), 125–139.

66 A Pramanick, M Sain. Application of the 'theory of mixtures' to temperature–stress equivalency in nonlinear creep of thermoplastic/agro-fibre composites. *Polymers & Polymer Composites*, 2006, **14**(5), 455–472.

67 A Pramanick, M Sain. Nonlinear viscoelastic creep characterization of HDPE – rice husk composites. *Polymers & Polymer Composites*, 2005, **13**(6), 581–598.

68 W S Lin, A K Pramanick, M Sain. Determination of material constants for nonlinear viscoelastic predictive model. *Journal of Composite Materials*, 2004, **38**(1), 19–29.

69 J George, M S Sreekala, S Thomas, S S Bhagwan, N R Neelakantan. Stress relaxation behaviour of short pineapple fibre reinforced polyethylene composites. *Journal of Reinforced Plastics and Composites*, 1998, **17**, 651–672.

70 T J George, S S Bhagwan, S Thomas. Improved interactions in chemically modified pineapple leaf fibre reinforced polyethylene composites. *Composites Interfaces*, 1997, **5**(3), 201–223.

71 T J George, S S Bhagwan, N Prabhakaran. Short pineapple leaf fibre reinforced low density polyethylene composites. *Journal of Applied Polymer Science*, 1995, **57**(7), 843–854.

72 T J George, S S Bhagwan. Effects of environment on the properties of low density polyethylene composites reinforced with pineapple leaf fibre. *Composites Science and Technology*, 1998, **58**, 1471–1485.

73 V G Geethamma, L A Pothen, B Rhao, N R Neelakantan, S Thomas. Tensile stress relaxation of short-coir-fibre-reinforced natural rubber composites. *Journal of Applied Polymer Science*, 2004, **94**(1), 96–104.

74 J Wortmann, K V Schulz. Stress relaxation and time temperature superposition of polypropylene fibres. *Polymer*, 1995, **36**, 315–321.

75 P A O'Connel, G B McKenna. Large deformation response of polycarbonate: time–temperature, time aging time, and time–strain superposition. *Polymer Engineering and Science*, 1997, **37**, 1485–1495.

76 J S Lai, W N Findley. Sress relaxation of non-linear viscoelastic material under uniaxial strain. *Transaction of the Society of Rheology*, **12**, 1968, 259–280.

77 A N Gent. Relaxation processes in vulcanized rubber. III. Relaxation at large strains and the effect of fillers. *Rubber Chemistry and Technology*, 1963, **36**(3), 697–708.

78 S S Bhagawan, D K Tripathy, S K De. Stress relaxation in short jute fibre-reinforced nitrile rubber composites. *Journal of Applied Polymer Science*, 1987, **33**(5), 1623–1639.

79 P Flink, B Stenberg. An indirect method which ranks the adhesion in natural rubber filled with different types of cellulose fibres by plots of $E(t)/E(t = 0)$ versus $\log(t)$. *British Polymer Journal*, 1990, **22**, 193–197.

80 S K N Kutty, G B Nando. Stress relaxation behaviour of short kevlar fibre reinforced thermoplastic polyurethane. *Journal of Applied Polymer Science*, 1991, **42**(7), 1835–1844.

81 A K Sircar, A Voet, F R Cook. Relaxation of stress and electrical resistivity in carbon-filled vulcanizates at moderate and high extensions. *Rubber Chemistry and Technology*, 1971, **44**(1), 185–198.

82 A N Gent. Crystallization and the relaxation of stress in stretched natural-rubber vulcanizates. *Transactions of the Faraday Society* (1954), **50**, 521–533.

83 S Varghese, B Kuriakose, S Thomas, T A Koshy. Dynamic mechanical properties of short sisal fibre reinforced natural rubber composites. *The Indian Journal of Natural Rubber Research*, 1991, **4**(1), 55–67.

84   S Varghese, B Kuriakose, S Thomas. Stress relaxation in short sisal-fiber-reinforced natural rubber composites. *Journal of Applied Polymer Science*, 1994, **53**(8), 1051–1060.

85   S Subramanian, K L Reifsnider, W W Stinchcomb. A cumulative damage model to predict the fatigue life of composite laminates including the effect of a fibre–matrix interphase. *International Journal of Fatigue*, 1995, **17**(5), 343–351.

86   K H Dharan. Interlaminar shear fatigue of pultruded graphite fibre–polyester composites. *Journal of Materials Science*, 1978, **13**, 1243–1248.

87   S A Hitchen, S L Ogin, P A Smith. Effect of fibre length on fatigue of short carbon fibre/epoxy composite. *Composites*, 1996, **26**(4), 303–310.

88   E K Gamstedt, R Talreja. Fatigue damage mechanisms in unidirectional carbon-fibre-reinforced plastics. *Journal of Materials Science*, 1999, **34**, 2535–2546.

89   B Harris (ed). *Fatigue in composites: Science and technology of the fatigue response of fibre-reinforced plastics*. Woodhead Publishing, Cambridge, 2003.

90   E K Gamstedt, L A Berglund, T Peijs. Fatigue mechanisms in unidirectional glass–fibre-reinforced polypropylene. *Composites Science and Technology*, 1999, **59**, 759–768.

91   M J Lamela, A F Canteli, H Reiter, B Harris. Comparative statistical analysis of the fatigue of composites under different modes of loading. *Journal of Materials Science*, 1997, **32**, 6495–6503.

92   N Otani, D Y Song. Fatigue life prediction of composites under two-stage loading. *Journal of Materials Science*, 1997, **32**, 755–760.

93   W V Paepegem, J Degrieck. Modelling damage and permanent strain in fibre-reinforced composites under in-plane fatigue loading. *Composites Science and Technology*, 2003, **63**, 677–694.

94   J Gassan, A K Bledzki. 7th International Techtextil Symposium, June 19–21, 1995.

95   A K Bledzki, S Reihmane, J Gassan. Properties and modification methods for vegetable fibres for natural fibre composites. *Journal of Applied Polymer Science*, 1996, **59**, 1329–1336.

96   S M Lee, R M Rowell. 4th ed., *International Encyclopedia of Composites*, **4**, VCH, New York, 1991.

97   H P Fink. Proceedings Akzo-Nobel Viscose Chemistry Seminar, Challenges in Cellulosic Man-Made Fibres, May 30–June 3, 1994.

98   H P Fink, J Ganster, J Fraatz, M Nywlt. Relations between structure and mechanical properties of cellulosic man-made fibres. Azko Nobel Viscose Chemistry Seminar 'Challenges in cellulosic man-made fibres', Stockholm, 1994.

99   L Mwaikambo. Plant-based resources for sustainable composites, PhD Thesis, University of Bath, 2002.

100   P K Ray, A C Chakravarty, S B Bandyopadhyay. Fine structure and mechanical properties of jute differently dried after retting. *Journal of Applied Polymer Science*, 1976, **20**(7), 1765–1767.

101   M A Semsazadeh. Fibre matrix interactions in jute reinforced polyester resin. *Polymer Composites*, 1986, **71**(1), 23–25.

102   E T N Bisanda, M P Ansell. The effect of silane treatment on the mechanical and physical properties of sisal–epoxy composites. *Composites Science and Technology*, 1991, **41**(2), 165–178.

103   S H Zeronian, H Kawabata. Factors affecting the tensile properties of nonmercerized and mercerized cotton fibres. *Textile Research Journal*, 1990, **60**(3), 179–183.

104   S S Zoolagud, T S Rangaraju. *Plant Fibres News*, 1997, **71**, 6.

105   S Mishra, J B Naik. Absorption of steam and water at ambient temperature in wood polymer composites prepared from agro-waste and Novolac. *Journal of Applied Polymer Science*, 1998, **68** (9), 1417–1421.

106   A F Millman. Strain limited design criteria for reinforced plastic process equipment, 34th Annual Technical Conference of the SPL, Sect. 3D, 1979.

107   J E Riccieri, A Vázquez, L Hecker, D Carvalho. Interfacial properties and initial step of the water sorption in unidirectional unsaturated polyester/vegetable fibre composites. *Polymer Composites*, 1999, **20**(1), 29–37.

108   R T Woodhams, G T Thomas, D K Rodgers. Wood fibres as reinforcing fillers for polyolefins, *Polymer Engineering and Science*, 1984, **24**(15), 1166–1171.

109   L Hua, P Flodin, T Rönnhult. Cellulose fibre-polyester composites with reduced water sensitivity. 2. Surface analysis. *Polymer Composites*, 1987, **8**(3), 203–207.

110   H F Mark, N C Gaylord and N M Bikales (eds). *Encyclopedia of polymer science and technology*, Wiley-Interscience, New York, 1970, 12, 1.

111   E T N Bisanda. The effect of alkali treatment on the adhesion characteristics of sisal fibres. *Applied Composite Materials*, 2000, **7**, 331–339.

112   N Chand, P K Rohatgi. Adhesion of sisal fibre–polyester system. *Polym Comm*, 1986, **27**, 157–160.

113   B N Cox, D B Marshall. Crack bridging in the fatigue of fibrous composites. *Fatigue & Fracture of Engineering Materials & Structures*, 1991, **14**(8), 847–861.

114   G Bao, Y Song. Crack bridging models for fibre composites with slip-dependent interfaces. *Journal of the Mechanics and Physics of Solids*, 1993, **41**(9), 1425–1444.

115   G M Newaz, N Bonora. Damage based fatigue life modeling for brittle composites. *ICCM-11*, Gold Coast, Australia, 1997.

116   J Gassan. A study of fibre and interface parameters affecting the fatigue behaviour of natural fibre composites. *Composites Part A: Applied Science and Manufacturing*, 2002, **33**(3), 369–374.

117   J Gassan. Naturfaserverstärkte Kunststoffe: Korrelation zwischen Struktur und Eigenschaften der Fasern und deren Composites. PhD Dissertation. Kassel: University of Kassel, 1997.

118   W Y Hamad. On the mechanisms of cumulative damage and fracture in native cellulose fibres. *Journal of Materials Science Letters*, 1998, **17**(5), 433–436.

119   K L Reifsnider, K Schulte, J C Duke. Long-term fatigue behaviour of composite materials. ed. T K O'Brien, *Long-term behaviour of composites*, ASTM, Philadelphia, 1983, 813.

120   G M Newaz. A quantitative assessment of debonding in unidirectional composites under long-term loading. *Journal of Reinforced Plastics and Composites*, 1985, **4**(4), 354–364.

121   V Ramakrishnan, N N Jayaraman. Mechanistically based fatigue-damage evolution model for brittle matrix fibre-reinforced composites. *Journal of Materials Science*, 1993, **28**(20), 5592–5602.

122   J Gassan. Fatigue behavior of cross-ply glass-fibre epoxy composites including the effect of fibre-matrix interphase. *Composite Interfaces*, 2000, **7**(4), 287–299.

123   J Gassan, A K Bledzki. Possibilities to improve the properties of natural fibre reinforced plastics by fibre modification – jute polypropylene composites. *Applied Composite Materials*, 2000, **7**(5–6), 373–385.

124   J Gassan, A K Bledzki. Dynamic–mechanical properties of natural fibre-reinforced

plastics: the effect of coupling agents. Fourth International Conference on Woodfibre–Plastic Composites, Madison, USA, 12–14 May, 1997.

125 J Gassan, A K Bledzki. Effect of moisture content on the properties of silanized jute-epoxy composites. *Polymer Composites*, 1997, **18**(2), 179–184.

126 C Lundemo. Fatigue of fibre-reinforced plastics: a literature survey. *Report of the Aeronautical Research Institute of Sweden (FFA-TN-Au-1499)*, Stockholm, 1979.

127 R Talreja. Fatigue of Composite Materials: damage mechanisms and fatigue-life diagrams, *Proceedings of the Royal Society, London*, 1981, **A378**, 461.

128 J Gassan, I Mildner, A K Bledzki. Influence of fibre structure modification on the mechanical properties of flax fibre-epoxy composites. *Mechanics of Composite Materials* (Translation of Mekhanika Kompozitnykh Materialov (Zinatne)), 2000, **35**(5), 435–440.

129 F W Zok, S M Spearing. Matrix crack spacing in brittle matrix composites. *Acta Metallurgica et Materialia*, 1992, **40**(8), 2033–2043.

130 J Gassan, A K Bledzki. Alkali treatment of jute fibres: Relationship between structure and mechanical properties. *Journal of Applied Polymer Science*, 1999, **71**, 623–629.

131 A Mukherjee, P K Ganguly, D Sur. Structural mechanics of jute: The effects of hemicellulose or lignin removal. *Journal of the Textile Institute Transactions*, 1993, **84**, 348–353.

132 D S Varma, M Varma, I K Varma. *Journal of the Textile Institute*, 1984, **54**, 349.

133 S Sreenivasan, B P Iyer, K K R Iyer. Influence of delignification and alkali treatment on the fine structure of coir fibres. *Journal of Materials Science*, 1996, **31**(3), 721–726.

134 H S Sharma, T W Fraser, D McCall, G Lyons. Fine structure of chemically modified flax fibre. *Journal of the Textile Institute*, 1995, **86**(4), 539–548.

135 B R Shelat, T Radhakrishnan, B V Iyer. The relation between crystallite orientation and mechanical properties of mercerized cottons. *Textile Research Journal*, 1960, **33**, 836–842.

136 J Gassan, A K Bledzki. The influence of fibre-surface treatment on the mechanical properties of jute–polypropylene composites. *Composites Part A: Applied Science and Manufacturing*, 1997, **28**(12), 1001–1005.

137 K P Mieck, A Nechwatal, C Knobelsdorf. Faser-Marix-Haftung in Kunstoffverbunden aus thermoplastischer Matrix und Flachs. *Angewandte Makromolekulare Chemie*, 1995, **224**(1), 73–88.

138 D Ray, B K Sarkar, N R Bose. Impact fatigue behaviour of vinylester resin matrix composites reinforced with alkali treated jute fibres. *Composites Part A: Applied Science and Manufacturing*, 2002, **33**(2), 233–241.

139 T Moe, L Kin. Durability of bamboo–glass fibre reinforced polymer matrix hybrid composites. *Composites Science and Technology*, 2003, **63**(3-4), 375–387.

140 I P Bond, M P Ansell. Fatigue properties of jointed wood composites. Part I. Statistical analysis, fatigue master curves and constant life diagrams. *Journal of Materials Science*, 1998, **33**, 2751–2762.

141 B Harris, N Gathercole, J A Lee, H Reiter, T Adam. Life-prediction for constant-stress fatigue in carbon–fibre composites. *Phil Trans R Soc Lond A*, 1997, **355**, 1259–1294.

142 A N Towo, M P Ansell. Fatigue evaluation and dynamic mechanical thermal

analysis of sisal fibre thermosetting resin composites. *Composites Science and Technology*, 2008, **68**(3–4), 925–932.

143   A N Towo, M P Ansell. Fatigue of sisal fibre reinforced composites: Constant-life diagrams and hysteresis loop capture. *Composites Science and Technology*, 2008, **68**(3-4), 915–924.

144   C L Hacker, M P Ansell. Fatigue damage and hysteresis in wood–epoxy laminates, *Journal of Materials Science*, 2001, **36**, 609–621.

145   Y. Tong, D H Isaac. Impact and fatigue behaviour of hemp fibre composites. *Composites Science and Technology*, 2007, **67**(15–16), 3300–3307.

146   K T Tsai, M P Ansell. Fatigue properties of wood in flexure. *Journal of Materials Science*, 1990, **25**, 865–878.

147   P W Bonfield, M P Ansell. Fatigue properties of wood in tension, compression and shear. *Journal of Materials Science*, 1991, **26**, 4765–4773.

148   M P Ansell. Fatigue of wood and wood panel products. In *Fatigue in composites*, ed. Bryan Harris, CRC press, 2003, 339–360.

149   G Kyanka. Fatigue properties of wood and wood composites. *International Journal of Fracture*, 1980, **16**(6), 609–616.

150   R J H Thompson, M P Ansell, P W Bonfield, J M Dinwoodie. Fatigue in wood-based panels. Part 1. The strength variability and fatigue performance of OSB, chipboard and MDF. *Wood Science and Technology*, 2002, **36**(3), 255–269.

12

Impact behavior of natural fiber composite laminates

C. SCARPONI, Sapienza University of Rome, Italy

Abstract: The behavior of natural fiber-reinforced composites under low-velocity impact is described, with particular emphasis on the phenomenon of impact, the sequence of damage, and the testing methods and instruments utilized. The ultrasonic nondestructive inspection (NDI) technique for the delamination evaluation of composite materials is illustrated. For each of these, both the 'classic' composites and the natural fiber composites are discussed. There are many images and tables, which show the behaviors and mechanical characteristics of the particular composite materials considered.

Key words: low-velocity impact, natural fiber, sequence of damage, nondestructive inspection, ultrasonic methods, delamination.

12.1 Introduction

An impact event involves relatively high contact forces acting on a small area over a period of short duration. In general, the composites absorb energy through fracture mechanisms such as indentation, matrix and fiber cracks and delaminations. In aircraft design, particularly for external structures, loads to be considered include impact events such as dropped tools, debris from runways, hailstones, bird strikes. This may result in a large internal damaged area in the laminate that is not detectable by visual observation. Variable service loads can continuously grow the damaged area, possibly resulting in complete structural collapse of the damaged part. In addition, although the use of natural fiber-reinforced plastics is continually increasing, there is still a poor understanding of certain aspects of their behavior, such as their response to impact and the influence of process parameters on mechanical performances.

In low-velocity impact loadings, the fiber–matrix adhesion can further affect the failure mode that occurs at a given load. The consequent damage is more critical in natural fiber composites (NFC) than in other composites because their fiber–matrix interface is far from regular (superficial roughness, defects, impurities). Some chemical or physical treatments can be carried out on NFC to enhance interfacial adhesion (removing weaker layers, creating an interphase with intermediate mechanical properties and/or creating chemical bonds between fibers and matrix), to reduce their hydrophilic nature and to improve wettability.

341

In the work of Valadez-Gonzalez et al.,[1] the effects of various effective chemical treatments (silane, NaOH and preimpregnation with polyethylene) on henequen/HPPE were evaluated using pull-out and SFFT. Wong et al.[2] made a comparative study of the effect of additives on interfacial strength of flax/PLA and flax/poly(3-hydroxybutyric acid) (PHB) composites and found that 4,4'-thiodiphenol and glycerol triacetate effectively improved adhesion for PHB and PLA matrices, respectively.

Many researchers, (such as Abrate, Anderson, Chang, Dhakal, Mueller and Krobjilowski, Scarponi et al., Santulli and Cantwell, Tita et al.[3–18]) have shown experimentally the damage status deriving from low velocity impact loads. In particular, studies on the same subject, but with the use of NF as a reinforcement, have been conducted by Dhakal, Mueller and Krobjilowski, Scarponi et al., Santulli and Cantwell, and Briotti.[9–12,19,20]

Scarponi et al.[17–20] investigated the impact behavior of jute/vinyl ester, glass/vinyl ester and hybrids, in order to verify the possible substitution/integration of NF instead of glass fibers. Santulli[9,10] studied the post-impact behavior of jute/polyester composites and pointed out that their interlaminar adhesion is sufficient to yield an impact damage pattern typical of stronger composites, often referred to as 'reversed-pine tree pattern'. Dhakal[11] dealt with polyester reinforced with needle punched random nonwoven hemp fibers and from impact test results it was shown that the total energy absorbed by 21% fiber volume (four layered) hemp-reinforced specimens was similar to the total energy absorbed by 21% fiber volume chopped strand mat E-glass-reinforced specimens.

Muller and Krobjilowski[12] described the effect of several material parameters and process conditions on impact strength of flax, hemp and kenaf composites. It can be seen from their work that increasing the share of reinforcing fiber in the composite, the maximum of impact strength moves to higher processing temperatures because, increasing the fiber volume, a better fiber impregnation owing to lower binder viscosity predominates over the weakening thermal decomposition.

12.2   Phenomenon description

The impact event involves relatively high contact forces acting on a small area over a period of short duration. The composites generally absorb energy through fracture mechanisms such as local indentation, matrix and fiber cracks and delaminations. Several authors[21–23] described how the delaminations and fiber breakage can often reduce the compressive strength. Thus the compression after impact (CAI) test has become important for the evaluation of composites in structural applications. Moreover, because of the failure mode, the impact damage is often not externally evident.

At relatively low impact energies, damage initiates with matrix cracking, fiber–matrix debonding and delaminations, whereas at higher energies the damage also occurs by fiber fracture and pull-outs. In general, impacts on composites with weak interfacial adhesion produce large areas of splitting and delaminations with strong effects on the compressive properties whereas localized impact loading on high-adhesion composites results in a smaller, more localized damage zone, with higher residual compressive properties of composite.[24]

To understand the low-velocity impact phenomenon Chang *et al.*[25–28] adopted a line-nose impactor in the investigation to study the impact damage in laminated composites, as show in Fig. 12.1. The results of line-loading impact from the experiment provided a considerably simplified two-dimensional impact damage pattern, instead of a complex three-dimensional one.

The sequence of damage is:

1. After the impact some cracks engage in the laminated composite between the plies (intraply matrix cracks).
2. Delaminations appear, caused by the intraply matrix cracks, and they spread through the interface.
3. New micro-cracks are generated.
4. A shear crack creates a delamination along the lower interface of the ply.
5. A bending crack creates a delamination along the upper interface of the ply.

This mechanism generates delamination areas that propagate conically through the thickness of the laminated composites, as shown in Fig. 12.2 and 12.3, in fact these areas increase far from the impacted zone.

Accordingly, based on the predictions and the experimental observations,[25] the physical process of impact damage mechanisms in laminated composites owing to low-velocity impact can be illustrated schematically in sequential steps as shown in Fig. 12.4. Some studies[29–33] have observed that the delamination, caused by an impactor with a hemispherical nose, has a 'peanut' shape with the major axis in the direction of the fibers of laminated composite (Fig.

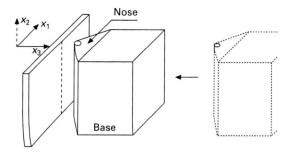

12.1 Schematic of the impactor used in the study.

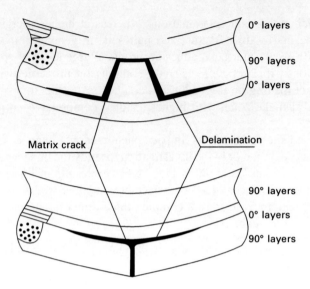

12.2 Matrix crack and delamination areas.

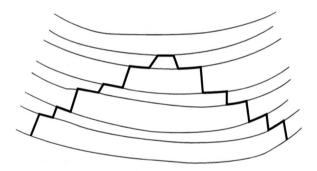

12.3 Mechanism of propagation of delamination areas.

12.5). An explanation of this particular shape is that when the delaminations start, caused by the intraply matrix cracks, they spread much more strongly in the same direction as the fibers.

12.3   Testing methods and instruments

The impact tests are usually performed with a falling weight machine, which allows the most important dynamic and kinetic parameters, such as the contact force, impactor velocity, displacement and perforation energy to be measured. Then, the delamination areas of each specimen can be measured using nondestructive techniques (NDT), such as ultrasonic, x-ray techniques and thermography. In order to determine the residual strength, CAI tests are

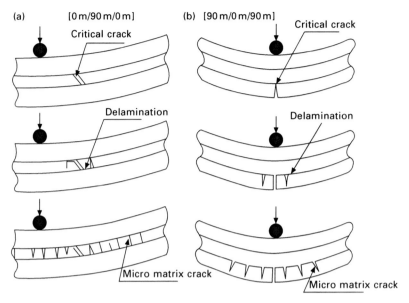

12.4 (a) and (b) Schematic illustration of two basic impact damage growth mechanisms of laminated composites.

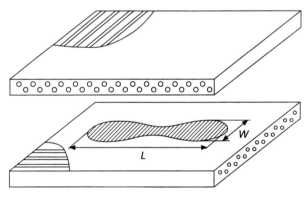

12.5 'Peanut' shape of delamination. *L*, delamination length; *W*, delamination width.

performed on specimens that have an indentation of approximately 0.3–0.4 mm according to the accepted concept of '*barely visible impact damage*'. Such an approach is usual in aeronautics.

12.3.1 Impact test

Scarponi *et al.*[7] studied the behavior of fiber-reinforced composites and sandwich panels for aeronautical applications under impact. Experimental tests were performed on several specimen configurations, based on different quasi-isotropic lay-up and three kinds of material: carbon/epoxy woven fabric, carbon/PEEK tape and syntactic foam. Impact tests were performed using a CEAST Fractovis Mk. 4 drop weight tester (Fig. 12.6). The specimens were clamped on all edges between two steel plates with a circular opening of 40 mm in diameter. The drop weight (4.801 kg) had an impactor with a hemispherical nose 12.7 mm in diameter (Fig. 12.7).

To measure the impact force, the projectile was provided with a contact force transducer (load cell). During each test, the drop weight was released at a predetermined height to impact the center of the specimen. Restrike of the drop weight was prevented by capturing it after the first impact. The impact tester had a photoelectric sensor in order to measure the impactor velocity at the beginning of contact. By means of this tester, the impact force F versus time t was recorded. Other graphs were computed by numerical integrations using suitable software. Thus, the contact force $F(t)$ was measured by a load cell, mounted on the impactor head, and the impactor acceleration was obtained as following:

$$a(t) = \frac{F(t) - P}{m} \qquad [12.1]$$

where F is the contact force and m and P are, respectively, the mass and

12.6 Low-velocity impact test machine.

12.7 Impactor head.

the weight of the falling impactor. By numerical integration the velocity during impact is:

$$v(t) = v_0 - \int_0^t \left(\frac{F(t) - P}{m} \right) dt \qquad [12.2]$$

where v_0 is the velocity at the beginning of contact ($t = 0$), with corresponding impactor kinetic energy τ_0. This velocity is measured by the photoelectric sensor. Further integration gives the displacement of the impactor during impact:

$$x(t) = x_0 + \int_0^t \left[v_0 - \int_0^t \left(\frac{F(t) - P}{m} \right) dt \right] dt \qquad [12.3]$$

where $x_0 = 0$. The energy exchanged at time t, owing to the work done by the contact force during the loading and unloading time, is defined as follows:

$$e(t) = \int_0^t F(t) \left[v_0 - \int_0^t \left(\frac{F(t) - P}{m} \right) dt \right] dt \qquad [12.4]$$

At time $t = t^*$, when the contact ends, the above defined energy results equal to e^* and, because no more energy exchange occurs, this value remains constant for $t > t^*$. The $e(t)$ reaches a maximum value e_{max}, which depends on the impactor kinetic energy (Fig. 12.8). Thus, for perfect elastic rebound:

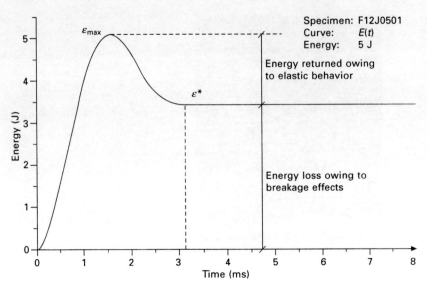

12.8 Typical energy versus time curve.

$$\tau_0 = e_{max} \qquad e_{max} > e^* = 0 \qquad\qquad [12.5]$$

Otherwise, when damage occurs, some of the energy transferred to the laminate is spent in damage phenomena:

$$\tau_0 = e_{max} \qquad e_{max} > e^* > 0 \qquad\qquad [12.6]$$

whereas, when the impactor passes through the laminate, the relations are the following:

$$\tau_0 > e_{max} \qquad e_{max} = e^* \qquad\qquad [12.7]$$

Thus, in this instance, just some of the impactor kinetic energy is transferred to the laminate during the contact; in this last instance, almost all of the energy transferred to the laminate is spent in damage phenomena.

12.3.2 Compression after impact test procedure

The CAI test for specimens in conditions of barely visible impact damage (BVID) is mandatory for composite laminates to be utilized on aeronautical applications (Scarponi et al.[7]). An antibuckling CAI test fixture (Fig. 12.9) was applied to the specimens to be subjected to a compressive load, in agreement with standard Boeing specifications BSS7620. The fixture imposed clamped boundary conditions on the loaded ends. Extreme care was taken to avoid misalignments and hence eccentricity of the load path which would increase the probability of failure owing to an overall geometric instability. The compression tests were conducted in a servo-hydraulic Instron testing

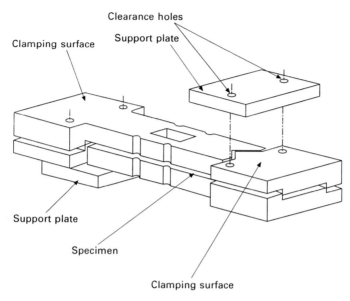

12.9 Compression after impact (CAI) test fixture. The nuts must be tightened by applying a 2 Nm torque using a dynamometric spanner.

machine at a constant crosshead speed of 1.27 mm min^{-1}. From this test, the compression modulus and failure load were obtained. For comparison, tests were also performed using undamaged specimens. For example, in Fig. 12.10 and Fig. 12.11, damaged and undamaged specimens after CAI test are shown. The failure for both of the specimens occurs in the test section.

For undamaged specimens, overestimated values of the ultimate loads were obtained because of the lateral support offered by the antibuckling fixture. On the contrary, for damaged specimens, the failure occurred in the damaged area, thus the previously defined values were correctly estimated. In either instance, it was possible to define the loss of compression properties owing to impact effects, with the described approximation.

12.3.3 Finite element method (FEM) analysis

Wu and Chang[34] presented a transient dynamic finite element analysis for studying the response of laminated composite plates to transverse foreign object impact. The analysis can be used to calculate displacements of composite plate during impact and the transient stress and strain distributions through the laminate thickness. A computer code (3DIMPACT) was based on the analysis. The result of calculations from the code were compared with the existing analytical solutions. An excellent agreement was found between the comparisons. In addition, the strain energy density distributions in the plate

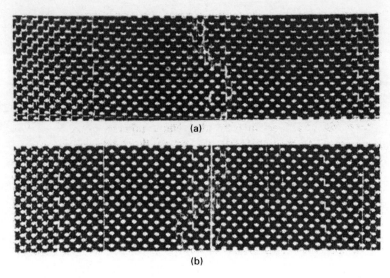

12.10 Specimen after compression test, top view: (a) undamaged and (b) impacted F16 specimens.

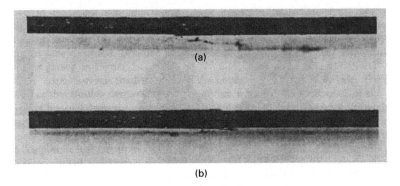

12.11 Specimen after compression test, side view: (a) undamaged and (b) impacted F16 specimens.

as a function of time were also calculated for different ply orientations. The results indicated that a correlation exists between the strain energy density distribution and impact damage. Such a code can be used to evaluate the distributions of all the stresses and strains through laminate thickness during transverse impact and to study the effect of laminate thickness and ply orientation on these distributions (Fig. 12.12–12.14)

In accordance with other authors,[25,34,35] the 3DIMPACT code simulation agreed fairly well with experimental tests at low-velocity impact. Thus, the maximum contact force and delaminated area can be calculated.

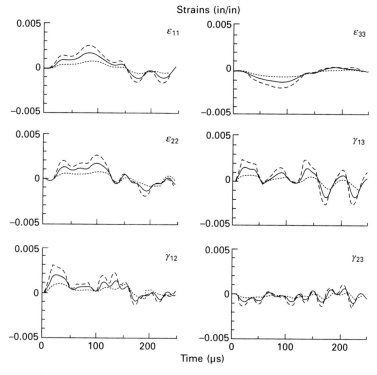

Strains (in/in)

12.12 Strains at a given point (0.3, –0.3, 0.040625) calculated by the present method on 3 in by 3 in T300/934 graphite/epoxy plates ([0/–45/45/90]2s) with clamped edges impacted by 0.5 in diameter aluminum spheres at 500 (– – –), 1000 (——), 1500 (·····) in/s.

12.4   Interpretation of the experimental data

A considerable number of studies have investigated damage in fiber-reinforced laminated composites caused by low-velocity impact. Choi, *et al.*[25] performed an investigation to understand impact damage mechanisms and the mechanics of laminated composites subjected to low-velocity impact and to determine essential parameters governing impact damage. In order to achieve this objective, a new impact tester was designed and built for the investigation. Different ply orientations and various thicknesses of specimens were selected for the tests. During impact, different weights and velocities of the impactor were used as additional test parameters and T300/976 graphite/epoxy preimpregnated samples (prepregs) were selected to fabricate specimen panels. One unique feature of the two-dimensional impact results, is that once damage occurred, delamination was always very extensive. Figure 12.15 presents the estimated delamination size in specimens of various ply orientations as a function of impact energy. There

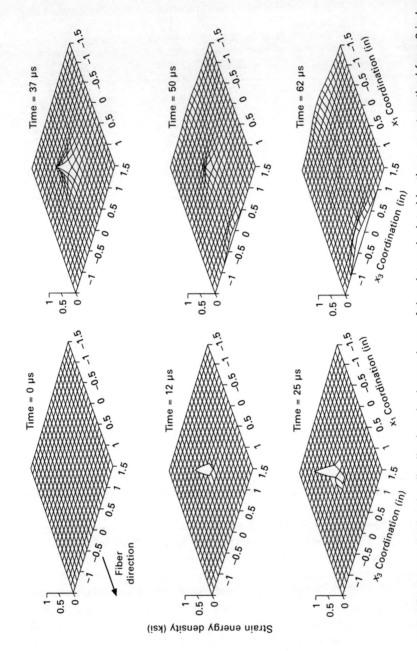

12.13 Strain energy density distributions in the top layer of the plate calculated by the present method for a 3 in by 3 in T300/934 graphite/epoxy plate ([0/−45/45/90]2s) with clamped edges impacted by 0.5 in diameter aluminum spheres at 1000 in/s. From Choi et al.[25]

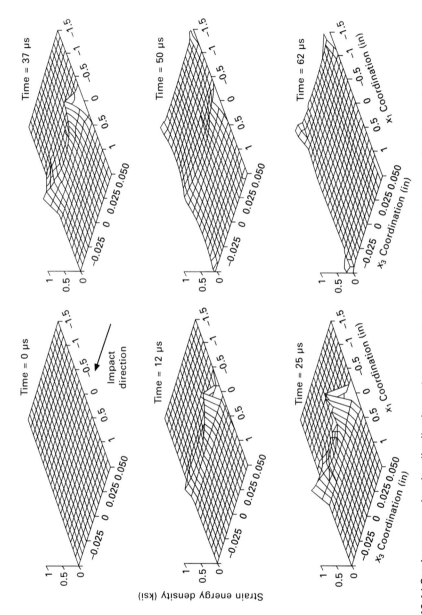

12.14 Strain energy density distributions along $x_2 = 0$ and through the thickness of the plate calculated by the present method for a 3 in by 3 in T300/934 graphite/epoxy plate ([0/−45/45/90]2s) with clamped edges impacted by 0.5 in diameter aluminum spheres at 1000 in/s. From Choi et al.[25]

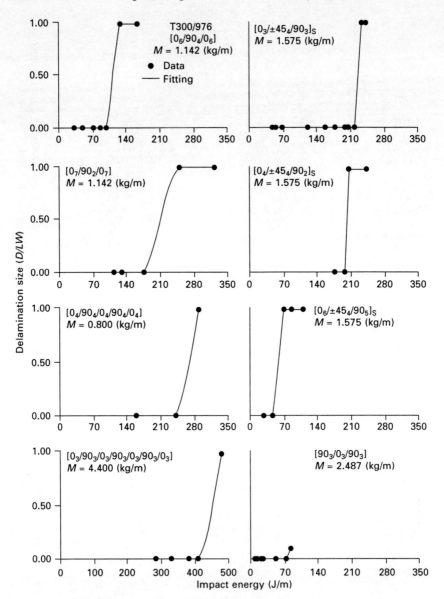

12.15 Illustration of the relationship between impact damage size and the impact for different ply orientations. *D*, effective delamination area; *L*, delamination length; *W*, delamination width.

apparently exists an impact energy threshold beyond which damage occurs. No damage, including matrix cracks and delaminations, was found in any of the specimens tested below that energy level. It was strongly implicated that the impact energy threshold was associated with the energy required to initiate the impact damage. The existence of an impact energy threshold for point-loading cases has also been reported by other investigators.[36–38]

Wu and Chang[34] developed a transient dynamic finite element model to analyze the mechanical response of a laminated composite plate subjected to a foreign object impact. The results of calculations showed that a correlation exists between the strain energy density distributions during impact and the resultant impact damage. Caprino et al.[15] performed low-velocity impact tests on graphite-fabric/epoxy specimens of various thicknesses, with an instrumented drop-weight apparatus. From the load/displacement graphs recorded during impact, the influence of material thickness on the main parameters involved in the impact phenomenon was evaluated. It was found that the force at delamination initiation can be accurately predicted by assuming a Hertzian contact law, coupled with a simple strength criterion based on the interlaminar shear stresses evaluated by strength-of-materials considerations. All the force/displacement graphs pertaining to the various thicknesses effectively converge to a single master graph, if a scaling parameter varying according to the power 1.5 is adopted for forces, whereas the displacements are held unchanged (Fig. 12.16–12.18). Consequently, the dependence of maximum force, energy at maximum force and penetration energy on the thickness can be established.

Gòmez–del Rio et al.[39] examined the response of carbon fiber-reinforced epoxy matrix (CFRP) laminates at low impact velocity and in low temperature conditions. Square specimens of carbon fiber/epoxy laminates with different stacking sequences (unidirectional, cross-ply, quasi-isotropic and woven

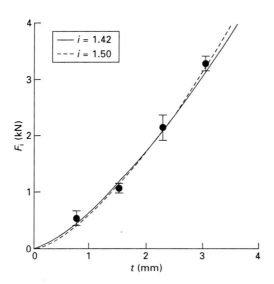

12.16 Force for delamination initiation F_i against specimen thickness t.

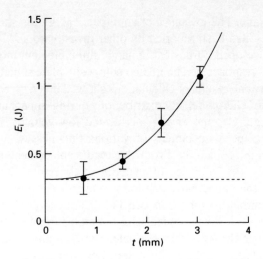

12.17 Energy for delamination initiation E_i against laminate thickness
t.

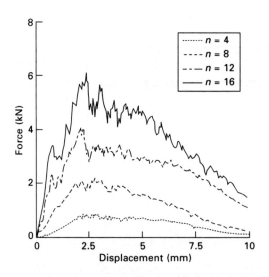

12.18 Typical force/displacement curves recorded during impact for
the different thicknesses (*n* = number of layers).

laminates) were tested using a drop weight tower device. The test temperature
ranged from 20 down to –150 °C. A wide destructive and nondestructive
evaluation (NDE) of the impact-induced damage showed how temperature has
an influence on the damage severity in carbon fiber composites. Figure 12.19
shows, for quasi-isotropic laminates, the mean values of absorbed energy,
resulting from laminate damage, versus impact energy at each test temperature.
This energy increases as temperature decreases. This observation has been

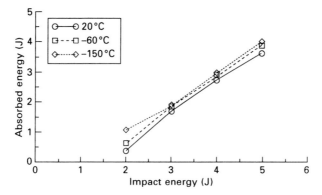

12.19 Mean values of absorbed energy versus impact energy for a quasi-isotropic laminate.

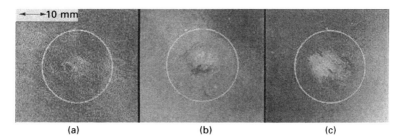

12.20 Effect of temperature on the impacted side damage for a quasi-isotropic laminate impacted at 5 J: (a) 20 °C; (b) –60 °C and (c) –150 °C.

applied to all tape laminates and impact energies. A visual inspection of the specimens shows a concave indentation caused by the edge of the striker bar. Indentation grows as the impact energy increases and as temperature decreases. Figure 12.20 and 12.21 show, in a quasi-isotropic laminate, the effect of low temperature on the external extension of the damage zone.

12.4.1 Impact tests on natural fiber-reinforced composite

Gassan and Bledzki[24] investigated the influence of fiber–matrix adhesion on the behavior of jute, maleic anhydride-polypropylene (MAPP) and untreated, reinforced polypropylene on fatigue and impact, showing that a strong interface is connected with a higher dynamic modulus and that, there was a reduction in stiffness degradation with increasing load cycle and applied maximum stresses. The stronger fiber–matrix adhesion owing to MAH-PP treatment reduced the loss-energy by 30% (Fig. 12.22). The post-impact dynamic modulus after five impact events was roughly 40%.

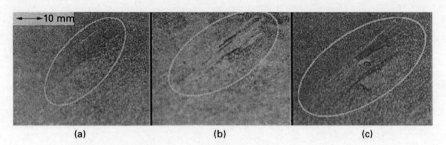

12.21 Effect of temperature on the rear side damage for a quasi-isotropic laminate impacted at 5 J: (a) 20 °C, (b) –60 °C and –150 °C.

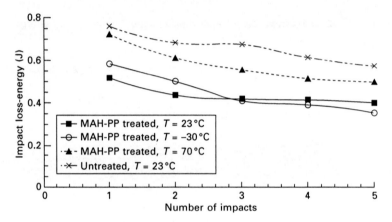

12.22 Influence of fiber surface treatment on nonpenetrating impact test (jute/PP; impact energy 1.5 J; fiber content 35% vol; thickness of specimens 4.4 mm).

Mueller and Krobjilowsky,[12] described the effects of several material parameters and process conditions on the impact strength of flax hemp and kenaf composites. It can be seen from this work that by increasing the share of reinforcing fiber in the composites, the maximum of impact strength moves to higher processing temperatures. An increasing share of NFCs requires a lower binder viscosity to achieve a reasonable fiber embedment; this explains why the peak value of impact strength is reached at higher processing temperatures and the strengthening influence of improved fiber embedment predominates over the weakening thermal decomposition owing to high temperatures. It can also be seen that increasing the preheating time decreases the impact strength.

Santulli[40] tested statically and dynamically (falling weight impact) unidirectional flax/epoxy laminates manufactured using different threads (0.2, 0.9 and 2.3 mm diameter) of untreated flax fibers. A basic study of the hysteresis graphs (Fig. 12.23) obtained from falling weight impact tests

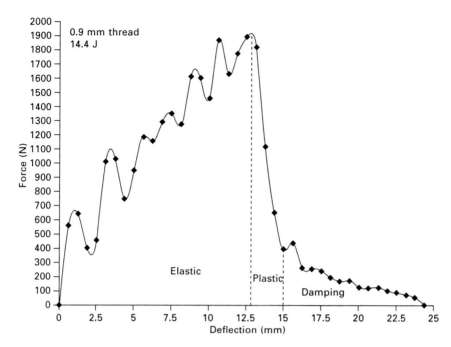

12.23 Impact hysteriesis graph for a flax/epoxy laminate impacted at 14.4 J.

would suggest that the energy generated in the impact event is divided into three parts, related to elastic, plastic and damping behavior of the laminate. The partition of typical hysteresis cycles on a flax/epoxy laminate in these three components is shown in Fig. 12.24.

Significant variations are observed among laminates manufactured with the three different flax threads, whereas those induced by the increased impact energy are only limited and not very consistent. The quasi-static bending tests results confirmed the better performance of the 0.9 mm thread laminates, as is indicated also by the higher value of their maximum load during impact tests. The not outstanding quality of the interface in the 2.3 mm flax thread samples is confirmed also from damage observation on the face opposite to impact, because of the very visible presence of spalling i.e., material loss at the rear (Fig. 12.25). Imperfect fiber–matrix adhesion was sometimes revealed in laminates reinforced with untreated flax fibers.[41] As a consequence, it is suggested that the smaller number of layers in the laminate reinforced with the 2.3 mm thread results in large debonded areas and therefore in easier spalling, in spite of its higher fiber content.

Scarponi *et al.*[42] determined experimentally some important mechanical characteristics of resin transfer moulding (RTM) hemp plain weave fabric/epoxy laminates; the impact performance is analyzed, in particular by comparing

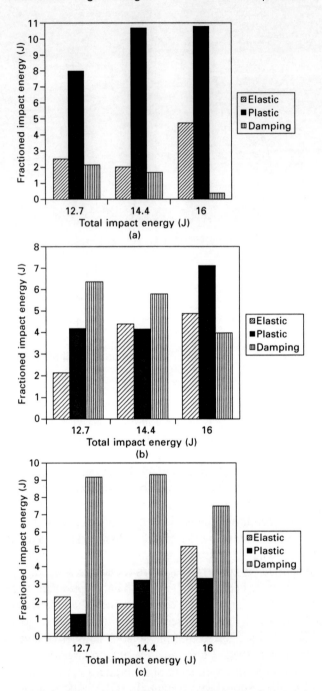

12.24 Partition of impact energy in laminates impacted at various energies: (a) 2.3 mm flax thread; (b) 0.9 mm flax thread; and (c) 0.2 mm flax thread.

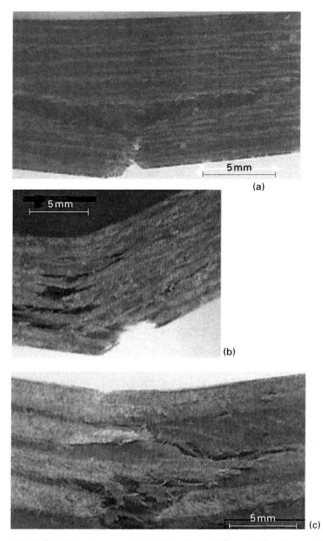

11.25 Spalling (rear material loss) on different flax/epoxy laminates (impact damage 16 J): (a) 0.2 mm thread, (b) 0.9 mm thread and (c) 2.3 mm thread.

data with other experimental results. In this study, the falling weight machine is equipped with an antirebound system and allowed variation of the mass, by means of removable weights (ASTM D5628-96, ASTM D5428-98).

Square specimens 100 mm × 100 mm have been utilized for the impact test. Specimens were held by a rectangular clamp. The test was conducted for three different levels of energy: 5, 10 and 15 J, with three samples for each level. The hemispherical head impactor had a mass of 3.966 kg and a diameter of 12.7 mm. The striker's speed was, respectively, 1.59, 2.25, and

2.75 m s^{-1}. The speed of the body at impact was measured by a photocell. From data elaboration, it was possible to obtain the impact speed v and the force $F(t)$ between the striker and the target.

Because the energies were not sufficient to produce the perforation of the laminates, the impact resulted in striker rebound. Information was obtained from the images of the impacted specimens, acquired also in backlight, and from force versus displacement, force versus time and velocity versus time curves recorded during the impact. From force versus time curves (Fig. 12.26) it is possible to see that:

- the shape of all of the force–displacement and velocity–time graphs (Fig. 12.27 and 12.28) show impact with rebound (displacement decreases after reaching the maximum load);
- for impact at 5 J, the graph is quasi-symmetrical, therefore the energy dissipation is reduced;
- the other two graphs show more energy dissipation, because of more marked asymmetry between the loading–unloading parts and because of rapid fluctuations owing to progressing damage.

From the force–time graphs it is also possible to measure the maximum load and the consequent onset of first damage. Results are reported in Table 12.1 together with the maximum displacement and the time at which maximum displacement is measured. The force–displacement graph (Fig. 12.29) shows clearly that loading and unloading phases do not coincide: rather a closed loop is apparent; its area (A1 + A2) represents the energy absorbed by the

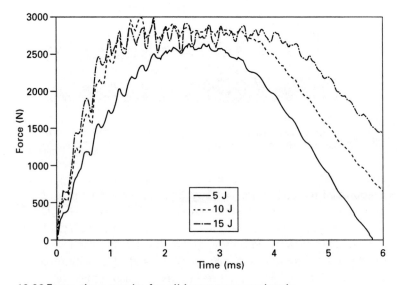

12.26 Force–time graphs for all impact energy levels.

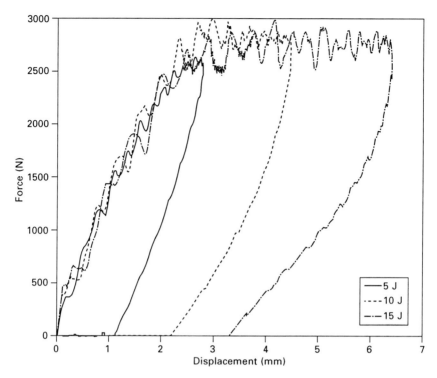

12.27 Force–displacement graphs for all impact energy levels.

specimen, dissipated by damage mechanisms, and is referred to as total hysteresis energy.

According also to Santulli,[43] and referring to Fig. 12.29, variables measured are: the slope of the quasi-elastic part of the impact graph, referred to as linear stiffness, maximum load, load drop (an indicator of damage severity), and damping ratio, defined as the ratio between the nonelastic energy (A2+A3) and the total hysteresis energy (A1+A2). These are all reported in Table 12.2. In particular, from the force–displacement graphs in Fig. 12.27 it can be seen that the stiffness, after reaching the first damage load, decreases continuously owing to material plasticity; for this reason the entity of the load drop is small and not very evident: in other words, progressive damage is observed.

When the striker stops without rebounding it is generally possible to divide the plot into three typical zones that have also been reported in Fig. 12.29:

A1 elastic response (with dynamical perturbations) until the yield force is reached;

A2 from the first sign of material damage until the peak force (with substantial

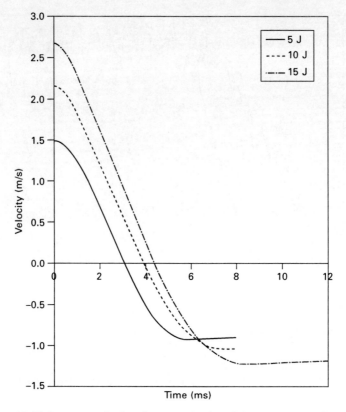

12.28 Average velocity–time graphs for all impact energy levels.

perturbations owing to damage development with formation of internal cracks and propagation of delaminations);

A3 force decrease, indicating the propagation of rupture phenomena.

Figure 12.30 shows a typical curve of the energy given from the striker to the specimen (for the 15 J impacted specimen). It can be assumed that the main contributions are represented by the energies of elastic deformation and energy dissipated during damage.[44] Whereas the first one is conserved, hence comes back to zero when the load is removed, the second one represents the proportion lost owing to irreversibility of the process. Referring to Fig. 12.31, it can be seen that after the maximum (E_{impact}) is reached, the trend is for the energy to decrease until a constant level of energy ($E_{absorbed}$) is reached at the end of the test.[44]

$$E_{impact} = E_{elastic} + E_{absorbed} \qquad [12.8]$$

where $E_{absorbed}$ is the asymptotic energy value, which comprises two terms: energy expended to generate the damage (E_{damage}) and energy absorbed by

Table 12.1 Impact results for hemp plain weave reinforced composites

Impact energy (J)	Incipient damage time (average) (ms)	Coefficient of variation (%)	Displacement (average) (mm)	Coefficient of variation (%)	Maximum displacement (average) (mm)	Coefficient of variation (%)	Maximum displacement time (average) (ms)	Maximum force (average) (N)	Coefficient of variation (%)
5	2.02	4.9	2.41	3.6	2.79	2.8	3.1	2689	1.9
10	1.15	1.4	2.31	1.3	4.58	2.7	3.64	2989	0.8
15	0.97	1.1	2.47	0.95	6.44	2.8	4.37	2996	4.1

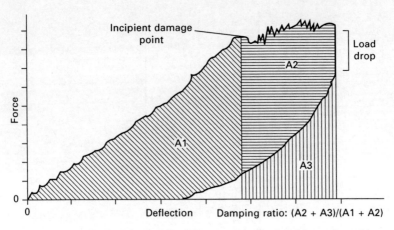

12.29 Typical force–displacement graph creating a closed loop.

the system by various means, such as vibrations, heat and anelastic behavior. (E_{disp}):[32]

$$E_{absorbed} = E_{damage} + E_{disp} \qquad [12.9]$$

The damage energy E_{damage} itself can be divided into three major components necessary for: indentation $E_{indentation}$, matrix damage E_{dm} and fiber rupture E_{df}.

A comparison between the three average curves obtained for each impact energy level is shown in Fig. 12.32.

In Fig. 12.33 are shown damaged surfaces on the side opposite to impact: for the 5 J impacted samples, damage is limited and no preferential directions of damage can be seen, whereas these are clearly seen in the 10 and 15 J samples. It is also apparent that the delamination propagation is not regular: in particular near the center, delaminations do not immediately head for 0°/90° directions (probably because of fabric defects and other irregularities), but that they begin at ±45° and then change to 0°/90°. At 15 J energy, from the damage extension, it can also be stated that a slightly higher energy would have probably led to penetration. For this reason higher energy impacts have not been performed.

Santulli[40] studied the falling weight impact properties of three different types of hemp/epoxy composites, in which hemp fibres are disposed in different ways: in particular, two configurations have been realised using loose hemp fibres, disposing them all with the same orientation (0°) in one instance (LU laminate), whereas for the other case layers are laminated with a 0°/90° configuration (LC laminate). The third configuration has been obtained using hemp mat (M laminate). The fiber volume introduced was limited only by the hand lay-up technique used for composite manufacturing, so that in every instance the maximum possible amount of fibres was introduced.

Table 12.2 Results for hemp plain weave/epoxy 5.1 mm thickness laminates

Impact energy (J)		A1 (J)	A2 (J)	A3 (J)	Damping ratio	Linear stiffness (kN/mm)	Load drop (N)	Load drop/maximum load (%)
5	Average	2547	0.211	0.755	0.351	1175	82	3.1
	CV (%)	2.1	25.3	19.3	20.6	4.7	17.3	17.6
10	Average	3409	3861	2224	0.837	1229	153	5.3
	CV (%)	2.6	1.7	1.1	1.4	3.2	6.6	8
15	Average	3748	7685	3020	0.936	1240	192	6.5
	CV (%)	1.5	1.1	5.1	1.9	1.2	5.5	7.5

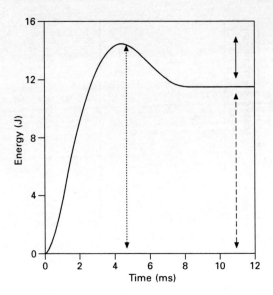

12.30 Energy–time graph; the dotted arrow indicates E_{impact}, the dashed arrow $E_{absorbed}$ and the solid line arrow $E_{elastic}$.

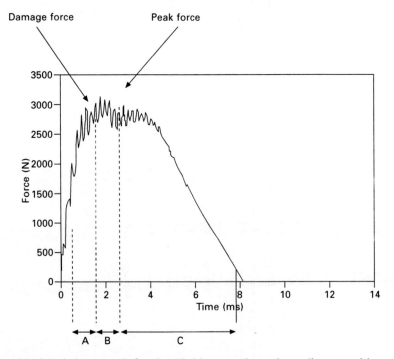

12.31 Load–time graph for the 15 J impacted specimen (impact with no perforation).

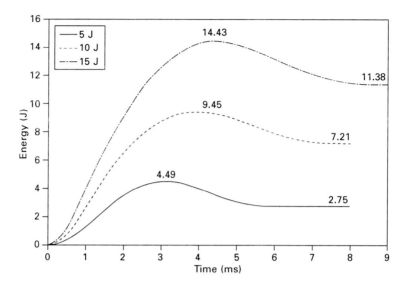

12.32 Energy–time graph for different energy levels.

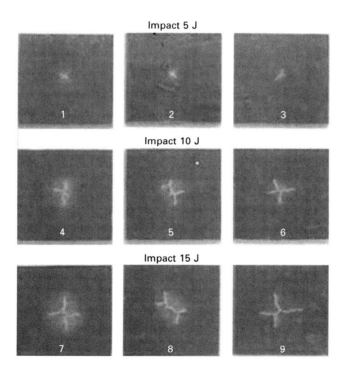

12.33 Back surfaces of impacted specimens (three for each energy level, one level per line, from 5 to 15 J).

In this way, the two configurations including loose hemp fibres, LU and LC, allowed 55 and 52% volume of fibres to be introduced, respectively, whereas hemp mat laminates M only included 43% vol. of fibres. In practice, this investigation evaluated whether disposing loose hemp fibres in a mat, an operation that results in the reduction of the volume of reinforcement introduced, could yield lower falling impact properties or not. Quasi-static test results show that the best performance, but also the largest scattering in properties, is obtained from the uni-directional composites (LU): this suggests the fundamental role played by fiber orientation in these materials. In contrast, the study of impact hysteresis cycles suggests that, despite the limitations owed to the presence of through thickness fibres and the lower amount of fibres introduced, hemp mat laminates (M) are superior in terms of impact properties.

In this instance, hysteresis cycles typically show a damping behaviour only for LU laminates, whereas, for LC and particularly M laminates, a large rebound energy after penetration, released at quasi-constant rate, is measured, with a limited plastic phase and a hardly detectable load drop, as it is shown by the examples of typical hysteresis cycles depicted in Fig. 12.34, for the three laminates.

12.5   Nondestructive inspection (NDI) ultrasonic techniques

It is well known that, owing to their brittle behavior and the manufacturing technique, composite materials are subjected to damage, such as delaminations, fiber and matrix crack and indentation. Moreover, in most cases, the damage develops inside the sample, such as for low-velocity impact load. Thus, several nondestructive inspection (NDI) techniques were developed. One of these comprises the use of ultrasonic waves that allows *in situ* inspection and does not need specimen pretreatment. NDI techniques based on ultrasonic waves have been utilized in the form of transmission and reflection for the detection of delaminations, cracks and other internal defects on anisotropic plates, but considerable operator skill is required to obtain reliable results.

Many studies[7,14,45–47] have utilized ultrasonic NDI systems for the detection of delaminations, and this is important in several applications involving fatigue and impact.

Scarponi et al.[48] described an ultrasonic NDI technique for the study of delamination in several composite materials. The methodology, based on an ultrasonic test apparatus in the form of reflection, allows the position to be determined along the thickness and extension of delaminations on several CFRP, GFRP and Kevlar fiber-reinforced plastic (KFRP) laminates subjected to low-velocity impact test.

The NDI system comprises (Fig. 12.35):

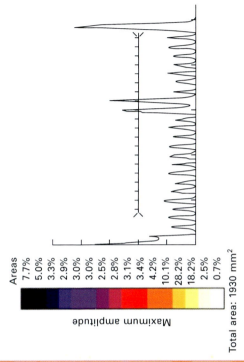

Areas	
7.7%	
5.0%	
3.3%	
2.9%	
3.0%	
3.0%	
2.5%	
2.8%	
3.1%	
3.4%	
4.2%	
10.1%	
28.2%	
18.2%	
2.5%	
0.7%	

Maximum amplitude

Total area: 1930 mm²

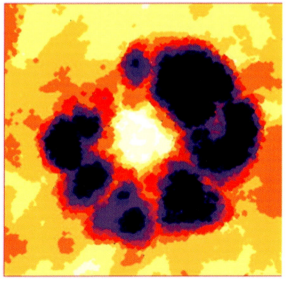

Plate I Amplitude mode.

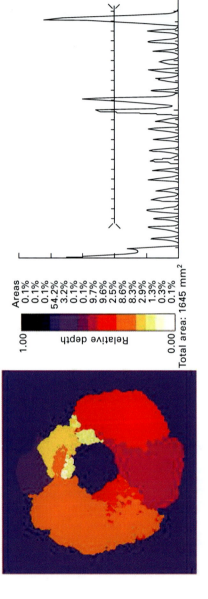

Total area: 1645 mm²

Plate II Relative depth mode.

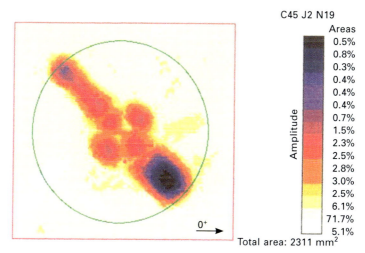

Plate III C-scan for ±45° laminate impacted at 2 J.

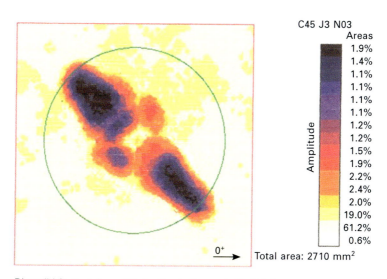

Plate IV C-scan for ±45° laminate impacted at 3 J.

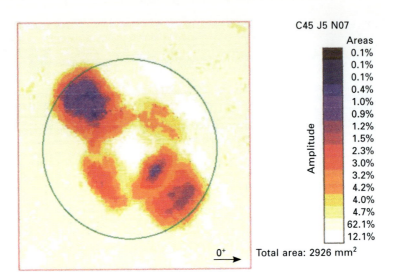

Plate V C-scan for ±45° laminate impacted at 5 J.

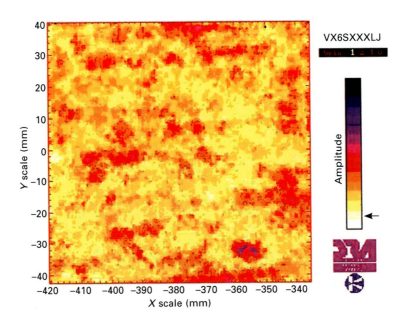

Plate VI Color map of sample before impact.

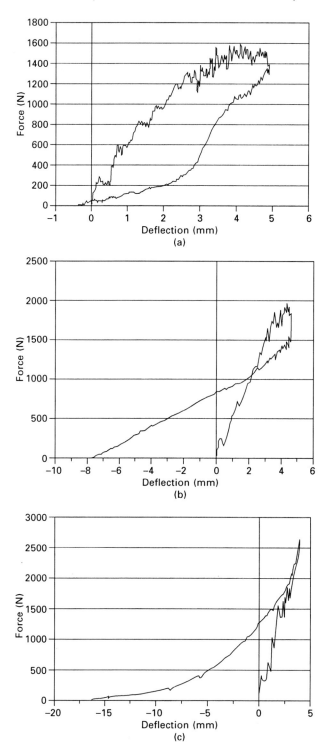

12.34 Typical falling weight impact hysteresis graph for the three hemp/epoxy laminates: (a) LU, (b) LC and (c) M.

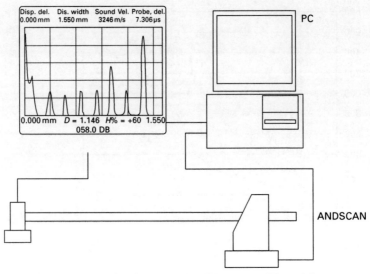

PC

ANDSCAN

12.35 Components of the ultrasonic NDI system. Disp. del., zero point; Dis. width, thickness of specimen; Sound Vel., sound velocity inside specimen; Probe, del., delay of the probe.

- An ultrasonic signal generator and an echo acquisition instrument, called USD 10, with a display and a user-oriented menu for the input/output operations;
- A 5 or 15 MHz probe for both the emission and reception of the ultrasonic beam;
- A probe arm, called ANDSCAN, fixed on a reference surface by a simple suction-cup system. The arm has two position encoders, in order to obtain the probe position with respect to the sample surface;
- A PC with dedicated software, that allows several operations such as scan map plotting in real-time and image post treatment.

The USD-10 display shows several echo peaks owing to the interlaminar surfaces; the first and the last peaks correspond to the top and the bottom surfaces of the sample and they are the highest in level. When a discontinuity increase (i.e. a delamination occurs) the corresponding peak rises, owing to the high reflection of ultrasonic wave. If the peak reaches the instrument gate, its position in depth is shown on the USD-10 display.

This tester allowed two working modes: the maximum amplitude mode and the relative depth mode. In the amplitude mode the maximum ultrasonic echo amplitude is measured, giving a precise damage location, because of the relationship between the damage extent and the echo amplitude. The USD-10 is arranged in order to measure and record the highest peak level. As a result of this inspection, a colorful map is obtained (Plate I between pages 370 and 371).

In the relative depth mode, the ultrasonic echo delay, that is the damage localization in terms of relative depth, is measured as a percentage of the specimen thickness (Plate II). In this instance, the USD-10 is arranged in order to measure and record the level of the peak above to the threshold limit and nearest to the specimen top surface (flank mode). In the depth map, the colors scale indicates the relative depth, and thus the damage position as a percentage of the specimen thickness. In Plates III to V, the C-scans for ±45° laminate impacted at 2, 3 and 5 J are shown.

Scarponi and Valente[18] described the experimental behaviour of very low-cost composite laminates, subjected to transverse impact loads at low velocity. The base materials consist of jute fibers and E-glass woven rovings of different specific area weights, embedded with a vinylester resin; the polymerization system was moulded by vacuum bag. Square specimens 100×100 mm between 4 and 8 mm thick for seven different lay-ups were subjected to impact loads on a horizontal axis machine at 5, 10, 15 and 20 J and the impact parameters were recorded. The impact tests were performed at the laboratories of the Aerospace Department of the Politechnic of Milan. The horizontal axis machine had an impactor with a hemispherical nose of 12.7 mm in diameter, accelerated by a precompressed spring. The material of the projectile was steel with various masses, from a minimum of 500 g up to a maximum of 5000 g.

The extension of the delaminations inside the specimens was determined by means of a new ultrasonic technique, developed by Scarponi and Valente[18]; ultrasonic maps were obtained before and after the impact load. The inspection was performed both from the impacted side and the rear side; in this way more information was obtained about the presence of delaminations. The samples were subjected to ultrasonic inspection before (following the maximum amplitude mode procedure on one side) and after the impact; in this way original defects, induced by the fabrication process, are evident.

In Plate VI, a typical colors map, obtained by the C-scan of a V5 specimen before impact is reported; it is possible to observe the presence of a spread of delaminations, the areas of which have been determined. After impact, the delaminated areas were evaluated (Fig. 12.36). The indentation zone, in accordance with other studies, was considered to be damaged along the whole thickness (Fig. 12.37 and 12.38).

12.6   Acknowledgements

The author thanks Dr Sara Iacobellis for her help with the literature search and assembly of the figures and text.

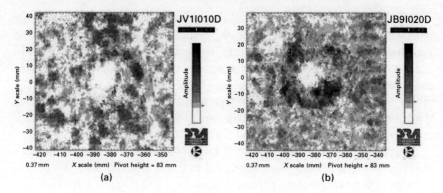

12.36 Evaluation of impacted delaminated area for a V5 glass fiber sample: (a) before impact and (b) post impact.

12.37 Rear part of the sample showing the prominence of the impacted zone.

12.38 Sample JV1-600 impacted at 20 J showing the impacted zone.

12.7   References

1 Valadez-Gonzalez, A., Cervantes-Uc, J.M., Olayo, R. and Herrera-Franco, P.J. 'Effect of fiber surface treatment on the fiber–matrix bond strength of natural fiber reinforced composites'. *Composites*, 1999, **B 30**, 309–320.

2 Wong, S., Shanks, R.A. and Hodzic, A., 'Effects of additives on the interfacial strength of poly(L-lactic acid) and poly(3-hydroxybutyric acid) – flax fibre composites'. *Composites Science and Technology*, 2007, **67**, 2478–2484.

3 Keršienė, N. and Žiliukas, A., 'Experimental investigation of low-velocity impact on woven glass-fibre-reinforced plastics composites'. *Mechanika*, 2005, **6**, 56.

4 Tita, V., de Carvalho, J. and Vandepitte, D., 'Failure analysis of low velocity impact on thin composite laminates: experimental and numerical approaches'. *Composite Structures*, 2008, **83**, 413–428.

5 Anderson, T.L., 'Fracture mechanics – fundamentals and applications'. New York: CRC Press; 1995.

6 Abrate, S., 'Impact on composite structures'. London: Cambridge University Press, 1998.

7 Scarponi, C., Briotti, G., Barboni, R., Marcone, A. and Iannone, M. 'Impact testing on composites laminates and sandwich panels'. *Journal of Composite Materials*, 1996, **30**, 17.

8 Chang, F.K. and Chang, K.Y. 'A progressive damage model for laminated composites containing stress concentration'. *Journal of Composite Materials*, 1987, **21**, 834–855.

9   Santulli, C. and Cantwell, W.J. 'Impact damage characterisation on jute/polyester composites'. *Journal of Materials Science Letters*, 2001, **20**, 477–479.

10 Santulli, C., 'Mechanical and impact properties of untreated jute fabric reinforced polyester laminates compared with different E–glass fibre reinforced laminates'. *Science and Engineering of Composite Materials*, 2001, **9**(4) 177–188.

11 Dhakal, H.N., Zhang, Z.Y., Richardson, M.O.W. and Errajhi, O.A.Z., 'The low velocity impact response of non-woven hemp fibre reinforced unsaturated polyester composites'. *Composite Structures*, 2007, **81**, 559–567.

12 Mueller, H.D. and Krobjilowski, A., 'Improving the impact strength of natural fiber reinforced composites by specifically designed material and process parameters'. *International Nonwovens Journal*, 2004, **13**(4), 31–38.

13 Scarponi, C., Maglione, D., Cosentino, G. and Lenzi, F., 'Hemp fabrics for natural fibres composites industrial applications'. *International Journal of Materials and Product Technology* (IJMPT), 2009, **36**(1–4), 261–277.

14 Scarponi, C. and Briotti, G., 'Ultrasonic detection of delaminations on composite materials'. *Journal of Reinforced Plastics and Composites*, 1997, **16**, 768–790.

15 Caprino, G., Lo Presto, V., Scarponi, C. and Briotti, G., 'Influence of material thickness on the response of carbon fabric/epoxy panels to low velocity impact'. *Composite Science and Technology*, 1999, **59**, 2279–2286.

16 Scarponi, C. and Briotti G., 'Ultrasonic techniques for the evaluation of delaminations on CFRP, KFRP and GFRP composite laminates'. *Composites Part B*, 2000, **31**, 237–243.

17 Briotti, G., Caneva, C., Valente, M. and Scarponi, C., 'Delamination on impacted natural fibers composite materials by means of a new ultrasonic methods', *Proceedings Structural Health Monitoring*, Stanford, 12–14 Septtember 2001.

18 Scarponi, C. and Valente, M., 'An application of a new ultrasonic technique to jute composite laminates subjected to low-velocity impact' *International Journal of Materials and Product Technology*, 2006, **26**(1–2), 6–18.

19 Briotti, G., Caneva, C., Valente, M. and Scarponi, C., 'Impact behaviour of jute fibers composites. *Proceedings International Conference Composite Engineering ICCE/8*, Tenerife, 5–11 August, 2001.

20 Briotti, G., Caneva, C., Valente, M. and Scarponi, C., 'Experimental evaluation of delaminations for low-cost composite laminates impacted at low-velocity'. *Proceedings International Conference Composite Engineering ICCE/7*, Denver, USA, 2–8 July, 2000.

21 Ghasemi Nejhad, M.N. and Parvizi-Majidi, A., 'Impact behavior and damage tolerance of women carbon fibre-reinforced thermoplastic composites'. *Composites*, 1990, **21**(2), 155–168.

22 Sjoblom, P.O., Hartness, J.T. and Cordell, T.M., 'On low-velocity impact testing of composite meterials'. *Journal of Composite Materials*, 1988, **22**, 30–52.

23 Iannone, M., Marcone, A., de Leo, M. and Ferraro, P., 'Impact damage in composites, controlling factors'. *Ninth International Conference on Composite Materials*, 12–16 July 1993, Madrid, Spain.

24 Gassan, J. and Bledzki, A.K., 'Possibilities to improve the properties of natural fiber reinforced plastics by fiber modification – jute polypropylene composites'. *Applied Composite Materials*, 2000; **7**, 373–385.

25 Choi, H.Y., Downs, R.J. and Chang, F.K., 'A new approach to toward understanding damage mechanisms and mechanics of laminated composites due to low-velocity impact: Part I – Analysis', *Journal of Composites Materials*, 1991, **25**, 992–1011.

26 Choi, H.-Y., Wu, T. and Chang, F.K., 'A new approach toward understanding damage mechanism and mechanics of laminated composites due to low-velocity impact. Part II. Analysis', *Journal of Composite Materials*, 1991, **25**, 1015.

27 Choi, H.Y., Wang, H.S. and Chang, F.K., 'Effect of laminate configuration and impactor's mass on the initial impact damage of composite plates due to line-loading impact', *Journal of Composite Materials*, 1992, **26**(6), 804–827.

28 Chang, F.K., Choi, H.Y. and Wang, H.S. 'Damage of laminated composites due to low velocity impact', *31st AIAA/ASME/ASCE/AHS/ASC Structures, Structural Dynamics and Materials (SDM) Conference*, April 2–4, Long Beach, CA.

29 Joshi, S.P. and Sun, C.T., 'Impact-induced fracture in quasi-isotropic laminate', *Journal of Composite Technology and Research*, 1986, **19**, 40–46.

30 Gosse, J.H. and Mori, P.B.Y. 'Impact damage characterization of graphite/epoxy laminates', *Proceedings of the Third Technical Conference of the American Society for Composites*, 1988, Seattle, WA, pp. 334–353.

31 Joshi, S.P., 'Impact-induced damage initiation analysis: an experimental study', *Proceedings of the Third Technical Conference of The American Society for Composites*, 1988, Seattle, WA, pp. 325–333.

32 Freeman, S.M., 'Characterization of lamina and interlaminar damage in graphite/epoxy composites by the deply technique', *Composite Materials; Testing and Design (Sixth Conference)*, 1982, ASTM STP 787, I.M. Daniel, ed., American Society for Testing and Materials, pp. 50–62.

33 Aggour, H. and Sun, C.T. 'Finite element analysis of a laminated composite plate subjected to circularly distributed central impact loading', *Computers and Structures*, 1988, **28**, 729–736.

34 Wu, H.Y.T. and Chang, F.K., 'Transient dynamic analysis of laminated composite plates subjected to transverse impact', *Computers and Structures*, 1989, **31**(3), 453–466.

35 Choi, H.Y. and Chang, F.K., 'A model for predicting damage in graphite/epoxy laminated composites resulting from low-velocity point impact', *Journal of Composites Materials*, 1992, **26**(14), 2134–2169.

36 Peter, O., Hartness, J.T. and Cordell, T.M., 'On low-velocity impact testing of composite materials', *Journal of Computers and Structures*, 1988, **22**, 30–52.

37 Reed, P.E. and Turner, S., 'Flexed plate impact, Part 7. Low energy and excess energy impacts on carbon fiber-reinforced polymer composites', *Composites*, 1988 **19**, 193–203.

38 Cantwell, W.J. and Morton, J., 'Geometrical effects in the low velocity impact response of CFRP', *Composite Structures*, 1989, **12**, 39–59.

39 Gòmez-del Rio, T., Zaera, R., Barbero, E. and Navarro, C., 'Damage in CFRPs due to low velocity impact at low temperature'. *Composites: Part B*, 2004, **36**, 41–50.

40 Santulli, C., 'Falling weight impact damage characterization on flax/epoxy laminates'. *International Journal of Materials and Product Technology*, 2009, **36**, 1/2/3/4.

41 Van De Weyenberg, I., Truong, T.C., Vangrimde, B. and Verpoest, I., 'Improving the properties of UD flax fibre reinforced composites by applying an alkaline fibre treatment', *Composites Part A – Applied Science and Manufacturing*, 2006, **37**, 1368–1376.

42 Scarponi, C., Pizzinelli, C.S., Sànchez-Sàez, S. and Barbero, E., 'Impact load behaviour of resin transfer moulding (RTM) hemp fibre composite laminates'. *Journal of Biobased Materials and Bioenergy*, 2009, **3**, 298–310.

43 Santulli, C. and Caruso, A.P., 'Effect of fiber architecture on the falling weight

impact properties of hemp/epoxy composites'. *Journal of Biobased Materials and Bioenergy*, 2009, **3**, 291–297.

44 Scarponi, C., Briotti, G., Barboni, R., Marcone, A. and Iannone, M., *Journal of Composites Materials*, 1996, **30**, 17.

45 Jeong, H., 'Effect of void on the mechanical strength and ultrasonic attenuation of laminated composites'. *Journal of Composite Materials*, 1997, **31**, 277–292.

46 Hsu, D.K., Hughes, M.S., 'Simultaneous ultrasonic velocity and sample thickness measurement and applications in composites'. *Journal of Acoustical Society of America*, 1992; **92**(2), 669.

47 Kaczmarek, H. 'Ultrasonic detection of damage in CFRPs'. *Journal of Composite Materials*, 1995, **29**, 59–95.

48 Scarponi, C. and Briotti, G. 'Ultrasonic technique for the evaluation of delaminations on CFRP, GFRP, KFRP composite materials'. *Composites: Part B*, 2000, **31**, 237–243.

13

Raman spectroscopy and x-ray scattering for assessing the interface in natural fibre composites

S. EICHHORN, University of Manchester, UK

Abstract: Raman spectroscopy and x-ray diffraction are applied in the study of interfaces in cellulose fibre-based composites. Detailed theory is given for the origins of Raman scattering by polymer molecules, as well as the changes in spectra as a consequence of mechanical deformation of fibres. The use of a model composite geometry of a cellulose fibre surrounded by a droplet of resin is introduced, showing how this can be used to obtain point-to-point information on the fibre–matrix interface. A similar treatment of the crystal deformation observed in deformed fibres using x-ray diffraction is also given. Particular emphasis is placed on the usefulness of synchrotron x-ray diffraction for this purpose. Specific examples relate to the use of these techniques for plant cellulose-based composites and also some more recent work conducted on nanocomposites.

Key words: cellulose, fibres, nanotechnology, interfaces, composites, Raman spectroscopy, x-ray diffraction.

13.1   Introduction to Raman spectroscopy

The Raman effect was first discovered in 1928 by the famous Indian scientist Chandrasekhara Venkata Raman (Raman and Krishnan 1928). It is alleged that Raman gained his inspiration for the effect whilst travelling by boat from India to the UK where he observed light scattering from the ripples on the surface of the sea. Raman won the Nobel Prize in 1930 for his discovery, which showed that when light is scattered by a material, it comes back with its original energy (elastic scattering) plus a small amount of energy associated with the vibration of bonds within the material. Infrared adsorption is different from Raman scattering in that in the former an oscillating dipole *absorbs* at a particular frequency whereas the latter phenomenon creates a small dipole, which then vibrates at a particular frequency emitting the additional photons observed in the scattered radiation. A theoretical description of the phenomenon can be given by first using Planck's equation to describe the energy of electromagnetic radiation (ΔE) as

$$\Delta E = h v_0 \tag{13.1}$$

379

where h is Planck's constant $(6.63 \times 10^{-34}$ J s$)$. The three components of the scattered radiation can be summarised as follows:

- **Rayleigh scattering**: the most intense component of the scattered light, having the same frequency as the incident radiation.
- **Stokes–Raman scattering**: very weak radiation (compared with Rayleigh scattering). In this instance, the incident photon excites a molecule to a higher energy state. The loss in energy of the photon is found to be equal to the gain in energy by the molecule and the scattered photon has a frequency $v_0 - v_m$; v_m is the vibrational frequency of the molecule.
- **Anti-Stokes–Raman scattering**: again a very weak radiation (compared with Rayleigh scattering) but it is the opposite to Stokes–Raman in that the molecules lose energy and the scattered photon has a higher frequency $v_0 + v_m$ than the incident photon.

These transitions are depicted pictorially in Fig. 13.1.

The interaction of the light with the material is between the electric field of the electromagnetic radiation and the electronic cloud of the molecules. If we consider an incident electromagnetic wave with an oscillating electric field of field strength E_s where:

$$E_s = E_{s0} \cos(2\pi v_0 t) \qquad [13.2]$$

where E_{s0} is the amplitude of the vibrating wave over time t. This wave then interacts with a molecule and induces an electric dipole moment P, which is given by the equation:

$$P = \alpha(q)E \qquad [13.3]$$

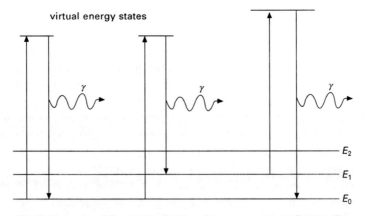

13.1 Schematic of Rayleigh, Stokes–Raman and Anti-Stokes–Raman scattering, where E_0, E_1 and E_2 are the ground, 1st and 2nd energy states of an arbitrary molecule and γ is a photon. Reproduced from Eichhorn *et al.* (2009) with permission.

where $\alpha(q)$ is the polarisability, which is itself a function of the nuclear displacement q. The molecule is also vibrating, owing to the interaction of the electromagnetic wave, with its own frequency v_m, and, therefore, the nuclear displacement q can be expressed using the equation:

$$q = q_0 \cos(2\pi v_m t) \qquad [13.4]$$

where q_0 is the amplitude of the vibration over time t. If the amplitudes are small then $\alpha(q)$ is best approximated from the first two terms of a power series, as

$$\alpha(q) = \alpha_0 + \left(\frac{\partial \alpha}{\partial q}\right)_0 q_0 + \dots \qquad [13.5]$$

where α_0 is the polarisability at the equilibrium position. A combination of equations [13.2]–[13.5] leads to the expression:

$$P = \alpha_0 E_0 \cos(2\pi v_0 t) + \left(\frac{\partial \alpha}{\partial q}\right)_0 q_0 E_0 \cos(2\pi v_0 t)\cos(2\pi v_{in} t) \quad [13.6]$$

which can be written, using a trigonometric identity, as:

$$P = \alpha_0 E_0 \cos(2\pi v_0 t) + \frac{1}{2}\left(\frac{\partial \alpha}{\partial q}\right)_0 q_0 E_0$$

$$\times \{\cos[2\pi(v_0 + v_m)t] + \cos[2\pi(v_0 - v_m)t]\} \qquad [13.7]$$

This expression separates out the contributions to the electric dipole moment from the Rayleigh scattering $\alpha_0 E_0 \cos(2\pi v_0 t)$ and the Raman scattering (with two frequencies; $v_0 + v_m$ (anti-Stokes) and $v_0 - v_m$ (Stokes). For a bond type to be Raman active, it must satisfy the condition that it has a non-zero rate of change of polarisability with respect to the nuclear displacement (*i.e.* $\partial \alpha/\partial q$ > 0). It is found that the intensity of a Raman vibration is proportional to the rate of change of polarisability.

Infrared spectroscopy is found to be more suited to polar groups, where a displacement of charge already exists between electronegative and electropositive atoms (*e.g.* O–H, N–H). For this reason, it is often more suited to side group vibrations of polymers, although C–O will give a reasonable infrared intensity. However, Raman spectroscopy occurs most intensely for symmetric vibrations (*e.g.* C–C, C=C), which are typically found along the backbone of polymers. It is for that reason that it is most suitable for following deformation, because changes in the vibrational state of the backbone bonds readily occur under these conditions.

13.2   Raman spectroscopy and measurements of molecular deformation in polymer fibres

13.2.1   Theoretical background to stress measurement in polymer fibres using Raman spectroscopy

The first report of using Raman spectroscopy to follow molecular deformation in polymer fibres was by Mitra, Risen and Baughman in 1977 (Mitra *et al.* 1977). They showed that Raman peaks for stressed single crystal fibres of polydiacetylene shifted towards a lower wavenumber position (see Fig. 13.2). These shifts were explained in terms of a change in the force constant of an anharmonic oscillator F_2', which itself is related to the vibrational frequency of the vibrational mode v_i via the equation:

$$v_i = (4\pi^2 c^2 \mu)^{-1/2}(F_2')^{1/2} \tag{13.8}$$

where c is the speed of light and μ is the reduced mass of the oscillator. This explanation was also confirmed experimentally and further explained theoretically by Batchelder and Bloor (1979), again for polydiacetylene single crystals. Penn and Milanovich (1979) initially found that they could not replicate these experiments on Kevlar fibres, as did Edwards and Hakiki (1989). In these studies, the high-power lasers used damaged the fibres leading

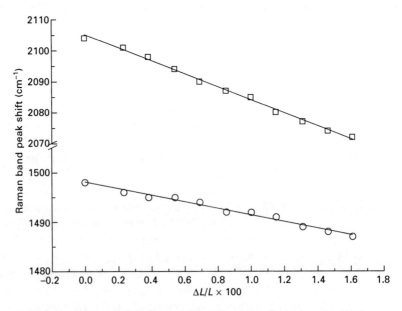

13.2 Shifts in Raman peaks for C=C (□) and C≡C (○) bonds within a polydiacetylene single crystal fibre as a function of percentage tensile strain ($\Delta L/L \times 100$). The results were replotted from Mitra *et al.* (1977) and fitted using a linear regression.

to a lack of a shift in their data (Young *et al.* 1991). Subsequent to the work by Penn and Milanovich (1979), Galiotis *et al.* (1984) showed that it was possible to follow the local strain state in a single aramid fibre embedded in an epoxy resin matrix. This work confirmed directly, for the first time, that the Cox shear lag theory worked for polymer fibre composite materials.

13.2.2 Stress measurements in oriented polymer fibres and composites using Raman spectroscopy

A large number of polymer fibres have been studied using the Raman spectroscopic technique. These have included aramid fibres (Young *et al.* 1991; Andrews and Young 1993), rigid-rod polymer fibres (Day *et al.* 1987; Young *et al.* 1990; Young and Ang 1992), high-performance polyethylene fibres (Kip *et al.* 1991; Moonen *et al.* 1992; Prasad and Grubb 1989; Wong and Young 1994), ceramic fibres (Day *et al.* 1989; Yang *et al.* 1992) and carbon fibres (Huang and Young 1993; Katagiri *et al.* 1988; Robinson *et al.* 1987; Huang and Young 1995). This is by no-means an exhaustive list and the reader is referred to a number of review articles on the subject (Young 1995; Young 1996; Young and Eichhorn 2007).

The experimental procedure for measuring Raman band shifts in single fibres is worth describing. Typically, for long fibres, cards can be made with windows cut into them, with a single filament passing through the centre of the window (much the same as for single-fibre tensile testing). A schematic of this set-up is shown in Fig. 13.3a. For short fibres and where end effects have to be eliminated, a beam bending experiment can be carried out. In this instance, the single fibre is placed on the surface of a thin polymer beam. A layer of resin is poured over the top of the fibre securing it to this surface. The beam can then be bent, by 4-point or 3-point bending to generate either a tensile or a compressive deformation of the fibre. To determine the strain, a small strain gauge is typically placed close to the fibre on the surface of the beam. A schematic of this experimental set-up is shown in Fig. 13.3b. The single fibre can then be deformed incrementally, and at each deformation step a Raman spectrum is recorded. By fitting the peaks, typically using Gaussian, Lorentzian or mixed functions, their positions can be found and plotted as a function of the deformation axis (either strain or stress).

Single fibres can also be embedded into a model composite material by surrounding the filament with a matrix material. To characterise the interface (particularly to determine the interfacial shear stress) in composites, a range of testing protocols has been developed over the years, including microbond, push-in, fragmentation and pull-out (Fig. 13.4) (Harris 1999). All of these tests rely on a number of assumptions, the most critical of which is that the *interfacial shear stress is constant along the interface*. The use of the Raman spectroscopic technique for analysing local strain/stress in a composite has been particularly instructive in showing that this assumption is not valid. All four of the testing protocols shown in Fig. 13.4 were analysed using the Raman technique, showing that the determination of the interfacial shear stress can be erroneous if local effects are not taken into account (Andrews

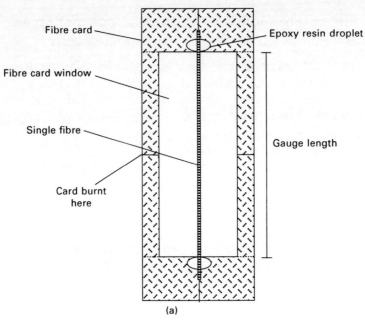

Fibre card

Epoxy resin droplet

Fibre card window

Single fibre

Gauge length

Card burnt
here

(a)

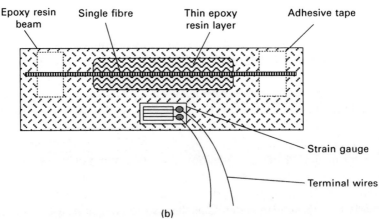

Epoxy resin
beam

Single fibre

Thin epoxy
resin layer

Adhesive tape

Strain gauge

Terminal wires

(b)

13.3 Schematics of (a) a single-fibre deformation card showing the positioning of the filament and (b) a resin beam fibre sample showing the position of a single fibre secured to the beam and the positioning of a strain gauge; the deformation modes for testing in tension (c) and compression (d) are also shown.

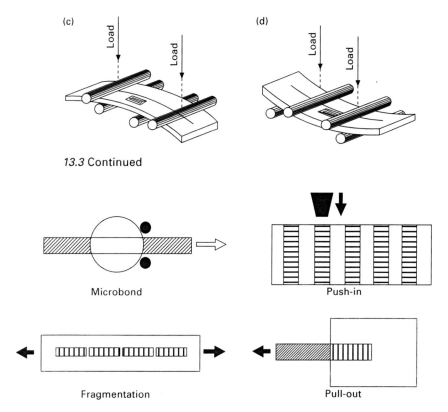

13.3 Continued

13.4 Schematics of common testing geometries for fibre–matrix interfacial shear strength analysis; namely microbond, push-in, fragmentation and pull-out.

et al. 1996; Melanitis and Galiotis 1993; Melanitis *et al.* 1992; Young *et al.* 2000).

13.3   X-ray diffraction and stress analysis in fibres and composites

13.3.1 Theoretical background to stress measurement in polymer fibres using x-ray diffraction

In oriented semicrystalline polymers there is a degree of perfection of the register of crystals along the fibre axis that gives rise to reflections, or layer lines, according to Bragg's law. Typically, x-ray diffraction patterns are obtained on flat films or, in more recent times, detectors. The layer lines are in the form of hyperbolae, and their distortion is caused by the flat geometry of the film or detector and not specifically by the material itself. A correction for the distortion owing to a flat detector plane can be made to such patterns based on the diffraction grating formula, whereby (Young and Lovell 1991):

$$c = \frac{l\lambda}{\sin \arctan(x/r)}$$ [13.9]

where c is the c-spacing of the crystal repeat (along the chain axis of the fibre for an oriented structure). A typical diffraction pattern from a cellulose fibre is shown in Fig. 13.5b indicating the position of the 002 layer line. It is often best to measure the position of this layer line at both the top and bottom of the pattern, and then divide this distance by two in order to obtain x in equation [13.9]. The crystal strain $\Delta\varepsilon_c$ can then be calculated by using the formula:

$$\Delta\varepsilon_c = \frac{c - c_0}{c_0}$$ [13.10]

where c_0 is the original c-spacing of the reflection 00l before deformation. It is typical that the crystal strain is then plotted as a function of the stress on the sample, be it bundles or single filaments.

13.3.2 Stress measurements in oriented polymer fibres and composites using x-ray diffraction

The first report of the measurement of crystal strain in fibres was by Sakurada $et\ al.$ (1962) on polyethylene, polyvinyl alcohol, polyvinylidene chloride, polypropylene, polyoxymethylene, and cellulose filament bundles. These bundles were stretched in a customised device and the change in the crystal repeat spacing c was recorded. Values of 240 GPa (polyethylene), 255 GPa (polyvinyl alcohol), 41.5 GPa (polyvinylidene chloride), 42 GPa (polypropylene), 54 GPa (polyoxymethylene) and 138 GPa (cellulose) were obtained for the elastic moduli of these materials. A simultaneous study of cellulose was conducted by Mann and Roldan-Gonzalez (1962). Subsequently a number of studies, particularly on aromatic polyesters, were published (Jakeways $et\ al.$ 1975; Nakamae $et\ al.$ 1987).

The most recent development in the study of polymer fibres and the measurement of the crystal modulus has been the use of microfocus beams, particularly at the ESRF (European Synchrotron Radiation Facility) (Engstrom $et\ al.$ 1995). This has allowed the study of single fibres of a whole range of materials, including Kevlar (Riekel $et\ al.$ 1999). Following this work a series of papers were then published on the determination of the physical and mechanical properties of rigid-rod fibres using the microfocus beam at the ESRF (Davies $et\ al.$ 2001; 2003; 2004; 2005; Montes-Moran $et\ al.$ 2002). The next few sections of this chapter deal with specific applications

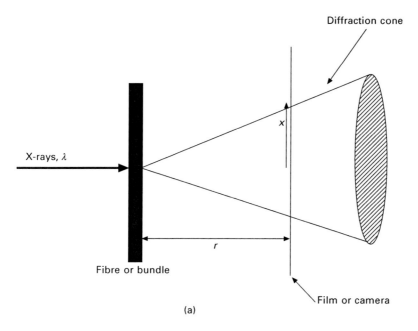

(a)

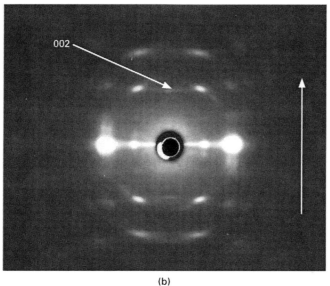

(b)

13.5 (a) Schematic of the experimental set-up for performing a fibre x-ray diffraction experiment where λ is the wavelength of the x-rays used, r is the sample (fibre) to camera or film separation and x is the vertical distance (from the centre of the pattern) of a meridional reflection (00l); (b) an example of an x-ray diffusion pattern for a cellulose fibre (Cordenka™, Acrodis) indicating the direction of the fibre axis and the position of the 002 reflection.

of both Raman spectroscopy and x-ray diffraction in the study of cellulose fibres and composites.

13.4   Raman spectroscopy and x-ray diffraction measurements of molecular and crystal deformation in cellulose fibres

The first ever report of the measurement of molecular deformation in cellulose fibres was by Hamad and Eichhorn (1997). They reported shifts in the positions of bands located both at 1095 cm^{-1} (corresponding to C–O stretching) and 895 cm^{-1} (corresponding to COH stretching) during tensile deformation of single filaments of lyocell and viscose fibres (cellulose II). This work was later built upon by Eichhorn *et al.* (2000) who showed that the same shifts could be observed from native cellulose fibres under tensile deformation (flax and hemp). Subsequently a large number of studies were conducted looking at a range of cellulose fibres (Eichhorn and Young 2001; Eichhorn *et al.* 2001a, 2001b; Eichhorn and Young 2003; Eichhorn *et al.* 2003). Other independent studies on the micromechanical deformation of cellulose fibres using Raman spectroscopy were carried out by Gierlinger *et al.* (2006) on single wood fibres and Peetla *et al.* (2006) on bast fibres (hemp, flax). The potential to use the local deformation micromechanics for the analysis of interfaces in natural fibre composites has been realised (see 13.4.1 and 13.4.2).

13.4.1 Analysis of interfaces in natural cellulose fibre composites

An example of a shift in a Raman peak with the application of tensile deformation is shown in Fig. 13.6a. Typical results for a shift in the position of a Raman band from a cellulose fibre as a function of deformation are shown in Fig. 13.6b. These results show that it is possible to determine the local stress state of a single fibre by interpolation (solid arrows). This can then be replicated on a single fibre, deformed by stress-transfer, in a composite specimen. A number of composite specimens have been evaluated using this approach, the first of which was a thin film of epoxy-resin surrounding a plant fibre (Eichhorn and Young 2003). A schematic of the specimen preparation and testing protocol is shown in Fig. 13.7a, and some typical results obtained by mapping along a fibre are shown in Fig. 13.7b (converted to fibre stress as shown in Fig. 13.6b). These results are fitted to the Cox model (Cox 1952) and a modified version which takes into account bonding across the fibre ends. Tze *et al.* (2006) showed that the interfacial shear stress obtained from interfacial mechanics experiments conducted using the Raman spectroscopic

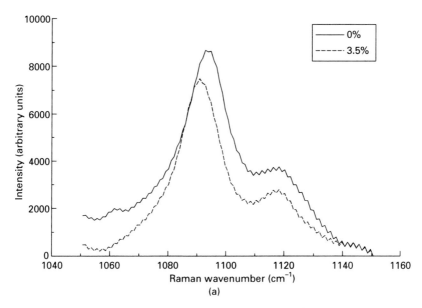

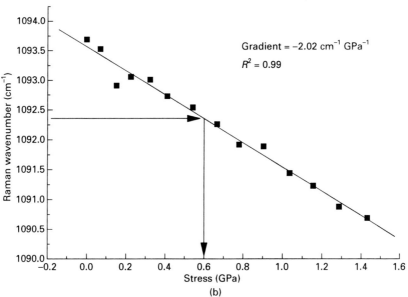

13.6 (a) Typical shift in the position of the Raman band initially located at ~1095 cm^{-1} for a flax fibre under tensile deformation (3.5% strain, ~0.6 GPa) and (b) the shift in the position of the Raman band initially located at ~1095 cm^{-1} as a function of fibre stress showing the interpolation of stress from the shift values (arrows).

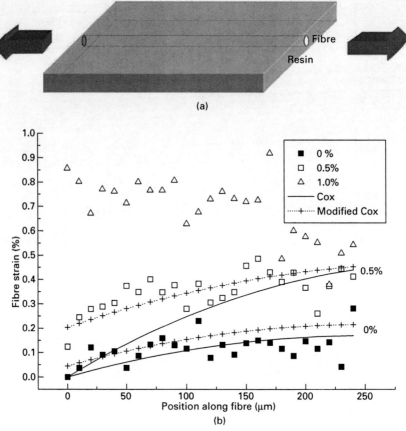

(a)

(b)

13.7 (a) Schematic of a thin film of epoxy resin surrounding a natural plant fibre for a micromechanical test using Raman spectroscopy showing the tensile stress direction (arrows) and (b) typical load fibre stress data obtained by tracking from a fibre end. Solid lines are the Cox model for stress transfer and crosses are for a modified Cox model where bonding across the ends is taken into account (Eichhorn and Young 2003). Panel (b) is reproduced with permission from Elsevier (2003).

technique can be correlated with the acid–base interaction parameter, which in turn depends on the fibre treatment. A plot of the data obtained by Tze *et al.* (2006) is shown in Fig. 13.8. Tze *et al.* (2006) used a microdroplet technique in order to obtain their interfacial shear stress measurements using the Raman spectrometer, based on the work of Eichhorn and Young (2004). An image of the set-up for the microdroplet test is shown in Fig. 13.9a. This model composite is similar in many respects to the microbond test shown in Fig. 13.4, although no restraint on the droplet is required. The laser of the Raman spectrometer can be mapped along the fibre, and

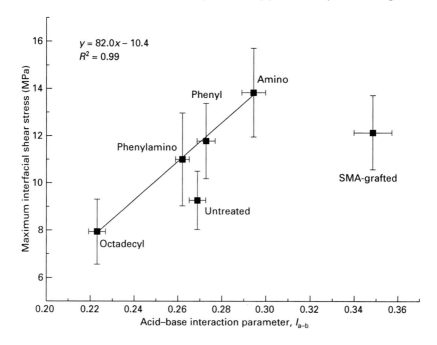

13.8 Effect of the acid–base interaction parameter I_{a-b} on the maximum interfacial shear stress, derived from microdroplet–fibre model composites scanned using the Raman spectroscopic technique. The solid line is a linear regression that does not include the data for 'Untreated' and 'SMA-grafted'. Replotted from Tze *et al.* (2006).

through the transparent resin droplet to obtain local stress data as a function of the position along the fibre. Example data are shown in Fig. 13.9b. These data can then be converted to interfacial shear stress values by fitting using a polynomial function, and then determining the first derivative of this fit ($d\sigma_f/dx$). This first derivative can then be inserted into:

$$\tau_i = \frac{r_f}{2}\frac{d\sigma_f}{dx} \qquad [13.11]$$

where r is the radius of the fibre and τ_i is the interfacial shear stress. These data can also be fitted using a model based on the pull-out test stress analysis [see Piggott (1980) for a full derivation]. However, most importantly, the fibre stress in the middle of the droplet is not zero, as it is for a fibre end in a pull-out test. The general differential equation describing the decay of fibre stress in this system is:

$$\frac{d^2\sigma_f}{dx^2} = \frac{n^2}{r_f^2}\sigma_f \qquad [13.12]$$

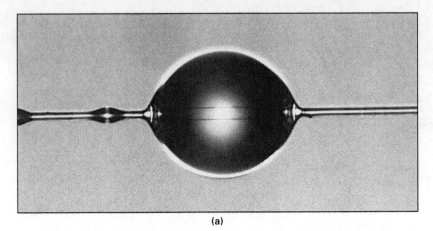

(a)

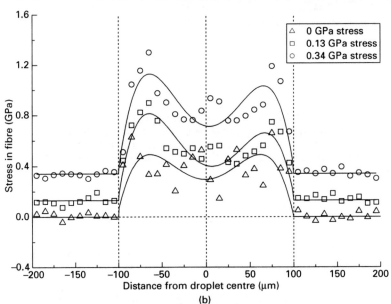

(b)

13.9 (a) A 200 μm polyester resin droplet (large droplet in the centre of image) cured on a regenerated cellulose fibre and (b) the fibre stress as a function of distance through the droplet at elevated levels of tensile deformation.

where

$$n^2 = \frac{E_m}{E_f} \frac{1}{\ln(R/r_f)} \frac{1}{(1 + v_m)}$$ [13.13]

E_m and E_f are the matrix and fibre moduli, respectively, and R is the radius of a solid cylinder of matrix around the fibre that is influenced by the applied stress.

When the boundary condition that the first derivative of the fibre stress with respect to the position along the fibre is zero at the centre of the droplet (*i.e.* $d\sigma_f/dx = 0$) is applied to the general solution to this equation, namely:

$$\sigma_f = A \sinh\left(\frac{n}{r_f}x\right) + B\cosh\left(\frac{n}{r_f}x\right) \qquad [13.14]$$

then the solution becomes:

$$\sigma_f = \sigma_{app}\frac{\cosh(nx/r_f)}{\cosh(nL/r_f)} \qquad [13.15]$$

where σ_{app} is the applied stress (typically $\sigma_{app} = \sigma_f$). This analysis has been applied to the regenerated cellulose fibre–droplet model composites (more specifically epoxy and polyester droplets on a liquid crystalline spun cellulose fibre) (Mottershead and Eichhorn 2007).

X-ray diffraction has also been used to map local deformation in cellulose fibre based composites. The first study to be reported using this technique, whereby the crystal strain of the fibre is used as a measure of stress-transfer was by Nishino *et al.* (2006) on a poly-(L-lactic acid) (PLLA)/kenaf model composite. They found that the stress on crystals within the fibres that were embedded in the resin was greater for silane-treated samples than for untreated fibres; this was a good indication of the effect of coupling on the interface between fibre and resin (Fig. 13.10). There have been surprisingly few studies on this particular use of x-ray diffraction to follow interfaces in cellulose-based composites and so it is clearly an area with potential. One major problem however, with the application of new generation synchrotron microfocus beams to cellulose is the damaging effect that they have on cellulose. If this can be overcome then a whole new field of research will be opened up.

13.4.2 Analysis of interfaces in cellulose nanocomposites

Raman spectroscopy has been recently used to follow interfaces in cellulose nanocomposites. Tunicate whisker-epoxy nanocomposites were the first to be investigated using this technique (Sturcova *et al.* 2005). In this study, the stiffness of a single whisker was estimated to be 143 GPa, close to a value determined by a theoretical calculation of the stiffness of cellulose Iβ (the crystal form of tunicate cellulose whiskers) (Sturcova *et al.* 2005). Subsequent work has shown that the stiffness of chemically derived cellulose whiskers is much smaller than that of tunicate (50–100 GPa) (Rusli and Eichhorn 2008). Both these experiments were carried out with cellulose nanowhiskers dispersed in a resin on a beam, similar to the set-up shown in Fig. 13.3b. Anisotropy in cellulose nanowhisker composites has also been measured using polarised Raman spectroscopy (Rusli *et al.* 2010). Hot pressed samples

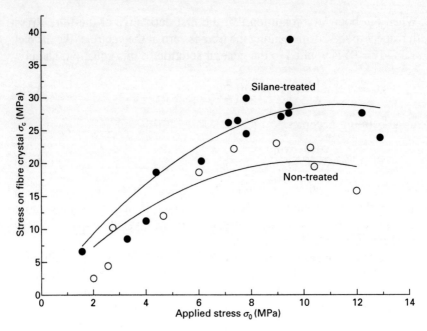

13.10 Relationship between the stress on the crystals within a kenaf fibre embedded in a PLLA matrix composite as a function of the applied stress on the sample; silane-treated (●) and non-treated (○) fibres. Replotted from Nishino *et al.* (2006). Solid lines are 2nd order polynomial fits to the data with intercepts at the origin. No information was given on the form of fit used for original data given by Nishino *et al.* (2006).

of poly(vinyl acetate) resin with 15 wt% of tunicate cellulose nanowhiskers were compared with solution cast samples (Rusli *et al.* 2010). It was shown that the former contained oriented domains of cellulose whiskers, whereas the latter only contained isotropic random networks. Both exhibited clearly different mechanical behaviour based on measurements of the local stress-transfer using Raman spectroscopy (Rusli *et al.* 2010). Images of the pressed and solution cast samples, viewed under a polarised light microscope, are shown in Fig. 13.11a. They show that the pressed sample consists of light and dark regions, corresponding to oriented and isotropic domains of whiskers respectively (Rusli *et al.* 2010). This approach therefore opens up the possibility of investigating orientation in cellulose whisker nanocomposites, and also the local mechanics of these interesting materials.

X-ray diffraction has also been used to follow the interfacial micromechanics of cellulose nanocomposites. Gindl *et al.* (2006) have shown that it is possible to independently follow the orientation of the matrix and reinforcing phases in a all-cellulose nanocomposite (where the matrix and reinforcement are both cellulosic). Orientation of the reinforcing phase is obtained from the diffraction

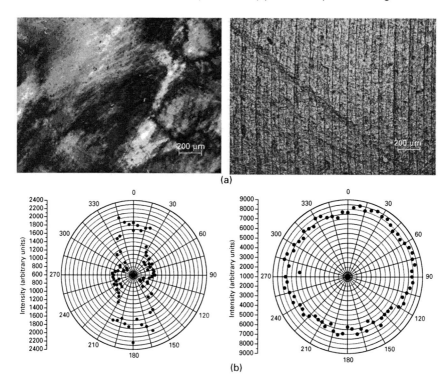

(a)

(b)

13.11 (a) Polarised light microscope images of (left) a pressed poly(vinyl acetate)–cellulose nanowhisker composite and (right) a solution cast sample. (b) The intenstiy of the Raman band located at 1095 cm^{-1} as a function of the rotation angle for a light and a dark region of a pressed sample [as in (a)]. Reproduced from Rusli *et al.* (2010), copyright ACS Publishing.

pattern of the cellulose, which is distinct for cellulose-I (reinforcement) and cellulose-II (matrix). Data from the experiment by Gindl *et al.* (2006) are reported in Fig. 13.12. This approach has not been developed for other cellulose nanocomposite systems and therefore an opportunity exists to study interfaces in these systems.

13.5   Discussion

The field of natural fibre-based composites is now in a mature stage of development, in that a large amount of technological know-how exists to enable these materials to become an industrial reality. In fact many products currently in the market-place exploit natural fibres in composites. However, there are many problems associated with their wider use in load-bearing applications. One of the key problems is that of variability of mechanical properties. Chemical modification of the fibre–matrix interface has gone

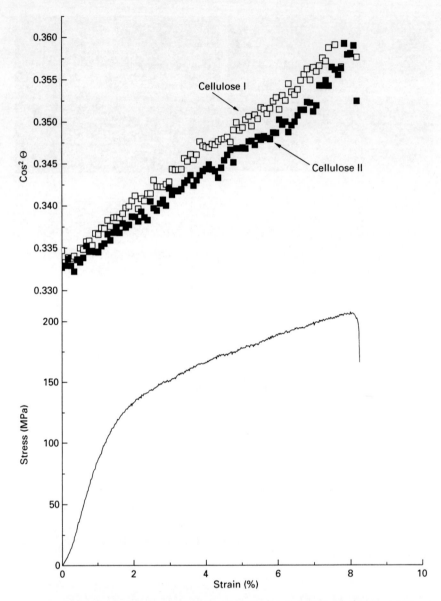

13.12 Line and scatter graphs showing stress (MPa) against strain (%) for cellulose I and cellulose II.

some way to alleviating this, as has the use of different processing routes for the fibres themselves. Advanced techniques that can probe the resultant effects on the interface are scarce. However, both Raman spectroscopy and x-ray diffraction (particularly using microfocus x-ray beams) have been, and are likely to be in the future, instructive for following the mechanics of these materials. The development of techniques such as these to characterise interfaces will require improvements in resolution (from the micron scale of beam sizes down to the nanoscale) in also detection (better and faster detection of spectra and x-ray diffraction patterns). One drive to overcome the issue of variability, and to provide a step-change in the mechanical properties of cellulose fibres has been the use of nanocellulose. By breaking down the cell wall into nanofibre components, either by chemical or mechanical means, or by generating them via bacterial means, can make materials that are useful for composite applications. However, these nanofibres do tend to cluster and so efforts to disperse them and cope with anisotropy issues will continue to occupy research efforts in this area. Raman spectroscopy offers a neat way of analysing such anisotropy and may serve as a diagnostic tool of nanocomposite performance.

13.6   Conclusions

This chapter has covered spectroscopic and diffraction techniques that can be used to investigate cellulose fibre-based composite materials. Raman spectroscopy has been shown to be a particularly powerful technique for following the local stress-state of natural and regenerated cellulose fibres in composites. This approach first takes advantage of the fact that the molecular deformation of a fibre can be followed, in air, at a spatial resolution of about 1–2 μm. Then, when a fibre is embedded in a transparent matrix, the stress that is transferred to the fibre can also be followed. This approach can also be adopted when using x-ray diffraction methods, and at the same (or smaller) spatial resolution. A combination of the two techniques could be a powerful way to examine composite materials in the future.

13.7   References

Andrews, M. C., D. J. Bannister, and R. J. Young (1996). 'The interfacial properties of aramid/epoxy model composites.' *Journal of Materials Science* **31**(15): 3893–3913.

Andrews, M. C. and R. J. Young (1993). 'Analysis of the deformation of aramid fibers and composites using raman spectroscopy.' *Journal of Raman Spectroscopy* **24**(8): 539–544.

Batchelder, D. N. and D. Bloor (1979). 'Strain dependence of the vibrational modes of a diacetylene crystal.' *Journal of Polymer Science Part B: Polymer Physics* **17**(4): 569–581.

Cox, H. L. (1952). 'The elasticity and strength of paper and other fibrous materials.' *British Journal of Applied Physics* **3**(MAR): 72–79.

Davies, R. J., S. J. Eichhorn, C. Riekel and R. J. Young (2004). 'Crystal lattice deformation in single poly(*p*-phenylene benzobisoxazole) fibres.' *Polymer* **45**(22): 7693–7704.

Davies, R. J., S. J. Eichhorn, C. Rukel and R. J. Young (2005). 'Crystallographic texturing in single poly(*p*-phenylene benzobisoxazole) fibres investigated using synchrotron radiation.' *Polymer* **46**(6): 1935–1942.

Davies, R. J., M. A. Montes-Moran, C. Riekel and R. J. Young (2001). 'Single fibre deformation studies of poly(*p*-phenylene benzobisoxazole) fibres. Part I. Determination of crystal modulus.' *Journal of Materials Science* **36**(13): 3079–3087.

Davies, R. J., M. A. Montes-Moran, C. Riekel and R. J. Young (2003). 'Single fibre deformation studies of poly(*p*-phenylene benzobisoxazole) fibres. Part II. Variation of crystal strain and crystallite orientation across the fibre.' *Journal of Materials Science* **38**(10): 2105–2115.

Day, R. J., V. Piddock, R. Taylor, R. J. Young and M. Zakikhani (1989). 'The distribution of graphitic microcrystals and the sensitivity of their Raman bands to strain in SiC fibers.' *Journal of Materials Science* **24**(8): 2898–2902.

Day, R. J., I. M. Robinson, M. Zakikhani and R. J. Young (1987). 'Raman-spectroscopy of stressed high modulus poly(*p*-phenylene benzobisthiazole) fibers.' *Polymer* **28**(11): 1833–1840.

Edwards, H. G. M. and S. Hakiki (1989). 'Raman-spectroscopic studies of Nomex and Kevlar fibers under stress.' *British Polymer Journal* **21**(6): 505–512.

Eichhorn, S. J. and R. J. Young (2001). 'The Young's modulus of a microcrystalline cellulose.' *Cellulose* **8**(3): 197–207.

Eichhorn, S. J. and R. J. Young (2003). 'Deformation micromechanics of natural cellulose fibre networks and composites.' *Composites Science and Technology* **63**(9): 1225–1230.

Eichhorn, S. J. and R. J. Young (2004). 'Composite micromechanics of hemp fibres and epoxy resin microdroplets.' *Composites Science and Technology* **64**(5): 767–772.

Eichhorn, S. J., J. W. S. Hearle, M. Jaffe and T. Kikutani Eds. (2009). *Handbook of textile fibre structure*, Woodhead Publishing Ltd.

Eichhorn, S. J., J. Sirichaisit and R. J. Young (2001a). 'Deformation mechanisms in cellulose fibres, paper and wood.' *Journal of Materials Science* **36**(13): 3129–3135.

Eichhorn, S. J., M. Hughes, R. Snell and L. Mott (2000). 'Strain induced shifts in the Raman spectra of natural cellulose fibers.' *Journal of Materials Science Letters* **19**(8): 721–723.

Eichhorn, S. J., R. J. Young and W.Y. Yeh (2001b). 'Deformation processes in regenerated cellulose fibers.' *Textile Research Journal* **71**(2): 121–129.

Eichhorn, S. J., R. J. Young, R. J. Davies and C. Riekel (2003). 'Characterisation of the microstructure and deformation of high modulus cellulose fibres.' *Polymer* **44**(19): 5901–5908.

Engstrom, P., S. Fiedler and C. Riekel (1995). 'Microdiffraction instrumentation and experiments on the microfocus beamline at the ESRF.' *Review of Scientific Instruments* **66**(2): 1348–1350.

Galiotis, C., R. J. Young, P. H. Yeung and D. N. Batchelder (1984). 'The study of model polydiacetylene epoxy composites. 1. The axial strain in the fiber.' *Journal of Materials Science* **19**(11): 3640–3648.

Gierlinger, N., M. Schwanninger, A. Reinecke and I. Burgert (2006). 'Molecular changes during tensile deformation of single wood fibers followed by Raman microscopy.' *Biomacromolecules* **7**(7): 2077–2081.

Gindl, W., K. J. Martinschitz, P. Boesecke and J. Keckes (2006). 'Structural changes during tensile testing of an all-cellulose composite by in situ synchrotron x-ray diffraction.' *Composites Science and Technology* **66**(15): 2639–2647.

Hamad, W. Y. and S. Eichhorn (1997). 'Deformation micromechanics of regenerated cellulose fibers using Raman spectroscopy.' *Journal of Engineering Materials and Technology: Transactions of the ASME* **119**(3): 309–313.

Harris, B. (1999). *Engineering composite materials*. London, IOM Communications Ltd.

Huang, Y. and R. J. Young (1995). 'Effect of fiber microstructure upon the modulus of pAN and pitch-based carbon-fibers.' *Carbon* **33**(2): 97–107.

Huang, Y. L. and R. J. Young (1993). 'Structure–property relationships in carbon-fibers.' *Institute of Physics Conference Series* (130): 319–322.

Jakeways, R., I. M. Ward, M. A. Wilding, I. H. Hall, I. J. Desborough and M. G. Pass (1975). 'Crystal deformation in aromatic polyesters.' *Journal of Polymer Science Part B: Polymer Physics* **13**(4): 799–813.

Katagiri, G., H. Ishida and A. Ishitani (1988). 'Raman spectra of graphite edge planes.' *Carbon* **26**(4): 565–571.

Kip, B. J., M. C. P. Van Eijk and R. J. Meier (1991). 'Molecular deformation of high-modulus polyethylene fibers studied by micro-Raman spectroscopy.' *Journal of Polymer Science Part B: Polymer Physics* **29**(1): 99–108.

Mann, J. and L. Roldan-Gonzalez (1962). 'X-ray measurements of the elastic modulus of cellulose crystals.' *Polymer* **3**(4): 549–553.

Melanitis, N. and C. Galiotis (1993). 'Interfacial micromechanics in model composites using laser Raman spectroscopy.' *Proceedings of the Royal Society of London Series A: Mathematical Physical and Engineering Sciences* **440**(1909): 379–398.

Melanitis, N., C. Galiotis, P. T. Tetlow and C. K. L. Davies (1992). 'Interfacial shear stress distribution in model composites. Part 2. Fragmentation studies on carbon fiber/epoxy systems.' *Journal of Composite Materials* **26**(4): 574–610.

Mitra, V. K., W. M. Risen, and R. H. Baughman (1977). 'A laser Raman study of stress dependence of vibrational frequencies of a monocrystalline polydiacetylene.' *Journal of Chemical Physics* **66**(6): 2731–2736.

Montes-Moran, M. A., R. J. Davies, C. Riekel and R. J. Young (2002). 'Deformation studies of single rigid-rod polymer-based fibres. Part 1. Determination of crystal modulus.' *Polymer* **43**(19): 5219–5226.

Moonen, J. A. H. M., W. A. C. Roovers, R. J. Meier and B. J. Kip (1992). 'Crystal and molecular deformation in strained high-performance polyethylene fibers studied by wide-angle x-ray-scattering and Raman spectroscopy.' *Journal of Polymer Science Part B. Polymer Physics* **30**(4): 361–372.

Mottershead, B. and S. J. Eichhorn (2007). 'Deformation micromechanics of model regenerated cellulose fibre-epoxy/polyester composites.' *Composites Science and Technology* **67**(10): 2150–2159.

Nakamae, K., T. Nishino, Y. Shimizu and T. Matsumoto (1987). 'Experimental determination of the elastic modulus of crystalline regions of some aromatic polyamides, aromatic polyesters, and aromatic polyether ketone.' *Polymer Journal* **19**(5): 451–459.

Nishino, T., K. Hirao and M. Kotera (2006). 'X-ray diffraction studies on stress transfer of kenaf reinforced poly(L-lactic acid) composite.' *Composites Part A: Applied Science and Manufacturing* **37**(12): 2269–2273.

Peetla, P., K. C. Schenzel and W. Diepenbrock (2006). 'Determination of mechanical strength properties of hemp fibers using near-infrared Fourier transform Raman microspectroscopy.' *Applied Spectroscopy* **60**(6): 682–691.

Penn, L. and F. Milanovich (1979). 'Raman spectroscopy of Kevlar-49 fiber.' *Polymer* **20**(1): 31–36.

Piggott, M. R. (1980). *Load bearing fibre composites*. Oxford, Pergamon Press.

Prasad, K. and D. T. Grubb (1989). 'Direct observation of taut tie molecules in high–strength polyethylene fibers by Raman spectroscopy.' *Journal of Polymer Science. Part B. Polymer Physics* **27**(2): 381–403.

Raman, C. V. and K. S. Krishnan (1928). 'A new type of secondary radiation.' *Nature* **121**: 501–502.

Riekel, C., T. Dieing, P. Engstrom, L. Vincze, C. Martin and A. Mahendrasingam (1999). 'X-ray microdiffraction study of chain orientation in poly(*p*-phenylene terephthalamide).' *Macromolecules* **32**(23): 7859–7865.

Robinson, I. M., M. Zakikhani, R. J. Day, R. J. Young and C. Galiotis (1987). 'Strain dependence of the Raman frequencies for different types of carbon fibers.' *Journal of Materials Science Letters* **6**(10): 1212–1214.

Rusli, R. and S. J. Eichhorn (2008). 'Determination of the stiffness of cellulose nanowhiskers and the fiber–matrix interface in a nanocomposite using Raman spectroscopy.' *Applied Physics Letters* **93**(3): 033111.

Rusli, R., K. Shanmuganathan, S. J. Rowan, C. Weder and S. J. Eichhorn (2010). 'Stress-transfer in anisotropic and environmentally adaptive cellulose whisker nanocomposites.' *Biomacromolecules* **11**(3): 762–768.

Sakurada, I., Y. Nukushina, and T. Ito (1962). 'Experimental determination of the elastic modulus of crystalline regions in oriented polymers.' *Journal of Polymer Science* **57**(165): 651–660.

Sturcova, A., G. R. Davies and S. J. Eichhorn (2005). 'Elastic modulus and stress–transfer properties of tunicate cellulose whiskers.' *Biomacromolecules* **6**(2): 1055–1061.

Tze, W. T. Y., D. J. Gardner, C. P. Tripp and S. C. O'Neill (2006). 'Cellulose fiber/polymer adhesion: effects of fiber/matrix interfacial chemistry on the micromechanics of the interphase.' *Journal of Adhesion Science and Technology* **20**(15): 1649–1668.

Wong, W. F. and R. J. Young (1994). 'Analysis of the deformation of gel-spun polyethylene fibers using Raman spectroscopy.' *Journal of Materials Science* **29**(2): 510–519.

Yang, X., X. Hu, R. J. Day and R. J. Young (1992). 'Structure and deformation of high-modulus alumina–zirconia fibers.' *Journal of Materials Science* **27**(5): 1409–1416.

Young, R. J. (1995). 'Monitoring deformation processes in high-performance fibres using Raman spectroscopy.' *Journal of the Textile Institute* **86**(2): 360–381.

Young, R. J. (1996). 'Evaluation of composite interfaces using Raman spectroscopy.' *Key Engineering Materials* **116–117**: 173–192.

Young, R. J. and P. P. Ang (1992). 'Relationship between structure and mechanical properties in high-modulus poly(2,5(6)-benzoxazole) (ABPBO) fibers.' *Polymer* **33**(5): 975–982.

Young, R. J., D. J. Bannister, A. J. Cervenka and I. Ahmad (2000). 'Effect of surface treatment upon the pull-out behaviour of aramid fibres from epoxy resins.' *Journal of Materials Science* **35**(8): 1939–1947.

Young, R. J., R. J. Day and M. Zakikhani (1990). 'The structure and deformation behavior of poly(*p*-phenylene benzobisoxazole) fibers.' *Journal of Materials Science* **25**(1A): 127–136.

Young, R. J. and S. J. Eichhorn (2007). 'Deformation mechanisms in polymer fibres and nanocomposites.' *Polymer* **48**(1): 2–18.

Young, R. J. and P. A. Lovell (1991). *Introduction to polymers*, Chapman & Hall.

Young, R. J., D. Lu, R. J. Day (1991). 'Raman spectroscopy of Kevlar fibers during deformation–*Caveat emptor*.' *Polymer International* **24**(2): 71–76.

Index